METHODS IN MOLECULAR BIOLOGY

Series Editor
John M. Walker
School of Life and Medical Sciences
University of Hertfordshire
Hatfield, Hertfordshire, AL10 9AB, UK

For further volumes:
http://www.springer.com/series/7651

Mammary Gland Development

Methods and Protocols

Edited by

Finian Martin

School of Biomolecular and Biomedical Science, University College Dublin, Belfield, Dublin, Ireland

Torsten Stein

Institute of Cancer Sciences, College of MVLS, University of Glasgow, Glasgow, UK

Jillian Howlin

Division of Oncology-Pathology, Lund University Cancer Center/Medicon Village, Lund, Sweden

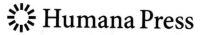

Editors
Finian Martin
School of Biomolecular and Biomedical Science
University College Dublin
Belfield, Dublin, Ireland

Torsten Stein
Institute of Cancer Sciences
College of MVLS
University of Glasgow
Glasgow, UK

Jillian Howlin
Division of Oncology-Pathology
Lund University Cancer Center/Medicon Village
Lund, Sweden

ISSN 1064-3745   ISSN 1940-6029   (electronic)
Methods in Molecular Biology
ISBN 978-1-4939-8212-7   ISBN 978-1-4939-6475-8   (eBook)
DOI 10.1007/978-1-4939-6475-8

Cover Illustration: Growing Mammary Ducts of a Pubertal Mouse - The image depicts the growing front of a pubertal mouse mammary ductal epithelium with its bulbous terminal end buds surrounded by the mammary fat pad.

Printed on acid-free paper

This Humana Press imprint is published by Springer Nature
The registered company is Springer Science+Business Media LLC New York

Preface

It is timely and exciting to compile some of the key molecular biological protocols and experimental strategies currently employed to study the various stages of mammary gland development. To this end we have invited leaders in the field to contribute detailed descriptions of their methodologies. We begin with a comprehensive overview of mouse mammary gland development drawing particular attention to comparative human biology and then present our selection of core methodologies in four parts.

Owing to the importance of transgenic and knock-out mouse models to the field, we begin Part I with two reviews of genetically modified mouse models that exhibit prenatal and pubertal mammary gland phenotypes. We discuss these phenotypes in the context of embryonic and postnatal gland development while emphasizing the study of the terminal end bud and the process of branching morphogenesis. We then present detailed descriptions of transplantation techniques, isolation and transcriptome analysis of the mammary terminal end bud during ductal morphogenesis, as well as transcriptome analysis of mammary fibroblasts isolated from sections taken at puberty. We finally describe how the post-lactational involuting gland can be used as a model to study epithelial cell death. In Part II we present a selection of 2D and 3D-model culture systems that have been employed to investigate a variety of mammary epithelial cell behaviors in vitro. We begin with a contractile assay for the study of myoepithelial cells and then present methods for in vitro recapitulation of mammary epithelial cell organization in the now infamous 3D acinus formation assay. In particular, we focus on the intrinsic molecular requirements for acini formation such as the role of MAP kinases and transient EMT, but also that of the microenvironment, examining the role of other cell types and how mechanical forces affect aggregated epithelial cells. Part III deals with stem cells and the mammary gland and we present methodologies for mammary stem cell isolation, reprogramming of progenitor populations, and a description of some strategies and methods for cell lineage tracing. Lastly, Part IV highlights some translational applications that provide a bridge between experimental studies of mammary gland development and the study of human breast cancer, with tissue microarrays for biomarker discovery and the increasingly popular practice of generating patient-derived xenographs for the study of cancer progression.

Our hope is that this volume will have a wide readership: researchers whose primary interest is in mammary gland development; developmental biologists interested in related internally branched epithelial organs, for instance the lung, kidney, and salivary gland; epithelial cell biologists and those with an interest in molecular mechanisms underlying breast cancer.

Belfield, Dublin, Ireland
Glasgow, UK
Lund, Sweden

Finian Martin
Torsten Stein
Jillian Howlin

Contents

Contributors

RENÉE VAN AMERONGEN • *Section of Molecular Cytology and Van Leeuwenhoek Centre for Advanced Microscopy, Swammerdam Institute for Life Sciences, University of Amsterdam, Amsterdam, The Netherlands*

CORINNE A. BOULANGER • *Mammary Stem Cell Section, Basic Research Laboratory, National Cancer Institute, National Institutes of Health, Bethesda, MD, USA*

CATHRIN BRISKEN • *The Swiss Institute for Experimental Cancer Research (ISREC), Swiss Federal Institute of Technology (EPFL), Lausanne, Switzerland*

DUJE BURIC • *The Swiss Institute for Experimental Cancer Research (ISREC), Swiss Federal Institute of Technology (EPFL), Lausanne, Switzerland*

CHRISTOPHER BYRNE • *Endocrine Oncology Research Group, Department of Surgery, Royal College of Surgeons in Ireland, Dublin, Ireland*

STÉPHANIE CAGNET • *Institut Curie, Centre de Recherche, Paris, France; CNRS, UMR144, Paris, France; ISREC-T Swiss Institute for Experimental Cancer Research, School of Life Sciences, Ecole Polytechnique Fédérale de Lausanne (EPFL), Lausanne, Switzerland*

CLAIRE CAIRNEY • *Institute of Cancer Sciences, College of MVLS, University of Glasgow, Glasgow, UK*

SINEAD COCCHIGLIA • *Endocrine Oncology Research Group, Department of Surgery, Royal College of Surgeons in Ireland, Dublin, Ireland*

NAI YANG FU • *The Walter and Eliza Hall Institute of Medical Research, Parkville, VIC, Australia; Department of Medical Biology, University of Melbourne, Parkville, VIC, Australia*

MALGORZATA GAJEWSKA • *Department of Physiological Sciences, Faculty of Veterinary Medicine, Warsaw University of Life Sciences (WULS – SGGW), Warsaw, Poland*

WILLIAM M. GALLAGHER • *UCD School of Biomolecular and Biomedical Science, UCD Conway Institute, University College Dublin, Belfield, Dublin, Ireland*

MARINA A. GLUKHOVA • *Institut Curie, Centre de Recherche, Paris, France; CNRS, UMR144, Paris, France*

JILLIAN HOWLIN • *Division of Oncology-Pathology, Canceromics Branch, Lund University Cancer Center/Medicon Village, Lund, Sweden*

AYMAN M. IBRAHIM • *Institute of Cancer Sciences, College of MVLS, University of Glasgow, Glasgow, UK; Zoology Department, Faculty of Science, Cairo University, Giza, Egypt*

WALTER KOLCH • *Systems Biology Ireland, University College Dublin, Belfield, Dublin, Ireland; UCD Conway Institute, University College Dublin, Belfield, Dublin, Ireland*

ZUZANA KOLEDOVA • *Wellcome Trust Centre for Cell Matrix Research, Faculty of Life Sciences, University of Manchester, Manchester, UK; Department of Histology and Embryology, Faculty of Medicine, Masaryk University, Masaryk, Czech Republic*

PETER A. KREUZALER • *Department of Pathology, University of Cambridge, Cambridge, UK*

BETHAN LLOYD-LEWIS • *Department of Pathology, University of Cambridge, Cambridge, UK*

PENGFEI LU • *Department of Anatomy and Program in Developmental and Stem Cell Biology, University of California, San Francisco, CA, USA; School of Life Science and Technology, Shanghai Tech University, Shanghai, China*

FINIAN MARTIN • *UCD School of Biomolecular and Biomedical Science, UCD Conway Institute, University College Dublin, Belfield, Dublin, Ireland*

JEAN McBRYAN • *Department of Molecular Medicine, Royal College of Surgeons in Ireland, Education and Research Centre, Beaumont Hospital, Dublin, Ireland*

SARA McNALLY • *UCD School of Biomolecular and Biomedical Science, UCD Conway Institute, University College Dublin, Belfield, Dublin, Ireland*

BENEDIKT MINKE • *Systems Biology Ireland, University College Dublin, Belfield, Dublin, Ireland; UCD School of Biomedical and Biological Science, University College Dublin, Belfield, Dublin, Ireland*

ANOESKA A.A. VAN DE MOOSDIJK • *Section of Molecular Cytology and Van Leeuwenhoek Centre for Advanced Microscopy, Swammerdam Institute for Life Sciences, University of Amsterdam, Amsterdam, The Netherlands*

JOANNA S. MORRIS • *School of Veterinary Medicine, University of Glasgow, Glasgow, UK*

LAOIGHSE MULRANE • *UCD School of Biomolecular and Biomedical Science, UCD Conway Institute, University College Dublin, Belfield, Dublin, Ireland*

CELESTE M. NELSON • *Departments of Chemical & Biological Engineering and Molecular Biology, Princeton University, Princeton, NJ, USA; Department of Molecular Biology, Princeton University, Princeton, NJ, USA*

DARRAN P. O'CONNOR • *RCSI Molecular & Cellular Therapeutics (MCT), Royal College of Surgeons in Ireland, Dublin, Ireland*

MEI FONG PANG • *Departments of Chemical & Biological Engineering and Molecular Biology, Princeton University, Princeton, NJ, USA*

SARA PENSA • *Department of Pathology, University of Cambridge, Cambridge, UK*

ALEXANDRA S. PIOTROWSKI • *Department of Chemical & Biological Engineering and Molecular Biology, Princeton University, Princeton, NJ, USA*

KARINE RAYMOND • *Institut Curie, Centre de Recherche, Paris, France; CNRS, UMR144, Paris, France; Saint Antoine Research Center, UPMC Université Paris, INSERM, Paris, France*

HENRIKE K. RESEMANN • *Department of Pathology, University of Cambridge, Cambridge, UK*

ANNE C. RIOS • *The Walter and Eliza Hall Institute of Medical Research, Parkville, VIC, Australia; Department of Medical Biology, University of Melbourne, Parkville, VIC, Australia*

TIMOTHY J. SARGEANT • *Department of Pathology, University of Cambridge, Cambridge, UK*

THOMAS SCHWARZL • *Systems Biology Ireland, University College Dublin, Belfield, Dublin, Ireland*

MONA SHEHATA • *Cancer Research UK Cambridge Institute, Li Ka Shing Centre, University of Cambridge, Cambridge, UK*

ALLISON K. SIMI • *Departments of Chemical & Biological Engineering and Molecular Biology, Princeton University, Princeton, NJ, USA*

GILBERT H. SMITH • *Mammary Stem Cell Section, Basic Research Laboratory, National Cancer Institute, National Institutes of Health, Bethesda, MD, USA*

TORSTEN STEIN • *Institute of Cancer Sciences, College of MVLS, University of Glasgow, Glasgow, UK*

JOHN STINGL • *Cancer Research UK Cambridge Institute, University of Cambridge, Cambridge, UK; STEMCELL Technologies Inc, Vancouver, British Columbia, Canada*

MARTA TERRILE • *UCD School of Biomolecular and Biomedical Science, UCD Conway Institute, University College Dublin, Belfield, Dublin, Ireland*

JOE TIEN • *Department of Biomedical Engineering, Boston University, Boston, MA, USA; Division of Materials Science and Engineering, Boston University, Boston, MA, USA*

DRIEKE VANDAMME • *Systems Biology Ireland, University College Dublin, Belfield, Dublin, Ireland*

DAMIR VAREŠLIJA • *Endocrine Oncology Research Group, Department of Surgery, Royal College of Surgeons in Ireland, Dublin, Ireland*

JACQUELINE M. VELTMAAT • *Institute of Molecular and Cell Biology, A*STAR (Agency for Science, Technology and Research), Singapore, Singapore*

JANE E. VISVADER • *The Walter and Eliza Hall Institute of Medical Research, Parkville, VIC, Australia; Department of Medical Biology, University of Melbourne, Parkville, VIC, Australia*

CHRISTINE J. WATSON • *Department of Pathology, University of Cambridge, Cambridge, UK*

LEONIE YOUNG • *Endocrine Oncology Research Group, Department of Surgery, Royal College of Surgeons in Ireland, Dublin, Ireland*

Chapter 1

Overview of Mammary Gland Development: A Comparison of Mouse and Human

Sara McNally and Torsten Stein

Abstract

The mouse mammary gland is widely used as a model for human breast cancer and has greatly added to our understanding of the molecular mechanisms involved in breast cancer development and progression. To fully appreciate the validity and limitations of the mouse model, it is essential to be aware of the similarities and also the differences that exist between the mouse mammary gland and the human breast. This introduction therefore describes the parallels and contrasts in mouse mammary gland and human breast morphogenesis from an early embryonic phase through to puberty, adulthood, pregnancy, parturition, and lactation, and finally the regressive stage of involution.

Key words Mammary gland, Breast, Development, Morphogenesis

1 The Mammary Gland

Mammals (class Mammalia) are uniquely characterized by the presence of mammary glands which form at the site of localized thickening of ventral epidermis or ectoderm. It has long been established that overlying ectoderm is induced and specified by the mammary mesenchyme to differentiate and form mammary buds [1], the development of which are required for milk production in order to nourish offspring. The broad term "mammary gland development" refers not only to growth but also the functional differentiation and regression of the mammary apparatus which occurs dynamically and in synchrony with the female reproductive cycle [2]. Indeed, distinct phases of development are driven by hormonal cues and most appealing as a model for molecular interrogation, the majority of mammary development is observed during postnatal life. This introduction describes mouse mammary morphogenesis from an early embryonic phase through to puberty, adulthood, pregnancy, parturition, and lactation, and finally the regressive stage of involution.

Finian Martin et al. (eds.), *Mammary Gland Development: Methods and Protocols*, Methods in Molecular Biology, vol. 1501, DOI 10.1007/978-1-4939-6475-8_1, © Springer Science+Business Media New York 2017

During embryonic/fetal development, the mammary "framework" is established. Prior to puberty and at the beginning of gonadal hormone release, growth of the mammary gland matches that of overall animal growth. At approximately 4 weeks of age in the mouse, the onset of puberty gives rise to accelerated ductal extension and branching and at this time, terminal end buds (TEBs) are observed as large club-shaped specialized structures at the tip of growing ducts. TEBs contain both body cells and cap cells, two histologically distinct cell types. Where body cells give rise to mammary epithelial cells, the cap cells are myoepithelial precursors [3]. Highly proliferative TEBs drive ductal tree extension where ducts comprise a single layer of luminal epithelial cells which will later form a transport channel for milk at lactation. Primary ducts are surrounded by a layer of myoepithelial cells which later become discontinuous around secondary and tertiary ducts and the TEBs.

Due to extensive lobulo-alveolar proliferation and additional ductal branching at pregnancy, the mouse fat pad is completely filled at parturition. During pregnancy and early lactation, alveolar and ductal cells undergo rounds of cell division [4]. And finally, functional differentiation is achieved at parturition with large amounts of milk being produced and secreted during lactation. Throughout pregnancy, lactation and weaning at involution, alveolar cells transition quickly through vigorous proliferation, differentiation, and eventual cell death [5]. Remodeling of the complete lobulo-alveolar epithelial compartment (involution) takes place after weaning of the pups, leaving a regressed gland that much resembles the virgin-like morphology. A new phase of lobulo-alveolar development occurs with each new pregnancy (see Fig. 1) [4, 6].

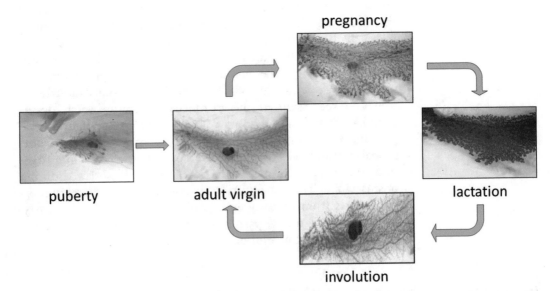

Fig. 1 Carmine alum-stained mouse mammary gland whole-mounts showing the changes in mammary ductal development during puberty, pregnancy, lactation, and involution

2   Morphology of the Mouse Mammary Gland

Located along the mammary line in the mouse, five pairs of mammary fat pads (containing specialized stroma of large amounts of fat and some fibrous connective tissue and a rudimentary ductal tree) are found just below the skin and between the forelimbs and hind limbs [7]. The gland itself is composed of two major cellular compartments, the epithelium and the surrounding stroma which derive from ectoderm and mesoderm of the embryo, respectively [8]. The mammary or "milk line" is simply a thickening of epidermis which runs on both sides of the mid-ventral embryo and in the mouse contains three thoracic fat pad pairs and two inguinal fat pad pairs [7]. Each fat pad is host to a lymph node and as these are readily identifiable, they are often a useful landmark for analysis of whole mounts [9]. Interestingly, the fifth inguinal gland is the most highly differentiated and the first thoracic gland is the least, highlighting a gradient of specification in the mouse [10]. An exterior nipple connected to the primary epithelial duct in each fat pad enables the release of milk at lactation [11, 12]. Where mice have five pairs of mammary glands, human development of one pair of mammary glands is observed although ducts from both species contain similar epithelial cells. A synopsis of the distinct stages of mammary gland development is provided below, beginning with development in utero.

3   Embryonic Mammary Gland Development

Embryonic mouse and human mammary development are documented to be very similar and in both species, underlying mesenchyme is invaded by the placode-derived mammary epithelial bud. However, human development is more complex than that of the mouse and is characterized by ten phases which are defined by fetal length [13].

An interplay of both mesenchymal and epithelial signaling is required for the early phase of mammary gland development in the mouse embryo. Between embryonic day 10 and 11 (E10 and E11), five pairs of ectodermal placodes begin to appear along two milk lines which run ventrally inside the limbs from the genital area to the neck. It is thought that ectodermal cells have in fact migrated to the points at which placodes are formed, and that they do not represent solely a point of heightened cell proliferation in the mouse embryo [7]. These placodes subsequently form buds which continue to increase in size up to E15 [12]. After formation of the mammary buds in female mice, further mammary development is temporarily halted. However, mammary epithelium in male embryos induces mesenchymal expression of androgen receptor and sensitivity to testosterone at E13 which leads to irreversible condensation of the mesenchyme around the bud and degeneration of bud epithelium during E13.5 and E15.5 [14]. This contrasts to human embryonic development where both male and female glands develop similarly in utero [15]. At E15.5–E.16, each individual bud elongates by proliferation at the bud tip and gives rise to a sprout/cord that invades the fat pad precursor (where "fat pad" refers to the mouse mesenchyme, a structure rich in adipose tissue) [7]. In the female mouse embryo, each sprout goes on to form a hollow lumen which opens on the skin surface to determine where the nipple will form by a process of epidermal invagination. The early signs of each sprout are evident by E16 and these very quickly (i.e., by E18.5) develop into the small glands that will be present at birth, by which time each mouse fat pad will host 15–20 branched ducts [16].

In the human gland primordium, the majority of epithelial cells stain positively for both luminal and basal markers (CK19 and CK14), though previous IHC analysis of human fetal tissue showed that the mammary bud was negative for CK14 [17]. The mammary epithelium cords become fully canalized near term and basal and luminal epithelial signatures are identifiable as either CK14+ or CK19+ cells (reviewed in ref. [18]). In contrast, only CK14+ cells have been detected in the mouse mammary gland during embryonic development (at E15.5) [19].

Canonical Wnt signaling is known to be required for specification of the mammary line. In fact, expression of a Wnt-responsive β-galactosidase (TOPGAL) transgene in cells between limb buds of E10.5 TOPGAL embryos is one of the earliest described markers of the mammary line. Later, during E11.25 and E11.5, expression of other Wnt family members (Wnt6, Wnt 10a, and Wnt10b for example) is detected in the mammary line and there is evidence that Wnt signaling may also contribute to mammary bud formation [20–22]. Embryonic mammary gland development also relies strongly on signaling from the fibroblast growth factor family and knockout mouse models have highlighted that four of the five mammary placodes are disrupted in the absence of FGF10 and FGFR2b [23]. Additionally, the FGF receptor FGFR1 along with FGF4, FGF8, FGF7, and FGF17 are all expressed in developing placodes [21].

Another important driver of mammary development in the embryo is parathyroid hormone-related protein (PTHrP) which is expressed by the mammary epithelial bud as it invaginates into the mesenchyme. Its receptor, PTH1R, is expressed in the mesenchyme underlying the developing bud. Deficiency of either PTHrP or its receptor generates normal mammary buds but these subsequently degenerate and as such, no ductal network is formed. It is the case that in order for mesenchyme to acquire specialized mammary fate, PTHrP is necessary. Moreover, in order for mammary mesenchyme to activate overlying epidermis to form the nipple, PTHrP signaling is required. Consequently, failure of nipple formation occurs in both PTHrP and PTH1R knockout mice [24].

The exact molecular events leading to the formation of a primitive ductal tree from the mammary bud are not clearly defined. However, there is relatively little demand for hormonal regulation of initial branching morphogenesis as mice deficient in growth hormone receptor, prolactin receptor, progesterone receptor, or estrogen receptor α or β display no obvious mammary defect at this developmental stage [8, 25]. Early bud outgrowth also occurs independently of growth factor signaling, for which there is a large requirement later during puberty [25].

4 The Prepubertal Mammary Gland

Between birth and puberty, the mammary gland exists in a relatively quiescent state (this is also the case for human glands) and at this time, the mouse mammary gland contains a stroma of fibroblasts and connective tissue along with a simple primary ductal network that fills only a fraction of the mammary fat pad. The rudimentary ductal tree observed in the parenchyma at this stage is characterized by each branch comprising a lumen surrounded by a single epithelial cell layer. Stromal and parenchymal compartments are separated by a laminin-containing basement membrane, on top of which is firstly, a layer on myoepithelial cells and secondly a layer of epithelial cells. In the mouse mammary gland, the presence of keratin 8, 11, 20, and 22 proteins (detected by immunohistochemistry) identifies epithelial cells [26]. Immunohistochemical techniques using an antibody against smooth muscle actin identifies myoepithelial cells [27]. However, human and mouse myoepithelial cells have both been found to express cytokeratin 5, 14, and p63 in addition to alpha smooth muscle actin [27]. A major difference between human and mouse mammary glands at birth is that several minor ductal networks are joined at the nipple in humans while a single network is found in mice [15]. Growth of this network keeps pace with normal body growth (is isomorphic) until puberty.

5 The Pubertal Mammary Gland

The pubertal mouse mammary gland is an ideal model for experimental morphogenesis. The primary glandular branching morphogenesis occurs at this time (approximately 5 weeks in mice, 9–12 years in humans), integrating epithelial cell proliferation, differentiation, and apoptosis. At the onset of puberty, rapid expansion of the preexisting rudimentary mammary epithelium generates an extensive ductal network by a process of branch initiation, elongation, and invasion of the mammary mesenchyme. Similarly in the human female, the onset of puberty sees a resumption in growth of both the mammary stroma (including fatty and fibrous tissue) and glandular tissue with ductal elongation and branching. It is this branching morphogenesis that characterizes pubertal mammary gland growth in both humans and rodent models. Tissue-specific molecular networks interpret signals from local cytokines/growth factors in both the epithelial and stromal microenvironments. This is largely orchestrated by secreted ovarian and pituitary hormones.

Drive from ovarian hormones such as estrogen stimulates the rapid proliferation and expansion of these simple structures (branching morphogenesis). This expansion is driven by terminal end buds TEBs (bulbous, club-shaped structures that encapsulate the tips of individual primary ducts) and in mice, for instance, as primary ducts elongate, bifurcation (or primary branching) of the TEBs generates additional primary ducts. These in turn are subjected to lateral secondary-branching [28]. Sexual maturity is reached at approximately 5 weeks in the mouse and 11–14 years in humans although the ductal network continues to grow beyond this point until it reaches its full dimension by around 8 weeks in the mouse and 18–24 years in the human mammary gland [29, 30]. When the extremities of the fat pad are reached, the end buds shrink in size and become mitotically inactive, then the pubertal growth phase is complete [31, 32]. The molecular regulators of terminal end bud formation and ductal morphogenesis are discussed in ref. [33].

Pubertal mammary gland development is both initiated and maintained by steroid hormones and pituitary hormones; and local growth factors and cytokines [34]. Ductal morphogenesis in the mammary gland is characterized by a dominant requirement for estrogen (reviewed in refs. [31, 35]). The mammary glands of mice ovariectomized at 5 weeks of age fail to develop a ductal network. This effect is rescued upon implantation of slow-release estrogen pellets into the mammary gland which stimulates ductal growth [11, 36]. The ductal network in mice deficient in ERα is severely stunted due to a complete failure to invade the stoma and adult mammary glands resemble glands of a newborn female [37, 38]. ERα$^{-/-}$ epithelial cells will not generate a mammary tree at puberty when transplanted into an ERα$^{+/+}$ fat pad; but ERα$^{+/+}$ epithelial cells will stimulate ductal outgrowth when transplanted into an ERα$^{-/-}$ fat pad; thus requirement is for ERα in epithelium, not stroma. In order for ductal morphogenesis to occur in the pubertal mammary gland, epithelial ERα is required to act in a paracrine fashion; proof is, ERα$^{-/-}$ epithelial cells will persist in a mammary tree when transplanted into fat pad, mixed with ERα$^{+/+}$ cells [39].

Ovarian steroids and pituitary hormones are not only necessary for ductal expansion in the pubertal mammary gland but also act as mediators of mammary stem-cell fate decisions. For a comprehensive review of molecular regulators of pubertal gland development, *see* [33]. The impact of regulators of pubertal mammary gland development on stem cell compartments and breast cancer progression is reviewed in ref. [40].

An extensive ductal tree with budding structures (mouse alveolar buds or human terminal duct lobular units (TDLUs)) is observed in the mature virgin gland [41]. TDLUs in humans are the functional unit of the breast and resemble a bunch of grapes and stem; one individual TDLU represents a collection of acini (from one single terminal duct) embedded in intralobular stroma [7]. Adequate space still remains in order for this network to further support tertiary lateral branches which will occur at each diestrus and during pregnancy [28, 42]. One important difference between the mouse and human mammary gland at puberty is that lobules form in the human breast at this stage, whereas in the mouse, lobules only appear with the onset of pregnancy [43].

The process of lateral secondary branching within the pubertal mammary gland receives less attention than ductal morphogenesis. Key regulators direct this molecular event which requires tight control; sufficient space must remain in the postpubertal gland to support lobuloalveolar development. This is achieved by restricting the numbers of branching events. Endogenous levels of inhibitory morphogens such as active TGFβ, act to control the spatial geometry of branching morphogenesis [44]. TGFβ has been found to complex with ECM molecules in areas where budding is "inhibited" and in this way, it prevents excessive branching and invasion of the fat pad by the advancing ductal network. It is also reported to have an epithelial-specific matrix-synthesis role: In those areas where morphogenesis is restricted there are concentrated chondroitin sulfate and type 1 collagen deposits [32]. More recently, the literature reports a strong upregulation of the chondroitin-sulfate proteoglycan versican in cap cells during puberty [45].

The cytokine hepatocyte growth factor (HGF)/Scatter factor (SF) is a positive regulator of branching morphogenesis and drives proliferation and EMT-like events in the mammary gland. In both, human and mouse mammary glands, the source of HGF/SF is the fibroblast population of cells [46, 47]. More specifically, HGF/SF is a stromal-derived paracrine mediator which promotes morphogenesis in-vitro and in-vivo and signals through the Ras pathway [48, 49]. The effects of TGFβ signaling are thought to compound the role of HGF/SF and both cytokines regulate the limits of branch spacing in the mammary gland. Growth factor and hormonal control of pubertal mouse mammary gland side-branching is reviewed in ref. [33].

6 Adult Mammary Gland Development

In the postpubertal/adult mouse mammary gland, secretion of ovarian hormones with every estrous cycle initiates the development of lateral and alveolar buds [50]. An estrous cycle in the mouse lasts for 4–5 days and as postpubertal development continues, rudimentary alveolar structures are formed from the subdivision of alveolar buds. After puberty, pregnancy represents the second postnatal mammary gland developmental stage and is a time of ductal branch expansion from the alveolar buds or human TDLUs. Rather than the alveolar buds found in the mouse mammary gland, the human mammary tree displays terminal ducts which end in clusters of ductules (these form the TDLUs) organized within loose intralobular connective tissue (*see* Fig. 2) [51]. Branching in the human gland is complex with a primary duct from the nipple connecting subsidiary ducts, segmental ducts and subsegmental ducts before reaching the TDLUs [15]. One other species difference to note is that the human mammary stroma is composed of less adipocytes than that of the mouse, but has more fibrous connective tissue [52]. In addition, the non-lactating adult human breast comprises as much as 80 % stroma [7]. However, mammary ducts from both species are made up of similar epithelial cells with ducts comprising epithelial cells surrounding a central lumen [6]. Luminal epithelial cells from both species have been shown express cytokeratin proteins 7, 8, 11, 15, 18, 19, 20, and 22 [26, 53].

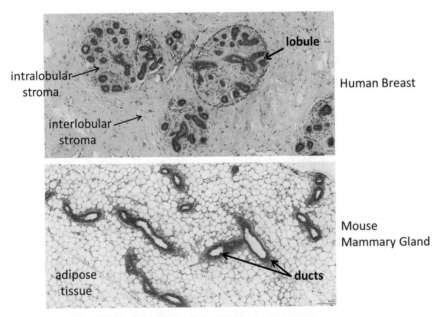

Fig. 2 Hematoxylin and eosin-stained sections of a human breast (*top*) and a mouse mammary gland (*bottom*). While the human breast section shows epithelial lobules with a collagenous intralobular and interlobular stroma, the epithelial ducts of the mouse mammary gland are surrounded by a collagen sheath within an adipocyte-rich tissue

7   The Pregnant Mammary Gland

Extensive tissue remodeling of the mammary gland occurs during pregnancy, the initial phase of which is defined by ductal branch proliferation, the development of alveolar buds (similar to those observed in the postpubertal gland) and remodeling of the ECM. By 19–21 days of pregnancy in the mouse, mammary differentiation peaks with the development of alveoli [54]. Pregnancy-induced growth of the mammary gland leads to differentiation of alveolar buds (at the tips of tertiary branches) into units capable of milk secretion (and a lactating gland at parturition) [6].

Hormonal control of mammary gland functional differentiation at this stage is well documented. Estrogen and progesterone are secreted from the corpus luteum. This is followed by estrogen, progesterone, and somatotropin from the placenta. Next are the pituitary hormones prolactin and then the adrenocorticoids from the adrenal gland. While estrogen signaling contributes indirectly to alveologenesis, alveolar proliferation can proceed in mammary epithelium lacking the estrogen receptor (its dominant requirement is during pubertal ductal elongation) [35].

A tightly controlled balance between progesterone and prolactin signaling is observed during the pregnancy-induced formation of milk secreting structures [55]. In early pregnancy, required progesterone levels are controlled by pituitary prolactin stimulation of ovarian progesterone. During day 2 to day 6 of pregnancy in the mouse, both of these hormones initiate a proliferation drive [56]. Epithelial recombination experiments and progesterone receptor knockout mice have shown that there is a requirement for progesterone during alveolar morphogenesis and that epithelial cell proliferation depends on epithelial progesterone receptor activity [57]. Specifically, reciprocal transplantation technology has highlighted a requirement for epithelial PR in lobuloalveolar development whereas the need for stromal PR may be restricted to ductal growth [57, 58]. Although tertiary side branching and lobuloalveolar development fail in pregnant and adult mice upon deletion of both progesterone receptor isoforms (PR-A and PR-B), selective knockout studies reveal that PR-B is the essential isoform at this time [59].

It is of note that ductal side-branching and alveolar morphogenesis in the cycling (nonpregnant adult) mouse is also dependent on progesterone signaling. The failure of a subset of epithelial cells to respond to progesterone directly, and the knowledge that some epithelial cells do not express the progesterone receptor point to a paracrine mode of progesterone signaling at this phase of mammary development [35]. Both receptor activator for nuclear factor kB ligand (RANKL) and Wnt4 (Wingless ligand) have been found to act as paracrine mediators of progesterone signaling in the mammary gland. In the absence of pregnancy, ectopic expression of RANKL or Wnt4 leads to tertiary side branching [60, 61]. In addition, pregnancy driven side branching and alveologenesis is impaired in mice upon RANKL or Wnt4 deletion [62, 63]. Expression in mice of RANKL and Wnt4 is regulated by progesterone and these proteins co-localize with PR-positive luminal epithelial cells [62] [64].

Epithelial PrlR has been shown to be required for normal lobuloalveolar differentiation in studies using prolactin receptor knockout mice (PrlKO); these studies also highlight that stromal PrlR plays a nonessential role at this point. The development of lobuloalveoli and the production of milk during pregnancy fails when PrlR−/− mammary fats pads are transplanted into wild-type fat pads, highlighting the essential requirement for PrlR in directing alveolar morphogenesis [65]. While alveolar development and lactogenesis during late pregnancy depend on PRL signaling, it appears nonessential for ductal outgrowth and side branching [65]. Members of the Suppressor of cytokine signaling (Socs) gene family are activated by prolactin signaling and, in turn, negatively regulate Prl action. Precocious mammary development at pregnancy is observed in Socs1−/− mice and it has been shown that the lobuloalveolar defect phenotype of Prlr+/− mice is rescued by Socs1+/− mice [66]. In addition, the lactation defect in Prlr+/− mice is ameliorated upon Socs2 knockout [67].

In the latter stage of pregnancy, individual alveoli differentiate from progressively cleaved alveolar buds and these alveoli will later form milk secreting lobules during lactation. A discontinuous layer of contractile myoepithelial cells surrounds each alveolus in the mammary gland [68, 69]. Complete lobuloalveolar differentiation requires direct contact between luminal cells and the extracellular matrix (ECM) of the basement membrane [70]. The transcription factor GATA3 has been shown to be required for complete alveolar maturation and its conditional deletion generates an alveolar/lactogenesis defect [71]. The onset of a secretory activation phase occurs by approximately day 18 of pregnancy when the majority of the mammary fat pad is composed of alveoli. At this point, lipids and milk proteins are produced by the alveolar epithelial cells in preparation for postpartum secretion [6, 72].

Overall, the extensive remodeling in the mouse mammary gland at pregnancy is comparable to the human morphogenesis at this time. However, lobules are present in the human female gland prior to pregnancy and the human female gland is not classed as fully differentiated until the first full-term pregnancy has reached its final stages [73]. Of note, the majority of human breast cancers arise from the TDLU. Hyperplastic enlarged lobular units (HELUs) evolve from normal TDLUs in the breast and this is the first histological sign of a potential disease state [74]. The conversion of a TDLU to a HELU is the most frequent growth abnormality in the human female breast and its identification represents the earliest possible chance of cancer precursor detection although not all hyperplasias will develop into cancer [74, 75]. Differential gene expression analysis of HELUs compared to normal TDLU's shows that estrogen receptor α (ERα) and amphiregulin (Areg) levels are elevated in HELUs [74]. A switch in preference of EGFR ligands (from EGF to Areg) may be mediated by estrogen and represents an early event in the progression to hyperplasia [74]. The role of ERα and Areg in regulating mammary gland epithelial expansion suggests a potential for two axes to contribute to breast tumor progression and Areg is found to be elevated in most ERα-positive breast tumors [74, 76].

8 The Lactating Mammary Gland

During lactation, milk-producing structures (the lobular alveoli) differentiate from alveolar buds or TDLUs and the provision of nutrition for offspring occurs in the fully differentiated mammary gland. PRL signaling is essential for the final events of lactogenic differentiation at this time in the mammary gland [55]. In response to a suckling stimulus from the infant, prolactin is released from the anterior pituitary and maintains the architecture of the "functional" gland. Mediators of PRL signaling at this time are signal transducer and activator of transcription 5 (Stat5) and Janus kinase 2 (Jak2) [77, 78]. It is also thought that cyclin D1 and IGF-2 may be upregulated in response to PRL activity [25, 79]. Large volumes of milk are produced by secretory cells which are fully active at this stage and in response to an increased milk burden, dilation of the alveoli occurs until they comprise the majority of the glandular volume. Suckling young stimulate the release of pituitary oxytocin which, in turn, induces contraction of the myoepithelial cells that form a continuous lining around each alveolus. This contraction forces milk into the ducts from the alveoli and milk production is stimulated further by removal of milk by suckling action [6, 80, 81]. When lactation is established, adipocyte fat is metabolized and the gland becomes completely filled with expanded alveoli. Mouse mammary gland lactation continues until pups are weaned after approximately 3 weeks and fully developed alveolar structures remain until this point [82].

9 The Involuting Mammary Gland

The (post-lactational) involution stage of mammary gland development is initiated by removal of the suckling stimulus (weaning) or the absence of breast feeding and milk stasis. A reduction in PRL levels in response to weaning results in milk secretion arrest, a process mediated by Akt and Stat5 signaling [83, 84]. Interestingly, early involution is a reversible (and possibly transiently quiescent) phase where suckling can reinitiate lactation [85]. However, an irreversible cascade of regressive remodeling and widespread programmed cell death of epithelial cells is initiated after approximately 2 days in the mouse mammary gland. Both the early and late phases of involution are controlled on a molecular level by the loss of survival factors and the accumulation of cell death mediators. In addition, suckling stimulus removal sees a reduction in systemic levels of lactogenic hormones and a buildup of milk in the alveolar lumina [86]. In both human and mouse mammary glands, post-lactational involution represents the removal of unwanted epithelial cells in a controlled fashion.

Early involution is evidenced by death of alveolar secretory epithelial cells where as much as 80% of glandular epithelial cells are lost by lysosome-mediated programmed cell death and removed by efferocytosis at this stage [87, 88]. Early programmed cell death has been detected within 24 h of pup removal in rodents [89] and involution-associated gene expression changes can be detected as early as 12 h after forced weaning [90]. As a result, alveolar distension occurs and alveoli collapse into epithelial clusters. In contrast, fat cells become more visible and adipocytes appear to refill and expand [6]. Ducts within a dense stroma are evident in early involution even though the epithelium appears disorganized. Gene expression analysis reveals that cell cycle control genes are upregulated in this phase [91, 92], as are caspases [92], acute-phase proteins [93], clusterin and tissue inhibitor of metalloproteinases-1 (TIMP-1) [94]. Constitutive expression of Akt/PKB, a mediator of cell survival, limits apoptosis in the involuting mammary gland [84]. The regulation of both survival and death signals in the mammary epithelium is tightly controlled and it has been shown that levels of activated Akt may be reduced in response to Stat3 mediated expression of PI3 kinase negative regulatory subunits [83]. Epithelial cell death in the mouse mammary gland at this time also relies on Fas ligand (FasL), transforming growth factor β (TGFβ) signaling and macrophages [95, 96].

Degradation of basement membrane and ECM proteins define the second phase of mammary gland involution whereby the reduction of lobulo-alveolar structures occurs concomitantly with an upregulation in genes associated with gland remodeling [6, 94]. In mice, the second phase of post-lactational involution occurs 48 h after weaning, and is accompanied by an influx of a number of immune cells, including plasma cells, macrophages, neutrophils, and eosinophils [93]. Upon completion of involution (which takes approximately 3 weeks in the mouse), the mammary gland resembles that of a mature, quiescent virgin gland albeit with an increased number of lobular alveoli. However, the parous gland does contain larger lobules than a nonparous human gland and so is composed of more glandular tissue [13]. In addition, altered gene expression profiles have been documented for the parous and nulliparous human breast [97]. A cyclic process following pregnancies will reinitiate alveolar bud (or TDLU) expansion, lobulo-alveolar differentiation, and eventual involution.

A second form of involution occurs within the mammary gland, independent of a prior lactational event but correlates with age. Age-related lobular involution in the human mammary gland refers to a gradual loss of breast epithelial tissue, a natural process which, when complete, may be physiologically protective against breast cancer incidence. Mammary gland complexity and function diminish in this time and in the human gland, the first signs of lobular involution arise at perimenopause [98]. Involution accelerates during menopause and this phase is characterized by a reduction in size and complexity of both the ductal network and TDLUs [98]. Clinical data reveals that postmenopausal women with delayed lobular involution have a higher risk of developing breast cancer when compared to menopausal woman experiencing partial or complete involution [99].

This introduction highlights the morphological lifespan of the mouse mammary gland from embryonic development, to puberty, pregnancy/lactation, and involution. Where possible, an effort is made to draw parallels (and highlight differences) between mammary gland development in the mouse and human. The remainder of this work serves to ascertain current knowledge regarding a wide range of mammary gland areas of active research, and transgenic mouse models displaying mammary gland phenotypes are described.

References

1. Propper A, Gomot L (1967) Tissue interactions during organogenesis of the mammary gland in the rabbit embryo. C R Acad Sci Hebd Seances Acad Sci D 264(22):2573–2575
2. Daniel CW, Smith GH (1999) The mammary gland: a model for development. J Mammary Gland Biol Neoplasia 4(1):3–8
3. Humphreys RC, Krajewska M, Krnacik S, Jaeger R, Weiher H, Krajewski S, Reed JC, Rosen JM (1996) Apoptosis in the terminal endbud of the murine mammary gland: a mechanism of ductal morphogenesis. Development 122(12):4013–4022
4. Hennighausen L, Robinson GW (1998) Think globally, act locally: the making of a mouse mammary gland. Genes Dev 12(4):449–455
5. Strange R, Li F, Saurer S, Burkhardt A, Friis RR (1992) Apoptotic cell death and tissue remodelling during mouse mammary gland involution. Development 115(1):49–58
6. Richert MM, Schwertfeger KL, Ryder JW, Anderson SM (2000) An atlas of mouse mammary gland development. J Mammary Gland Biol Neoplasia 5(2):227–241
7. Parmar H, Cunha GR (2004) Epithelial-stromal interactions in the mouse and human mammary gland in vivo. Endocr Relat Cancer 11(3):437–458
8. Hennighausen L, Robinson GW (2001) Signaling pathways in mammary gland development. Dev Cell 1(4):467–475
9. Russo IH, Russo J (1996) Mammary gland neoplasia in long-term rodent studies. Environ Health Perspect 104(9):938–967
10. Bolander FF Jr (1990) Differential characteristics of the thoracic and abdominal mammary glands from mice. Exp Cell Res 189(1):142–144
11. Daniel CW, Silberstein GB, Strickland P (1987) Direct action of 17 beta-estradiol on mouse mammary ducts analyzed by sustained release

implants and steroid autoradiography. Cancer Res 47(22):6052–6057

12. Sakakura T, Kusano I, Kusakabe M, Inaguma Y, Nishizuka Y (1987) Biology of mammary fat pad in fetal mouse: capacity to support development of various fetal epithelia in vivo. Development 100(3):421–430

13. Russo J, Russo IH (2004) Development of the human breast. Maturitas 49(1):2–15. doi:10.1016/j.maturitas.2004.04.011

14. Kratochwil K (1971) In vitro analysis of the hormonal basis for the sexual dimorphism in the embryonic development of the mouse mammary gland. J Embryol Exp Morphol 25(1):141–153

15. Howard BA, Gusterson BA (2000) Human breast development. J Mammary Gland Biol Neoplasia 5(2):119–137

16. Veltmaat JM, Mailleux AA, Thiery JP, Bellusci S (2003) Mouse embryonic mammogenesis as a model for the molecular regulation of pattern formation. Differentiation 71(1):1–17. doi:10.1046/j.1432-0436.2003.700601.x

17. Anbazhagan R, Osin PP, Bartkova J, Nathan B, Lane EB, Gusterson BA (1998) The development of epithelial phenotypes in the human fetal and infant breast. J Pathol 184(2):197–206. doi:10.1002/(SICI)1096-9896(199802)184:2<197::AID-PATH992>3.0.CO;2-J

18. Petersen OW, Polyak K (2010) Stem cells in the human breast. Cold Spring Harb Perspect Biol 2(5):a003160. doi:10.1101/cshperspect.a003160

19. Sun P, Yuan Y, Li A, Li B, Dai X (2010) Cytokeratin expression during mouse embryonic and early postnatal mammary gland development. Histochem Cell Biol 133(2):213–221. doi:10.1007/s00418-009-0662-5

20. Chu EY, Hens J, Andl T, Kairo A, Yamaguchi TP, Brisken C, Glick A, Wysolmerski JJ, Millar SE (2004) Canonical WNT signaling promotes mammary placode development and is essential for initiation of mammary gland morphogenesis. Development 131(19):4819–4829. doi:10.1242/dev.01347

21. Eblaghie MC, Song SJ, Kim JY, Akita K, Tickle C, Jung HS (2004) Interactions between FGF and Wnt signals and Tbx3 gene expression in mammary gland initiation in mouse embryos. J Anat 205(1):1–13. doi:10.1111/j.0021-8782.2004.00309.x

22. Veltmaat JM, Van Veelen W, Thiery JP, Bellusci S (2004) Identification of the mammary line in mouse by Wnt10b expression. Dev Dyn 229(2):349–356. doi:10.1002/dvdy.10441

23. Mailleux AA, Spencer-Dene B, Dillon C, Ndiaye D, Savona-Baron C, Itoh N, Kato S, Dickson C, Thiery JP, Bellusci S (2002) Role of FGF10/FGFR2b signaling during mam-

mary gland development in the mouse embryo. Development 129(1):53–60

24. Foley J, Dann P, Hong J, Cosgrove J, Dreyer B, Rimm D, Dunbar M, Philbrick W, Wysolmerski J (2001) Parathyroid hormone-related protein maintains mammary epithelial fate and triggers nipple skin differentiation during embryonic breast development. Development 128(4):513–525

25. Hovey RC, Harris J, Hadsell DL, Lee AV, Ormandy CJ, Vonderhaar BK (2003) Local insulin-like growth factor-II mediates prolactin-induced mammary gland development. Mol Endocrinol 17(3):460–471. doi:10.1210/me.2002-0214

26. Asch HL, Asch BB (1985) Expression of keratins and other cytoskeletal proteins in mouse mammary epithelium during the normal developmental cycle and primary culture. Dev Biol 107(2):470–482

27. Radice GL, Ferreira-Cornwell MC, Robinson SD, Rayburn H, Chodosh LA, Takeichi M, Hynes RO (1997) Precocious mammary gland development in P-cadherin-deficient mice. J Cell Biol 139(4):1025–1032

28. Ingman WV, Robertson SA (2008) Mammary gland development in transforming growth factor beta1 null mutant mice: systemic and epithelial effects. Biol Reprod 79(4):711–717

29. Russo IH, Russo J (2011) Pregnancy-induced changes in breast cancer risk. J Mammary Gland Biol Neoplasia 16(3):221–233. doi:10.1007/s10911-011-9228-y

30. Brisken C, Duss S (2007) Stem cells and the stem cell niche in the breast: an integrated hormonal and developmental perspective. Stem Cell Rev 3(2):147–156

31. Sternlicht MD, Kouros-Mehr H, Lu P, Werb Z (2006) Hormonal and local control of mammary branching morphogenesis. Differentiation 74(7):365–381

32. Silberstein GB, Strickland P, Coleman S, Daniel CW (1990) Epithelium-dependent extracellular matrix synthesis in transforming growth factor-beta 1-growth-inhibited mouse mammary gland. J Cell Biol 110(6):2209–2219

33. McNally S, Martin F (2011) Molecular regulators of pubertal mammary gland development. Ann Med 43(3):212–234. doi:10.3109/07853890.2011.554425

34. Topper YJ, Freeman CS (1980) Multiple hormone interactions in the developmental biology of the mammary gland. Physiol Rev 60:1049–1056

35. Brisken C (2002) Hormonal control of alveolar development and its implications for breast carcinogenesis. J Mammary Gland Biol Neoplasia 7(1):39–48

36. Deroo BJ, Hewitt SC, Collins JB, Grissom SF, Hamilton KJ, Korach KS (2009) Profile of estrogen-responsive genes in an estrogen-specific mammary gland outgrowth model. Mol Reprod Dev 76(8):733–750

37. Couse JF, Korach KS (1999) Estrogen receptor null mice: what have we learned and where will they lead us? Endocr Rev 20(3):358–417

38. Bocchinfuso WP, Korach KS (1997) Mammary gland development and tumorigenesis in estrogen receptor knockout mice. J Mammary Gland Biol Neoplasia 2(4):323–334

39. Mallepell S, Krust A, Chambon P, Brisken C (2006) Paracrine signaling through the epithelial estrogen receptor alpha is required for proliferation and morphogenesis in the mammary gland. Proc Natl Acad Sci U S A 103(7):2196–2201

40. LaMarca HL, Rosen JM (2008) Minireview: hormones and mammary cell fate – what will I become when I grow up? Endocrinology 149(9):4317–4321

41. Britt K, Ashworth A, Smalley M (2007) Pregnancy and the risk of breast cancer. Endocr Relat Cancer 14(4):907–933. doi:10.1677/ERC-07-0137

42. Atwood CS, Hovey RC, Glover JP, Chepko G, Ginsburg E, Robison WG, Vonderhaar BK (2000) Progesterone induces side-branching of the ductal epithelium in the mammary glands of peripubertal mice. J Endocrinol 167(1):39–52

43. Silberstein GB (2001) Postnatal mammary gland morphogenesis. Microsc Res Tech 52(2):155–162. doi:10.1002/1097-0029(20010115)52:2<155::AID-JEMT1001>3.0.CO;2-P

44. Nelson CM, Vanduijn MM, Inman JL, Fletcher DA, Bissell MJ (2006) Tissue geometry determines sites of mammary branching morphogenesis in organotypic cultures. Science (New York, NY) 314(5797):298–300

45. Olijnyk D, Ibrahim AM, Ferrier RK, Tsuda T, Chu ML, Gusterson BA, Stein T, Morris JS (2014) Fibulin-2 is involved in early extracellular matrix development of the outgrowing mouse mammary epithelium. Cell Mol Life Sci 71(19):3811–3828. doi:10.1007/s00018-014-1577-4

46. Kamalati T, Niranjan B, Yant J, Buluwela L (1999) HGF/SF in mammary epithelial growth and morphogenesis: in vitro and in vivo models. J Mammary Gland Biol Neoplasia 4(1):69–77

47. Niranjan B, Buluwela L, Yant J, Perusinghe N, Atherton A, Phippard D, Dale T, Gusterson B, Kamalati T (1995) HGF/SF: a potent cytokine for mammary growth, morpho-genesis and development. Development 121(9):2897–2908

48. Soriano JV, Pepper MS, Orci L, Montesano R (1998) Roles of hepatocyte growth factor/scatter factor and transforming growth factor-beta1 in mammary gland ductal morphogenesis. J Mammary Gland Biol Neoplasia 3(2):133–150

49. Pollard JW (2001) Tumour-stromal interactions. Transforming growth factor-beta isoforms and hepatocyte growth factor/scatter factor in mammary gland ductal morphogenesis. Breast Cancer Res 3(4):230–237

50. Andres AC, Strange R (1999) Apoptosis in the estrous and menstrual cycles. J Mammary Gland Biol Neoplasia 4(2):221–228

51. Russo J, Gusterson BA, Rogers AE, Russo IH, Wellings SR, van Zwieten MJ (1990) Comparative study of human and rat mammary tumorigenesis. Lab Invest 62(3):244–278

52. Visvader JE (2009) Keeping abreast of the mammary epithelial hierarchy and breast tumorigenesis. Genes Dev 23(22):2563–2577. doi:10.1101/gad.1849509

53. Moll R, Franke WW, Schiller DL, Geiger B, Krepler R (1982) The catalog of human cytokeratins: patterns of expression in normal epithelia, tumors and cultured cells. Cell 31(1):11–24

54. Nandi S (1958) Endocrine control of mammary-gland development and function in the C3H/He Crgl mouse. J Natl Cancer Inst 21(6):1039–1063

55. Neville MC, McFadden TB, Forsyth I (2002) Hormonal regulation of mammary differentiation and milk secretion. J Mammary Gland Biol Neoplasia 7(1):49–66

56. Traurig HH (1967) A radioautographic study of cell proliferation in the mammary gland of the pregnant mouse. Anat Rec 159(2):239–247. doi:10.1002/ar.1091590213

57. Brisken C, Park S, Vass T, Lydon JP, O'Malley BW, Weinberg RA (1998) A paracrine role for the epithelial progesterone receptor in mammary gland development. Proc Natl Acad Sci U S A 95(9):5076–5081

58. Humphreys RC, Lydon J, O'Malley BW, Rosen JM (1997) Mammary gland development is mediated by both stromal and epithelial progesterone receptors. Mol Endocrinol 11(6):801–811. doi:10.1210/mend.11.6.9891

59. Conneely OM, Mulac-Jericevic B, Lydon JP (2003) Progesterone-dependent regulation of female reproductive activity by two distinct progesterone receptor isoforms. Steroids 68(10-13):771–778

60. Bradbury JM, Edwards PA, Niemeyer CC, Dale TC (1995) Wnt-4 expression induces a preg-

nancy-like growth pattern in reconstituted mammary glands in virgin mice. Dev Biol 170(2):553–563. doi:10.1006/dbio.1995.1236

61. Fernandez-Valdivia R, Mukherjee A, Ying Y, Li J, Paquet M, DeMayo FJ, Lydon JP (2009) The RANKL signaling axis is sufficient to elicit ductal side-branching and alveologenesis in the mammary gland of the virgin mouse. Dev Biol 328(1):127–139. doi:10.1016/j.ydbio.2009.01.019

62. Brisken C, Heineman A, Chavarria T, Elenbaas B, Tan J, Dey SK, McMahon JA, McMahon AP, Weinberg RA (2000) Essential function of Wnt-4 in mammary gland development downstream of progesterone signaling. Genes Dev 14(6):650–654

63. Fata JE, Kong YY, Li J, Sasaki T, Irie-Sasaki J, Moorehead RA, Elliott R, Scully S, Voura EB, Lacey DL, Boyle WJ, Khokha R, Penninger JM (2000) The osteoclast differentiation factor osteoprotegerin-ligand is essential for mammary gland development. Cell 103(1):41–50

64. Mulac-Jericevic B, Lydon JP, DeMayo FJ, Conneely OM (2003) Defective mammary gland morphogenesis in mice lacking the progesterone receptor B isoform. Proc Natl Acad Sci U S A 100(17):9744–9749. doi:10.1073/pnas.1732707100

65. Brisken C, Kaur S, Chavarria TE, Binart N, Sutherland RL, Weinberg RA, Kelly PA, Ormandy CJ (1999) Prolactin controls mammary gland development via direct and indirect mechanisms. Dev Biol 210(1):96–106. doi:10.1006/dbio.1999.9271

66. Ormandy CJ, Naylor M, Harris J, Robertson F, Horseman ND, Lindeman GJ, Visvader J, Kelly PA (2003) Investigation of the transcriptional changes underlying functional defects in the mammary glands of prolactin receptor knockout mice. Recent Prog Horm Res 58:297–323

67. Harris J, Stanford PM, Sutherland K, Oakes SR, Naylor MJ, Robertson FG, Blazek KD, Kazlauskas M, Hilton HN, Wittlin S, Alexander WS, Lindeman GJ, Visvader JE, Ormandy CJ (2006) Socs2 and elf5 mediate prolactin-induced mammary gland development. Mol Endocrinol 20(5):1177–1187. doi:10.1210/me.2005-0473

68. Adams JC, Watt FM (1993) Regulation of development and differentiation by the extracellular matrix. Development 117(4):1183–1198

69. Howlett AR, Bissell MJ (1993) The influence of tissue microenvironment (stroma and extracellular matrix) on the development and function of mammary epithelium. Epithelial Cell Biol 2(2):79–89

70. Fata JE, Werb Z, Bissell MJ (2004) Regulation of mammary gland branching morphogenesis by the extracellular matrix and its remodeling enzymes. Breast Cancer Res 6(1):1–11. doi:10.1186/bcr634

71. Kouros-Mehr H, Slorach EM, Sternlicht MD, Werb Z (2006) GATA-3 maintains the differentiation of the luminal cell fate in the mammary gland. Cell 127(5):1041–1055. doi:10.1016/j.cell.2006.09.048

72. Nguyen DA, Parlow AF, Neville MC (2001) Hormonal regulation of tight junction closure in the mouse mammary epithelium during the transition from pregnancy to lactation. J Endocrinol 170(2):347–356

73. Russo J, Russo IH (1987) The mammary gland: development, regulation and function. Plenum, New York, NY

74. Lee S, Medina D, Tsimelzon A, Mohsin SK, Mao S, Wu Y, Allred DC (2007) Alterations of gene expression in the development of early hyperplastic precursors of breast cancer. Am J Pathol 171(1):252–262

75. LaMarca HL, Rosen JM (2007) Estrogen regulation of mammary gland development and breast cancer: amphiregulin takes center stage. Breast Cancer Res 9(4):304

76. Johnston SR (2006) Clinical efforts to combine endocrine agents with targeted therapies against epidermal growth factor receptor/human epidermal growth factor receptor 2 and mammalian target of rapamycin in breast cancer. Clin Cancer Res 12(3):1061–1068

77. Chen CC, Stairs DB, Boxer RB, Belka GK, Horseman ND, Alvarez JV, Chodosh LA (2012) Autocrine prolactin induced by the Pten-Akt pathway is required for lactation initiation and provides a direct link between the Akt and Stat5 pathways. Genes Dev 26(19):2154–2168. doi:10.1101/gad.197343.112

78. Wagner KU, Krempler A, Triplett AA, Qi Y, George NM, Zhu J, Rui H (2004) Impaired alveologenesis and maintenance of secretory mammary epithelial cells in Jak2 conditional knockout mice. Mol Cell Biol 24(12):5510–5520. doi:10.1128/MCB.24.12.5510-5520.2004

79. Brisken C, Ayyannan A, Nguyen C, Heineman A, Reinhardt F, Tan J, Dey SK, Dotto GP, Weinberg RA (2002) IGF-2 is a mediator of prolactin-induced morphogenesis in the breast. Dev Cell 3(6):877–887

80. Prilusky J, Deis RP (1975) Effect of L-dopa on milk ejection and prolactin release in lactating rats. J Endocrinol 67(3):397–401

81. Mather IH, Keenan TW (1998) Origin and secretion of milk lipids. J Mammary Gland Biol Neoplasia 3(3):259–273

82. Neville MC (1999) Physiology of lactation. Clin Perinatol 26(2):251–279, v

83. Abell K, Bilancio A, Clarkson RW, Tiffen PG, Altaparmakov AI, Burdon TG, Asano T, Vanhaesebroeck B, Watson CJ (2005) Stat3-induced apoptosis requires a molecular switch in PI(3)K subunit composition. Nat Cell Biol 7(4):392–398. doi:10.1038/ncb1242

84. Schwertfeger KL, Richert MM, Anderson SM (2001) Mammary gland involution is delayed by activated Akt in transgenic mice. Mol Endocrinol 15(6):867–881. doi:10.1210/mend.15.6.0663

85. Li M, Liu X, Robinson G, Bar-Peled U, Wagner KU, Young WS, Hennighausen L, Furth PA (1997) Mammary-derived signals activate programmed cell death during the first stage of mammary gland involution. Proc Natl Acad Sci U S A 94(7):3425–3430

86. Feng Z, Marti A, Jehn B, Altermatt HJ, Chicaiza G, Jaggi R (1995) Glucocorticoid and progesterone inhibit involution and programmed cell death in the mouse mammary gland. J Cell Biol 131(4):1095–1103

87. Kreuzaler PA, Staniszewska AD, Li W, Omidvar N, Kedjouar B, Turkson J, Poli V, Flavell RA, Clarkson RW, Watson CJ (2011) Stat3 controls lysosomal-mediated cell death in vivo. Nat Cell Biol 13(3):303–309. doi:10.1038/ncb2171

88. Monks J, Rosner D, Geske FJ, Lehman L, Hanson L, Neville MC, Fadok VA (2005) Epithelial cells as phagocytes: apoptotic epithelial cells are engulfed by mammary alveolar epithelial cells and repress inflammatory mediator release. Cell Death Differ 12(2):107–114. doi:10.1038/sj.cdd.4401517

89. Monks J, Smith-Steinhart C, Kruk ER, Fadok VA, Henson PM (2008) Epithelial cells remove apoptotic epithelial cells during post-lactation involution of the mouse mammary gland. Biol Reprod 78(4):586–594. doi:10.1095/biolreprod.107.065045

90. Clarkson RW, Wayland MT, Lee J, Freeman T, Watson CJ (2004) Gene expression profiling of mammary gland development reveals putative roles for death receptors and immune mediators in post-lactational regression. Breast Cancer Res 6(2):R92–R109. doi:10.1186/bcr754

91. Marti A, Jehn B, Costello E, Keon N, Ke G, Martin F, Jaggi R (1994) Protein kinase A and AP-1 (c-Fos/JunD) are induced during apoptosis of mouse mammary epithelial cells. Oncogene 9(4):1213–1223

92. Marti A, Lazar H, Ritter P, Jaggi R (1999) Transcription factor activities and gene expression during mouse mammary gland involution. J Mammary Gland Biol Neoplasia 4(2):145–152

93. Stein T, Morris JS, Davies CR, Weber-Hall SJ, Duffy MA, Heath VJ, Bell AK, Ferrier RK, Sandilands GP, Gusterson BA (2004) Involution of the mouse mammary gland is associated with an immune cascade and an acute-phase response, involving LBP, CD14 and STAT3. Breast Cancer Res 6(2):R75–R91. doi:10.1186/bcr753

94. Lund LR, Romer J, Thomasset N, Solberg H, Pyke C, Bissell MJ, Dano K, Werb Z (1996) Two distinct phases of apoptosis in mammary gland involution: proteinase-independent and -dependent pathways. Development 122(1):181–193

95. Watson CJ, Kreuzaler PA (2011) Remodeling mechanisms of the mammary gland during involution. Int J Dev Biol 55(7-9):757–762. doi:10.1387/ijdb.113414cw

96. O'Brien J, Martinson H, Durand-Rougely C, Schedin P (2012) Macrophages are crucial for epithelial cell death and adipocyte repopulation during mammary gland involution. Development 139(2):269–275. doi:10.1242/dev.071696

97. Balogh GA, Heulings R, Mailo DA, Russo PA, Sheriff F, Russo IH, Moral R, Russo J (2006) Genomic signature induced by pregnancy in the human breast. Int J Oncol 28(2):399–410

98. Hutson SW, Cowen PN, Bird CC (1985) Morphometric studies of age related changes in normal human breast and their significance for evolution of mammary cancer. J Clin Pathol 38(3):281–287

99. Milanese TR, Hartmann LC, Sellers TA, Frost MH, Vierkant RA, Maloney SD, Pankratz VS, Degnim AC, Vachon CM, Reynolds CA, Thompson RA, Melton LJ 3rd, Goode EL, Visscher DW (2006) Age-related lobular involution and risk of breast cancer. J Natl Cancer Inst 98(22):1600–1607. doi:10.1093/jnci/djj439

Part I

In Vivo Models

Chapter 2

Prenatal Mammary Gland Development in the Mouse: Research Models and Techniques for Its Study from Past to Present

Jacqueline M. Veltmaat

Abstract

Mammary gland development starts during prenatal life, when at designated positions along the ventrolateral boundary of the embryonic or fetal trunk, surface ectodermal cells coalesce to form primordia for mammary glands, instead of differentiating into epidermis. With the wealth of genetically engineered mice available as research models, our understanding of the prenatal phase of mammary development has recently greatly advanced. This understanding includes the recognition of molecular and mechanistic parallels between prenatal and postnatal mammary morphogenesis and even tumorigenesis, much of which can moreover be extrapolated to human. This makes the murine embryonic mammary gland a useful model for a myriad of questions pertaining to normal and pathological breast development. Hence, unless indicated otherwise, this review describes embryonic mammary gland development in mouse only, and lists mouse models that have been examined for defects in embryonic mammary development. Techniques that originated in the field of developmental biology, such as explant culture and tissue recombination, were adapted specifically to research on the embryonic mammary gland. Detailed protocols for these techniques have recently been published elsewhere. This review describes how the development and adaptation of these techniques moved the field forward from insights on (comparative) morphogenesis of the embryonic mammary gland to the understanding of tissue and molecular interactions and their regulation of morphogenesis and functional development of the embryonic mammary gland. It is here furthermore illustrated how generic molecular biology and biochemistry techniques can be combined with these older, developmental biology techniques, to address relevant research questions. As such, this review should provide a solid starting point for those wishing to familiarize themselves with this fascinating and important subdomain of mammary gland biology, and guide them in designing a relevant research strategy.

Key words Mouse embryo, Mammary gland development, Techniques, Mouse models, Explant cultures, Tissue recombination, Tissue interactions, Molecular interactions, Morphogenesis

1 Introduction

Already around 350 BCE Aristotle had documented that some but not all terrestrial and marine animal species have special milk-producing glands, usually with a teat or nipple as outlet, to which

Finian Martin et al. (eds.), *Mammary Gland Development: Methods and Protocols*, Methods in Molecular Biology, vol. 1501, DOI 10.1007/978-1-4939-6475-8_2, © Springer Science+Business Media New York 2017

the newborn can latch on for its feeding [1]. In some species he observed those glands only in females, in other species in both males and females [1], but even though males may lactate, e.g., in bats [2, 3] only females were observed to nurse the young. Perhaps this explains why these glands are called "mammary glands," as a referral to the word "mama" or "mamma" for mother. Over 2000 years later, Linnaeus used the possession of mammary glands as the defining feature for a separate Class of animals, named Mammalia after the mammary gland [4].

Mammary glands are apocrine glands that reside on the ventral side of the trunk of adult mammals; most often they are present in pairs of which the singletons are displaced more or less symmetrically away from the ventral midline. In monotremes (platypus and echidna), each gland exists as one lobule budding off a single duct which is connected to a hair shaft [5]. Due to its small size, its milk producing capacity is low. Furthermore, in the absence of nipple or teat, the milk seeps out along the hair to be licked up by the newborn [6]. This type of gland and mode of excretion may reflect ancestral glands that birds used to moisten their eggs [5, 7], but it is relatively inefficient for nursing newborns. The low milk production per gland and wastage of milk is compensated by a high number (between 100 and 150 pairs) of glands in monotremes.

Mammary glands of marsupial (e.g., kangaroos) and placental (e.g., humans, whales) mammals have a large internal surface of secretory cells, owing to reiterated branching of the primary duct. Moreover, as all the milk of one gland drains to one teat or nipple from which the newborn can suckle, milk spillage is minimized. This generally ensures sufficient milk production per gland to feed one newborn. Compared to monotremes, marsupials and placentals can therefore do with fewer mammary glands. Indeed, their number of pairs of mammary glands ranges between 1 and 25 [8], in a correlation close to 1 for "average litter size" to "number of mammary gland pairs" across species [9]. With their maximum litter size seldom exceeding twice the average litter size, this ratio generally still leaves one gland available per newborn.

Interestingly, even if the number of mammary gland pairs is the same between some species, the location of these glands may differ between these species. For example, elephants, humans, and horses each have one pair of mammary glands, which is located at the chest in elephants and humans, but near the hind leg in the horse. This variation in position of the glands along the anteroposterior body axis seems to correspond to habitat, method of rearing, and degree of maturity of the offspring at birth [6].

Why would it be relevant to study the prenatal phase of mammary gland development? First of all, the mother's milk is the only source of nutrients for the newborn, and provides antibodies and other immune support as well until the newborn's own immune system becomes active [10–12]. Though humans may substitute their own breast milk by formula, most formula is still a dairy product. As such, mammary glands are directly crucial to the survival of mammalian species; and indirectly as well, through the close bond that nursing forges between the newborn and its mother. Even though the gland's milk-producing function is not required before adulthood, almost all aspects of mammary morphogenesis and functional differentiation already take place before birth. It is therefore not surprising that throughout the centuries, zoologists found the prenatal phase of mammary gland development important for study.

Moreover, almost all aspects of mammary morphogenesis and functional differentiation already take place before birth, only to be reiterated or enhanced postnatally under the influence of puberty and pregnancy hormones. Downstream of these hormones seem to act many of the signaling cascades that regulate prenatal mammary development [13–16]. Even stem cells, which are required to regenerate the mammary gland with each pregnancy, are already present in the prenatal gland [17–19]. As the prenatal mammary gland is relatively accessible for experimentation and is less complex in tissue composition than the adult mammary gland, it may be a practical additional or alternative research model for research questions pertaining to development of the postnatal mammary gland.

The regulation of the variation in number and position of the mammary glands raises additional interesting questions for developmental biologists about regulatory mechanisms creating this variation. For the high degree of similarity in shape and function between the multiple pairs of mammary glands in for example cats or pigs would suggest these glands are mere copies of each other. Yet the variation in number and position of glands between and even within species, and the heritable propensity for having too few or many mammary glands in for example sheep, pigs, humans, and macaques [20–26] indicates that each mammary gland must have some unique genetic component or protein activity that determines whether its development will be initiated and continued or not. Insights in these differences between the pairs and even between the left and right counterparts of each pair [27] may affect our thinking about the extrapolation of results obtained with one gland to other glands.

Of particular interest are the parallels in tissue interactions and molecular activity between prenatal mammogenesis and mammary tumorigenesis and metastasis [28–33]. Although better screening, care and treatment options for breast cancer have improved survival chances for patients with breast cancer over the past twenty years, this cancer is still the second leading cancer-related cause of death for women worldwide [34, 35]. Progress in finding even better therapies is impeded by the wide heterogeneity in the molecular mechanisms of the wide variety of breast cancer types, only 2–10 % of which seems to have a familial component [36, 37]. As embryonic mammary glands are less complex and heterogeneous in tissue composition than adult mammary glands and tumors, and are easily accessible and available, new candidates for nonfamilial forms of breast cancer may be identified through the study of prenatal mammary gland development [28, 38].

For obvious reasons of ethics, human fetuses are insufficiently available for such studies. Comparative studies from the past have revealed that prenatal mammary gland development in rabbit embryos closely resembles that in human fetuses [39]. Nonetheless, currently most research on prenatal mammary glands is done in mice, and some of the techniques are optimized for use on his research model in particular, despite a few morphogenic differences in mammary development between men, rabbit, and mice [40, 41]. The choice for mice is largely based on the wealth of genetically engineered mice becoming available since 1989 [42]. Several of the genes that have so far been identified as regulators of early mammary gland development in the mouse embryo are known to also underlie defects in prenatal mammary development in humans [33, 43, 44]. Those findings validate the use of mouse embryos as a model for human prenatal mammary development.

Therefore, this review focuses primarily on mammary development in mouse embryos. It takes the approximate chronological order in which techniques were developed and used to study developmental biology, as a basis to describe how insights were gained in the different aspects of mammary gland development in mouse embryos.

2 Macroscopic and Microscopic Aspects of Prenatal Morphogenesis of the Mammary Gland in Mouse

From the mid-nineteenth century onwards there has been a steady stream of publications pertaining to embryonic mammary gland development in a broad variety of mammalian species. The earliest studies were based on macroscopic analysis of embryos to assess the number, positions, and external morphology of mammary glands, and microscopic analysis to study tissue composition and internal morphology of mammary glands at different embryonic ages.

Determination of embryonic age: For many species, embryos were obtained by chance without knowledge of the onset of pregnancy and age of the embryo. Size (e.g., crown–rump length) or weight measurements of embryos of different mothers were used to assess the relative chronological age between embryos of different pregnancies. Although this is a helpful method in the absence of knowledge of the onset of pregnancy, size and weight are not precise determinants of (relative) chronological age, due to the normal variation in size and weight of embryos at any given developmental age.

Already in the early nineteenth century, rats, rabbits, and mice were kept in captivity for research purposes [45]. In captivity, the onset of pregnancy can be controlled. If the day–night (light–dark) cycle is kept regular, female mice in estrous will ovulate at around the middle of dark time, and produce more pheromones that entice the male to copulate. Copulation results in production of a sturdy white vaginal plug in the female that remains present for about half a day. Nowadays, in a laboratory setting, the middle of the dark time is often conveniently set to be around midnight. Therefore, noon of the day a vaginal plug is observed, is usually considered embryonic day 0.5 (E0.5), assuming copulation resulted in a pregnancy. The female is then separated from the male, and monitored for signs of pregnancy. Embryos are collected at the desired age for study. Embryos of the same pregnancy, thus same chronological age, will differ in their true developmental age. The relative developmental stage of embryos within one batch can be assessed by their progress in a developmental process that is particular for that chronological age, e.g., the number of somites between E8 and E12, and number of branches of the salivary gland at E13, unless one compares wild type embryos with littermates that carry a genetic mutation that disturbs the developmental process that is used for staging. Note that in the older literature, and occasionally in current publications, the progress of pregnancy is counted only in full days, and some may consider the day a plug is observed as embryonic day 0 (E0), while others consider it day 1 (E1). This may lead to small discrepancies in the literature regarding the timing of morphogenetic events of mammary morphogenesis. The different speeds of embryonic development between different mouse strains may be another source of small discrepancies in the literature regarding the timing of morphogenetic events.

Histology: Early descriptions of the murine mammary glands were based on microscopic analysis of histology, for which embryos were treated with a fixative, dehydrated, embedded in wax, and sliced into sections with a minimum thickness of 4 μm, and stained with a variety of chemical solutions to facilitate the recognition of different cell or tissue components (nucleus, cytoplasma, extracellular matrix fibers, etc.) [45–48]. From around the 1970s–1980s, histology was also performed on frozen sections, or specimens were embedded in a plastic or epoxy resin, to cut semi-thin (1 μm) sections which provide a higher resolution of intracellular structures [49, 50]. Such histological studies led to most of the insights about morphogenesis as described further below.

Electron microscopy: From the 1970s onwards a few studies incorporated scanning electron microscopy of whole embryos to analyze changes on the surface of the embryo associated with mammary development [51–54].

Microscopy of whole glands: Embryonic skins can also be peeled off the embryo and mounted on a microscope slide for examination of gross morphology of the rudimentary glands under bright field stereoscopy. When the skin is peeled off sufficiently thin, transmitted light allows recognition of the rudimentary gland without further treatment of the specimen. After E16.5 the mouse epidermis becomes keratinized and subdermal fat develops. Visualization of the mammary rudiments (MRs) can then be enhanced by defatting and staining the skins with carmine alum [52, 53], according to a protocol routinely used for adult mammary glands [55].

3D-reconstruction of mammary rudiments: Recently, the application of bioinformatics and image analysis to digital images of histological preparations, or optical sectioning of intact fluorescently labeled MRs has allowed to generate 3D-constructions of complete series of (optical) serial sections through mammary rudiments [54, 56–58]. Different tissue components or differently labeled cell types can be identified manually or automatically, allowing measurements of volume, and proportions of different cell populations as well as recognition of regionalized distribution of specific cell populations within the MR [54].

2.1 Brief Overview of Macroscopic and Microscopic Aspects of Embryonic Morphogenesis

This section will only briefly describe the morphogenetic stages in mouse embryos, just to introduce the terminology and concepts of the field and facilitate the understanding of the subsequent passages of this current review. For more details on morphogenesis, the reader is referred to previously published reviews [41, 47, 53, 59].

Mammary gland development takes place along the ventrolateral boundaries in the surface ectoderm (i.e., the prospective epidermis) of the embryonic trunk. One could draw an imaginary line called *mammary line* or milk line (ML) extending from axilla (armpit) and inguen (groin) along both boundaries (Fig. 1). These boundaries are histologically detectable in the surface ectoderm as the junctions between squamous cells on the ventrum (belly) and cuboidal cells on both flanks.

In the course of the tenth day of mouse embryogenesis (E10.5), cuboidal cells along the two MLs first elongate to a columnar shape, rapidly followed by multilayering [45, 60, 61]. This cell elongation and multilayering occurs in three separate *mammary streaks* per ML: One extends between the forelimb and hindlimb and is approximately 30 cells wide, while separate streaks develop in the

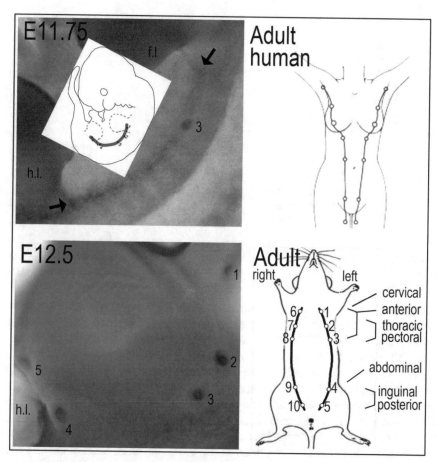

Fig. 1 Position of the mammary line and rudiments in embryo and adult. The *left two panels* show mouse embryos at E11.75 and E12.5, hybridized with a *Wnt10b* probe which visualizes the mammary line (between *arrows*) and rudiments (*numbered*). The *inset cartoon* shows how the mammary line extends from axilla, along the flank, to inguin. In adult mice and humans imaginary mammary lines can still be drawn more ventrally, connecting all sites where mammary glands reside, normally five pairs in mouse and one pair in human, but occasionally supernumerary mammary glands develop at other sites as indicated with *open circles*. In mouse, the embryonic mammary rudiments are usually numbered as pairs 1 through 5 in anteroposterior order, but in adult mice the glands may be indicated by individual number (1–10), or position on the trunk. *fl* forelimb, *hl* hindlimb. Adapted from ref. [27], with permission

axilla and inguen [61]. This marks the onset of mammogenesis. These streaks extend towards each other, and ultimately represent one continuous histologically detectable *mammary line* on each flank (Fig. 1). In species like rabbit, the MLs rapidly become elevated above the surface ectodermal landscape and are therefore called mammary ridges [41, 51]. At designated positions along the left and right ML, mammary glands will develop as symmetrically located pairs, of which the number varies in a species-dependent manner. They undergo a series of morphological changes or stages with each their own name as described below and depicted in Fig. 2.

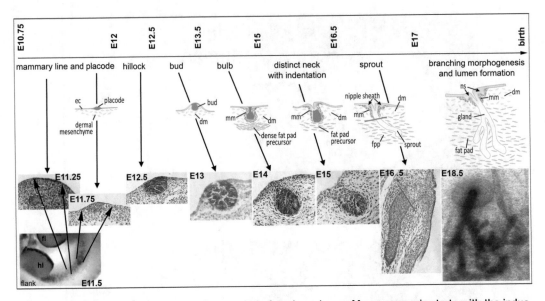

Fig. 2 Stages of mouse mammary morphogenesis in female embryos. Mammogenesis starts with the induction of mammary streaks which fuse into a continuous line from E10.75 onwards (shown as a lateral view on the flank of a TOPGAL-F stained E11.5 embryo) while at designated positions placodes are formed asynchronously before E12. Histological sections of TOPGAL-F stained embryos and cartoons with the blue TOPGAL-positive domains, illustrate how the epithelial mammary placodes transform to hillocks and spherical buds that are first raised above the landscape of surface ectoderm (ec), but by E13.5 they subside below the surface. By then, some mammary rudiments have acquired a bulb-shape, and a few layers of contiguous dermal mesenchyme (dm) condense around all buds/bulbs to become mammary mesenchyme (mm). By E14.5, subdermal mesenchyme differentiates into the dense fat pad precursor. Then the neck area of the mammary epithelium begins to differentiate and forms a funnel-shaped indentation as the future outlet of the milk canal. Around E16, the tip of the bud/bulb breaks through the mammary mesenchyme and invades the fat pad precursor, while a nipple sheath develops at the neck area. Within a day, branching morphogenesis and canalization occur, such that the mammary gland resembles a miniature mammary gland before birth, as shown with a carmine-red stained fragment of an E18.5 skin with gland. Adapted from refs. [41, 53], with permission

The embryonic mammary gland(s) may be called *mammary primordium* (primordia), *mammary anlage(n)*, or *mammary rudiment(s)* (MRs) in reference to any developmental stage or none in particular. They include the mammary epithelium (ME), mammary mesenchyme (MM), and fat pad (FP), as these tissues develop in an interdependent manner.

In mouse embryos, multilayering is advanced in one subdomain per mammary streak and at the subaxillary and suprainguinal junction of these streaks. Between E11 and E12, five pairs of lentil-shaped mammary *placodes* arise in the axillae (MR1), at a subaxillary position (MR2), at the level of the diaphragm (MR3), at a suprainguinal position (MR4), and in the inguinae (MR5). Intriguingly, they arise asynchronously, not in numerical order, and independently of each other [41, 52, 53, 61, 62]. By increased multilayering, each placode becomes a *hillock* within half a day, slightly elevated in the ectodermal landscape [45]. Each hillock grows larger and changes shape into a spherical *bud* between E12 and E13, still elevated above the adjacent ectoderm [53]. During that day, the ML disappears as a histologically and molecularly detectable entity [60], but even at later stages the name *mammary line* may still be used to refer to the imaginary line that connects all mammary glands on one flank.

Subsequently the buds invaginate deeper into the underlying dermal mesenchyme, such that they are no longer elevated above the ectodermal landscape [45, 47, 53, 63]. From bud-stage, some MRs simply elongate (MR2, MR4, and MR5) while in MR1 and MR3, the proximal part which connects to the overlying ectoderm, takes on the shape of a narrow neck, such that these MRs each resemble a *bulb* (MR1 and MR3) [54]. Meanwhile, the contiguous layers of dermal mesenchyme condense concentrically and differentiate into a specialized fibroblastic mesenchyme called *dense* or *primary mammary mesenchyme* or simply *mammary mesenchyme* (MM) by E13.5 [64]. Between then and E15.5 the mesenchyme around the neck of the MRs in male embryos condenses. The spherical part of most/all MRs becomes disconnected from the epidermis and nipples fail to form in males [45, 47]. In most mammalian species including human such drastic sex-specific differences do not occur.

In E13.5 female mouse embryos the MRs remain intact and continue to grow, though slowly, over the next two days. Meanwhile, around E14.5, a subdermal layer of mesenchyme condenses and differentiates into the *secondary mammary mesenchyme* or *dense fat pad precursor* (FP) consisting of presumptive adipocytes, fibroblast, endothelial cells, nerve cells, and perhaps other cell types [65]. By E15.5 rapid proliferation of ME cells provides a growth spurt particularly at the distal end of each bulb [46], which elongates and breaks through its surrounding basal lamina and *primary mammary mesenchyme*. At that time, the position of the MR is outwardly visible as a funnel-shaped depression in the skin; the position of the future nipple [53]. By E16.5, the bulb has elongated further into a solid cord of epithelial cells. This so-called *sprout* penetrates the *fat pad precursor*, which has now a much lower cell density than at E14.5. While the *sprout* undergoes bifurcation [66] and side-branching by E17 [41], small internal cavities appear and join each other to generate a canal [66]. Meanwhile, the skin adjacent to the neck of the sprout differentiates into a *nipple* [45, 47, 67], which becomes the outlet for the milk canal. By E18.5, most MRs have undergone several rounds of reiterated branching and resemble a miniature *mammary ductal tree* by E18.5. MR2 and MR3 have the most branches, and MR5 may just show one bifurcation [53]. Birth is expected between E19.5 and E21.5 depending on the strain of mice. By then, and in the context of differential growth speeds of different regions of the body, the imaginary mammary lines have acquired a more ventral position than at E11.5, and the MRs are spaced differently along those lines, such that MR1, MR2, and MR3 attain pectoral/thoracic positions, while MR4 and MR5 reach a low abdominal respectively inguinal position as seen in adults (Fig. 1).

2.2 Overinpretation of Static Histological Data as if Tell-Tales of Kinetic Events

Although histological data only provide static information, they were in some cases used to make unfounded conclusions about kinetic events, such as the histological ontogeny of the ME and mechanisms of its morphogenesis. For example, Bresslau concluded that the ME must be of ectodermal origin, as in the several species he had examined so far, he had found no apparent boundary between the mammary placodes and the surface ectoderm, while these two tissues are separated from the underlying dermal mesenchyme by a continuous basement membrane [8]. Moreover, without measuring proliferative activity or pressures, Charles Turner and Elisio Gomez attributed the multilayering of the epithelium in the ML and MRs to rapid proliferation of the basal cell layer of these structures. They also considered the condensation of the underlying dermal mesenchyme a consequence of an increased pressure on the dermis by the multiple layers of epithelium [45]. Decades later, Albert Raynaud argued, equally without proof, that mesenchymal condensation was a result of local fluid extraction from the dermis by the ectoderm, which also led to enlargement (elongation) of cells in the mammary line [47].

However, Boris Balinsky challenged the presumed role of cell proliferation in multilayering as he observed too few mitotic cells (which he scored by the absence of a nuclear membrane) in the ML and MRs of E11–E14 mouse embryos to account for the rapid increase in ME cell number. He suggested that growth must be provided by surrounding ectodermal/epidermal cells streaming towards the places where the MRs are forming, but had no technique in place to demonstrate such cell movements [46, 68].

2.3 Scanning Electron Microscopy (SEM)

Propper used SEM to scan the surface of rabbit embryos at an age when their mammary ridge was clearly elevated. On the apex of the ridge, he observed occasional cells with a rounded cell body and filopodia-like extensions along the length of the ridge. He proposed these cells as "wandering" cells migrating towards sites of mammary placode formation [51]. At that time, it was assumed that the MLs are complete and continuous between axilla and inguen prior to placode formation, and that MRs will subsequently derive from the ML by localized enhanced cell proliferation ([45] and references therein). Interestingly, Propper had already called for more nuanced thinking about that dogma, as the MRs in the axilla and inguen of rat and rabbit seemed to develop without apparent connection to the region of the mammary line on the flank between forelimb and hind limb ([69] and references therein). Nonetheless, his SEM data were extrapolated to mammogenesis in the mouse embryo, and the dogma now became that the MRs derive from the ML by cell migration, still implying the ML is complete before MR formation starts [60]. However, although Propper's SEM data may suggest cell migration, it is still static data, and no formal proof for migration. Moreover, contrary to rabbit embryos, mouse embryos do not form an elevated mammary ridge, and the ML in mouse embryos most likely only resembles the apex of the ridge in rabbit. In hindsight, the supposedly migratory cells at the apex of the ridge were detected at an embryonic age when the placodes are already present and transitioning to the hillock stage [41]. Thus one can also question the relevance of these supposedly migratory cells for the initiation of placode formation, as well as the validity of the extrapolation of the SEM data from rabbit to mouse.

Notably, like Bresslau, also Balinsky, Propper, and Sakakura assumed that the epithelial compartment of the MR is of pure ectodermal/epidermal origin. Nonetheless, one could for example also argue that dermal cells may locally traverse the basement membrane and contribute to the emerging mammary placodes, but that the sections may have been too thick, or not examined in sufficient numbers, to observe examples suggesting such events. It took another 45 years and development of tissue recombination techniques to unequivocally confirm the pure ectodermal origin of the mammary gland epithelium [70], see Subheading 4.4.

3 Combining Microscopy with Cell Labeling Techniques to Explore Whether Cell Migration Contributes to Mammary Placode Formation in the Surface Ectoderm

Meanwhile, the possible involvement of cell migration in early mammogenesis was studied more aptly by labeling cells in a defined region, and verifying their position after a certain period of time.

3.1 Charcoal Depositions

Alain Propper deposited charcoal on explanted flanks of rabbit embryos and cultured the flanks for several days before harvesting them and determining the location of the particles in histological sections of the flanks. Charcoal deposited *outside* the mammary ridge never ended up in the MRs, but charcoal deposited *on* the slopes of the mammary ridge around the time of placode formation, was incorporated in the MRs within 24–48 h

[39]. Although these experiments demonstrate the involvement of cell migration, the significance of these data for mammogenesis in mouse was not clear, as mouse embryos do not form an elevated mammary ridge. The ML in the mouse embryo more closely resembles the apex than the whole width of the mammary ridge. If one wants to extrapolate Propper's charcoal data to mouse, one has to consider the possibility that ectodermal cells flanking the ML in mouse embryos may also contribute to mammary placode formation.

3.2 DiI-Injections

DiI can be injected in embryonic flanks in explant cultures. When injected near the presumptive ML at E10.5, the labeled domain expands in the course of 3 days, suggesting that cell migration occurs in that time [71]. Shortcomings of this technique are that the precise location of the prospective ML is undetectable, and the relevant site of injection can only be estimated; cells are not labeled individually but as a cluster; the explant undergoes extensive growth in 3 days, which on the one hand leads to the loss of focal plane due to 3D growth and makes live or time-lapse recording of the culture impossible, and on the other hand allows for expansion of the labeled domain simply by cell proliferation and passing on of the dye to daughter cells. A comparison of start point and endpoint of a cluster of labeled cells does not differentiate between expansion of the domain by cell proliferation or migration.

3.3 Labeled Markers of Cell Proliferation Offer Opportunities for Cell Tracing in Early Mammogenesis

Balinsky's low count of mitotic cells in the ML and MRs [46] was confirmed by injecting pregnant female mice with tritiated thymidine (^{3}H-TdR) and analyzing the incorporation of ^{3}H-TdR in the skin and developing mammary tissues by autoradiography of histological sections of embryos that were harvested several hours after injection. No ^{3}H-TdR was incorporated in the ME of the embryonic MR3 when females were injected at different time points at the 13th day of pregnancy, indicating a proliferative arrest in this ME between E13 and E14. This was in stark contrast to the high ^{3}H-TdR incorporation, thus high proliferative activity, of cells in the adjacent ectoderm/epidermis and mesenchyme [13]. However, when embryos were harvested and analyzed 24 h after injection, the neck of MR3 contained labeled cells. As ^{3}H-TdR rapidly degrades when not incorporated in cells, these cells must have been labeled about a day earlier. Given that no ^{3}H-TdR was incorporated in the ME proper at the preceding day, these positive cells must represent cells that resided in the adjacent epidermis at their time of labeling, 24 h earlier. The authors also labeled and harvested embryos at E14.5, and observed a high proportion of positive cells in the ME, indicating a resumption of cell proliferation [13].

Somehow, the results of the previous study were later referred to as if ME undergoes a 24 h proliferative arrest between E12 and E13.5, even though this study did not include time points before E12, and not all MRs. Therefore Lee et al. elaborated on this study, widening the time range from E11.5 to E13.5 and including all MRs separately [54]. They replaced the ^{3}H-TdR by the thymidine analog BrdU—which also rapidly degrades if it is not incorporated in cells—and used immunohistochemistry to detect BrdU incorporation in histological sections of embryos harvested 2 or 24 h after injection of the mother. They found almost no BrdU positive cells in the epithelium of the ML and all MRs. Although they found small differences between the MRs, their overall data confirmed Balinsky's low counts of mitotic cells, thus little to no proliferative activity in the epithelium of the ML and MRs between E11.5 and E13.5 [54]. However, if embryos were harvested 24 h after injection of the mother, the embryonic ME contained a high number of BrdU-positive cells. Their number was too high to be explained by proliferation of the initially rare BrdU-labeled cells present at 2 h after labeling. As such, cell proliferation was excluded as a significant contributor of the initiation and growth of MR formation, while cell migration was identified as a major contributor to the initiation and early growth of the ME up to E13.5 [54].

The disadvantage of the ^{3}H-TdR or BrdU labeling technique is that cells are still not individually traced; it does not reveal the exact directionality (e.g., along the DV axis, along the AP axis/ML or centripetal aggregation) and distance of migration, nor does it distinguish between the peridermal and basal cell layers of the ectoderm/epidermis as putative contributors to the ME.

Regardless and importantly, the contention that the mammary placodes are (solely) derived from the ML [45, 65] was contested by these data, as the ML itself would contribute mostly unlabeled cells. Moreover, ME growth was mostly explained by the influx of labeled ectodermal cells [54].

4 Organ Explant Culture and Tissue Recombination Techniques Uncover Continuous Reciprocal Tissue Interactions That Drive the Induction and Morphogenesis of Embryonic Mammary Glands

4.1 Ex Vivo Explant Cultures

In order to facilitate the manipulation of mammary gland development and to address questions concerning regulatory mechanisms of mammary development, an existing in vitro organ explant culture technique [72] was modified to support the growth of embryonic mammary glands ex vivo. With this purpose, Margaret Hardy cut out the ventral and lateral body wall including the ML region of E10, E12, and E13 mouse embryos. She cultured them in adult cock plasma and chicken embryo extract in a watch-glass [73]. Boris Balinsky reduced the explants to a smaller strip of tissue encompassing the ML region, modified the medium, and also tried to culture explants of E8 and E9 embryos [68]. While these younger explants necrotized, explants from E10 and older embryos survived in both Hardy's and Balinsky's experiments. These were examined directly under the stereoscope, or prepared for sectioning and histological analysis. Both Hardy and Balinsky observed MRs in a small percentage of cultured E10 embryos, even though these embryos had no MRs at the time of explantation. Later also Etienne Lasfargues and Margaret Murray [74] successfully grew mammary glands in explants of E10 embryos. While explants of E10 embryos yielded MRs at different stages of morphogenesis within the same E10 explant after 18 days of culture, MR development was more successful and at a more consistent speed in E12 and E13 explants. In such explants, development was only slightly delayed to in vivo development and even progressed to branching morphogenesis [73].

4.2 Tissue Recombination

More than a decade later, Alain Propper and coworkers successfully modified the culture technique for rabbit embryos, albeit that the explants did not attain branching morphogenesis [75]. At the time, developmental biologists were discovering important roles for mesenchymal tissues in organ development. In that context, Propper wanted to assess whether MR formation is an intrinsic property of the ectoderm, the mesoderm, or induced in the ectoderm by the mesoderm. He dissected embryonic flanks, separated the mesenchyme from the ectoderm/epidermis by a mild trypsin digestion, and put them in culture. The mesenchyme or epidermis alone did not give rise to MRs, and often degenerated. He also separated the mesenchyme and epidermis from the head region, and then recombined flank mesenchyme with head epidermis and vice versa (the so-called heterotopic tissue recombinations). Head mesenchyme did not induce a ML or MRs in E12 flank epithelium, although it would sustain MRs present in E13 and E14 flank epithelium. By contrast, flank mesenchyme from E12 (no mammary line/ridge yet) to E14 (hillock stage) embryos did induce a mammary ridge and subsequently MRs in head epithelium [76, 77]. Propper went on to recombine flank mesenchyme of E12 (pre-ML) or E13 (ML) rabbit embryos with chick or duck epidermis just prior to (E6, E7) or after (E8, E9) feather bud induction, and even with chick amnion or chorion (the so-called heterospecific recombinations). In all cases he observed spherical buds resembling mammary buds. In recombinants with bird epidermis he observed concentrically condensed mesoderm around these buds, and upon longer culture periods, these buds developed deep invaginations with a lumen, thus morphologically closely resembling mammary sprouts [78,

79]. These experiments showed that the initiation of mammogenesis is not intrinsic to the ectoderm/epidermis, but induced by local factors in the flank mesenchyme underlying the ML in rabbit embryos. Moreover, the flank mesenchyme exerts an *inductive* role, and can even induce mammary morphogenesis in epithelium that normally does not form mammary glands, even from other species as long as this epithelium is not yet committed to a particular fate. Similar heterotopic and heterospecific experiments at slightly different embryonic ages revealed that once the ML is formed, it needs mesenchyme for its fractionation into MRs. However, this mesenchyme need not be the flank mesenchyme, thus any mesenchyme can take over this *permissive* role.

While Propper was working on the rabbit, Klaus Kratochwil aimed to improve morphogenesis of mouse embryonic mammary glands in culture. He replaced the watch-glass used by Hardy, Balinsky, and Lasfargues and Murray with Grobstein's special glass organ culture dishes [80] that have a central depression containing 0.7–0.9 ml of nutrient medium. He placed a thin (22 ± 3 μm) filter with an average pore size of 0.35 μm on the depression such that it was in contact with, but not submerged in the medium. At the air–liquid interface on these filters, he cultured either intact MRs with a fair amount of subjacent mesenchyme and a small piece of epidermis, or he separated the ME from its subjacent mesenchyme and cultured the two tissues in isolation or recombined them with each other [81]. With these techniques, he was able to achieve normal mammary morphogenesis in organ culture, including the formation of a nipple with nipple sheath, a ramifying ductal system based on monopodial branching as is typical for mammary glands, and adipose tissue. However, when he recombined E12 and E16 ME with E13 salivary mesenchyme, he observed a dichotomous branching pattern that is typical for a salivary instead of mammary gland. From his experiments, he concluded that ME requires any mesenchyme to continue growth and morphogenesis; that the organ-specific morphology is induced by the mesenchyme; and that at E16, the ME is not yet committed to this mammary-specific morphology [82].

4.3 Applications of the Explant Culture Technique

Kratochwil used his culture technique mostly for recombinant explants to study aspects of the sexual dimorphism of mammary development observed in mouse, as described below. But even nowadays, the technique of culturing explants in the air-liquid is still frequently used with individual MRs, tissue recombinants, or whole flanks. It is very amenable to the introduction of experimental variables that also address fundamental questions about the nature and role of tissue-interactions in organ development in a very precise and elegant manner, as will become clear in the course of this review. It is a practical method to monitor daily progress of mammary development, especially in cases when for example a prenatally lethal mutation would prevent mammary development in vivo. It facilitates the study of the roles of genes or proteins of interest in tissue-interactions by electroporation of expression constructs [71, 83], creating heterogenic (female/male or wt/mutant) tissue recombinants at developmental stages of interest [84], or by manipulating the levels of soluble proteins by adding them to the medium or implanting slow-release beads coated with proteins in flanks in culture [57, 60, 62, 71, 85]. A detailed protocol for dissection of flanks and individual MRs and tissues has been published recently [29, 56, 58] and is illustrated with movies as well [86]. Even if in the latter protocol, tissues were treated with RNALater™ or a fixative for gene expression or protein analysis, the general steps of dissection are similar for cases where tissues are harvested for culture. Additional protocols describe variations on the culture protocol to analyze branching morphogenesis or perform tissue recombination [83, 87–89].

4.4 Transplantation or Grafting of Explant Cultures

One drawback of the in vitro explant culture technique is that the medium needs to be daily replaced, and does not contain the maternally derived or self-produced hormones that may circulate through the bloodstream of mammalian embryos. To test the morphogenic effect of pregnancy hormones on embryonic MRs, Teruyo Sakakura and colleagues repeated Kratochwil's recombination experiments of E16 ME with E13 salivary mesenchyme, but subsequently grafted the recombinants under the kidney capsule of syngeneic female mice, which were then made pregnant. Similar to Kratochwil, Sakakura observed a salivary gland morphology in her transplanted recombinants, and in addition she found that this epithelium produced milk proteins. Thus, morphological development and functional differentiation of the ME are not coupled, and commitment to the lineage-specific differentiation program is established in the ME before E16 [90].

Two decades earlier, K.B. DeOme and colleagues had published the successful grafting and growth of ME of an adult donor mouse into the mammary fat pads of 3-week-old female mice that was cleared of its own mammary epithelium [91]. After a desired period of growth of such grafts, the fat pads are dissected, fixed, dehydrated, defatted, and stained with hematoxylin/eosin or carmine alum for stereoscopic analysis of the outgrowth [55]. As the mammary fat pad is the natural environment for ME from around E16 onwards, Sakakura next tried if E16 embryonic ME could also thrive in such cleared prepubertal fat pads. Indeed this was the case, and even MRs from E13 donors developed rigorously and with normal branching patterns in such cleared fat pads [92]. She observed that the fat pad also sustains the growth of embryonic primary (dense) mammary mesenchyme (MM) and secondary mammary mesenchyme or fat pad precursor (FP) and studied their effect on adult ME by not clearing the host fat pad prior to grafting. She identified different effects on adult ME morphogenesis: Where adult ME was in contact with MM, it underwent hyperplastic branching in a monopodial pattern without ductal elongation, whereas adult ME in contact with FP underwent monopodial branching and ductal elongation, and as such was indistinguishable from a normal adult gland [92]. She observed a close resemblance between the MM-induced hyperplastic nodules and hyperplastic nodules that were already at the time considered preneoplastic lesions [91, 93], and recognized that it was of importance to study whether the MM has a tumor-enhancing potential and if so, how this potential was suppressed in the embryo [30, 92].

Building on the works of Kratochwil and Sakakura, and with a similar interest in the role of mesenchyme in organ development, Cunha and coworkers combined recombined E13 mouse mesenchyme underlying the ML with E13 rat ectoderm from the dorsal or ventral region (thus not from the ML) and transplanted these heterospecific, heterotypic recombinants in lactating female mice. The developing ME in such recombinants was entirely rat-derived, finally confirming the ectodermal origin of mammary gland epithelium [70] as suggested decades earlier by Bresslau [8], Turner and Gomez [45] and Balinsky [68].

Currently, the technique of transplanting embryonic mammary tissues in the cleared prepubertal fat pad is still used regularly, e.g., when embryos of mutant mice do not survive long enough to monitor mammary development, or to test whether observed mammary phenotypes in mutant embryos are due to the altered gene function in the ME, in the MM or in the FP [14, 52, 94–97].

5 Techniques to Study the Role of Steroid Hormones in Prenatal Sexual Dimorphism of Mammogenesis

Observations of sexual dimorphism.

5.1 Histological Analysis

In 1933, Turner and Gomez already mentioned that in male mouse (and rat) embryos, contrary to other species they knew, the MRs become detached from the epidermis and do not form nipples [45]. Albert Raynaud studied this in more detail and observed no notable differences in MRs between male and female embryos of E12 to E14 [98] and Raynaud (1947) cited in ref. [47] though Kratochwil observed a slightly smaller size of MRs in E14 males compared to females [63]. At E15, the MM around the neck of the bud/bulb is in males much more condensed than in females and pyknotic cells are present in the neck epithelium at E15. Soon the bulb of the MR detaches from the epidermis, likely due to this mesenchymal constriction and epithelial cell death ([47] and Raynaud (1947) cited therein). Notably, not all five pairs of MRs in males undergo this process: Raynaud observed that the fifth pair of MRs apparently regresses without prior separation from the epidermis [99] and considerable variations were observed between strains [63, 100].

5.2 Manipulation of Embryonic Mammary Development In Utero

Albert Raynaud and Marcel Frilley hypothesized that the differences in mammary development between male and female embryos may be due to functional differentiation of the gonads occurring before that time. To test this, they performed a fetal gonadectomy by X-ray irradiation of the gonads of E13 mouse fetuses of both sexes in utero, which they then allowed to develop in utero until E18.5. In both gonadectomized sexes, the MRs developed as in untreated female embryos, indicating that by default, mammary development proceeds along a female program, which does not require embryonic gonadal function. The perturbed mammary development in males is due to gonadal function in male embryos (Raynaud and Frilley (1947, 1949), cited in ref. [47]).

Unraveling the actions of testosterone.

In other experiments, pregnant females were injected with male steroid hormones. This led to involution of the MRs in female as well as in male embryos (Raynaud (1947a, 1949) cited in ref. [47]; [101]), whereas injection of a synthetic antiandrogenic steroid prevented the regression of MRs in male embryos [102]. Together, these experimental data demonstrated that the MRs need no embryonic gonadal secretions for their development, and that the embryonic testes are responsible for perturbed mammary and nipple development in male embryos [47].

5.3 Explant Culture and Organ Coculture

Kratochwil argued that gonadectomy may affect other endocrine organs in the embryo, and the injections may create a hormonal imbalance in the pregnant mother. Therefore, the abovementioned experiments could not answer the question whether the steroid hormones act directly or indirectly on the MRs, whereas explant culture experiments could. He observed a female developmental program in E12 and E13 mammary explants of both male and female embryos. However, of E14 male explants, MR2, MR3, and M5 were very susceptible to regression, while MRs that survived (50 % of MR1 and MR4 and some MR2 and MR3), resumed growth along a female developmental program albeit with a 2-day delay. In explants of E12–E15 females that were cocultured with E13 testes, all E12–E14 MRs regressed, while 75 % of the E15 MRs survived. Kratochwil obtained similar results when he replaced the testes by testosterone. He therefore concluded that testosterone acts directly on MRs, without necessary involvement of other endocrine organs, although the speed and nature of the morphological response to testosterone may differ between MR pairs. Moreover, the arrest or degeneration of MRs as observed in males is not dependent on the genetic sex of the MRs proper, but on the presence of androgenic hormones in the embryo. Importantly, these androgenic hormones can exert their effect only during the limited time-window between E13 and E14 [50, 63].

5.4 Analysis of (Spontaneous) Mutant Mice

At the time, Lyon and Hawkes had just recovered spontaneous mutant mice carrying an X-linked mutation, X^{Tfm}, leading to testicular feminization [103] that was attributed to a nonfunctional androgen receptor [104, 105]. Kratochwil and Schwartz used these mice to uncover whether the androgen response of male MRs occurs in the ME, MM, or both [106]. They made heterogenic (wild type/mutant) recombinations of ME and mesenchyme of male embryos only and cultured them ex vivo in the presence of testosterone. While all recombinants with mutant mesenchyme underwent female morphogenesis, approximately 60% of the recombinants with wild type mesenchyme underwent the typical developmental arrest or regression normally seen in wild type males. Kratochwil and Schwartz therefore concluded that in male embryos, testosterone only acts on the MM and not the ME, and that the observed epithelial cell death in male MRs is mediated by the mesenchyme.

5.5 Radioactive Cell Labeling and Autoradiography

To test whether cell proliferation contributed to the higher density of MM compared to dermal mesenchyme, Kratochwil and colleagues cultured explants several hours in the presence of tritiated thymidine to label cells in S-phase prior to harvesting the explants for histology combined with autoradiography. The virtual absence of radioactivity in the MM indicated that the higher density of this mesenchyme compared to the dermal mesenchyme is not due to increased proliferation [50]. Later they immersed skin strips with mammary glands from freshly dissected embryos in radiolabeled testosterone, and processed them for histological sectioning and autoradiography or for radioactivity measurements in tissue extracts [107, 108]. They such established that the greatest testosterone-binding capacity is localized in the dense MM adjacent to the ME.

5.6 Heterogenic wt/ Mutant Explant Cultures

Because the higher cell density of the MM could not be attributed to locally enhanced proliferation [50] Kratochwil and colleagues wanted to investigate whether mesenchymal cell migration towards the bud contributes to the condensation of the MM. They made heterogenic with *wt* ME with adhering MM and a large mass of X^{Tfm}/Y MM and dermal mesenchyme, and vice versa. In this case, all recombinants of wt epithelium and adhering MM responded to testosterone despite their environment of androgen-insensitive *Tfm* mesenchyme, whereas recombinants of *Tfm* epithelium and MM with a mass of wt mesenchyme showed no androgen response. Thus it seemed that the mesenchymal response is initiated at the epithelial–mesenchymal interface only, and does not involve migration of distant mesenchymal cells toward the ME. This was further supported by experiments with recombinants of wt epithelium with X^{Tfm}/X mesenchyme, i.e., from heterozygous females, instead of from mutant males. In cells of females at an early embryonic age, one of both X-chromosomes is randomly inactivated and remains inactive in daughter cells. This random X-inactivation resulted in clusters of androgen-responsive cells with an active wt X chromosome, and clusters of androgen-insensitive cells with an active X^{Tfm} chromosome in the MM of X^{Tfm}/X females. The mesenchyme of such recombinants cultured in the presence of testosterone showed similar clusters of mesenchymal condensation representing clones of cells with an active X chromosome, and clusters of loose mesenchyme represented clones of cells with an active X^{Tfm} chromosome. This heterogeneity also indicated that condensation, once initiated, does not spread across the mesenchyme independently of the hormone [49, 50].

Recombinants of wt male MM with wt epithelia of other organs did not show this androgen response, indicating that an interaction with specifically the ME is required for the mesenchyme to pack densely in response to testosterone. Moreover, wt ME induces this testosterone response even in wt mesenchyme that is normally not in contact with ME, e.g., the mesenchyme that is situated in between positions where MRs form along the ML [49]. This was later attributed to the localized induction of a testosterone-binding capacity by the ME in the adjacent mesenchyme [107, 108], provided by androgen receptors [109]. In addition, heterochronic recombinants (different in age) of wt ME and mesenchyme showed that the developmental age of the MM, but not of the epithelium is key to this response [50].

Notably, this strong androgen-response in males is specific for rats and mice, as in other species under study, mammary gland development in male embryos proceeds the same as that in females embryos. In correspondence, testosterone binding was not observed in MM of rabbit embryos, and heterospecific recombinants of mouse ME with rabbit mesenchyme did not exhibit any condensation in response to testosterone [49, 108].

5.7 Androgen Receptor Activation in Females In Utero

However, low concentrations of testosterone have been found in female mouse embryos [110], and the MM of female mouse embryos also expresses androgen receptors [109]. While androgen receptor activation was long considered to be nonexistent or too low in females to affect their mammary development, E18 females with an intrauterine position in between two males (2 M females) have smaller mammary glands than females flank by two females (0 M females), which is likely attributable to androgen receptor activation in 2 M females by testosterone diffusing from their flanking males [111].

Unraveling the actions of estrogens.

Mammary development in gonadectomized male and female embryos proceeds as in normal female embryos. Although Raynaud therefore concluded that MRs need no embryonic ovarian secretions for their development (Raynaud and Frilley (1949), cited in ref. [47]), he did nonetheless consider the possibility that maternal hormones may be present in the amniotic fluid or traverse the placental barrier, and as such may contribute do the default, female, developmental program for MRs (Raynaud (1947), cited in ref. [47]).

5.8 In Utero Manipulation of Mouse Embryos

Indeed, MRs are able to respond to estrogenic compounds, as the injection of high doses of estrogenic compounds in pregnant females stimulated nipple development [112, 113] and led to failure of the sprout to elongate and branch [114–120]. When 16-day pregnant females were subcutaneously injected with the radiolabeled estrogenic compound diethylstilbestrol, followed several hours later by dissection and cryosectioning of the embryos for histology and autoradiography, these estrogens were traced back in the nuclei of E16 MM, but not ME [121]. This location corresponded nicely with the aforementioned phenotypes caused by exposure to high levels of estrogenic compounds.

5.9 Gene and Protein Expression Analysis

Meanwhile, molecular cloning techniques had resulted in the identification of two (α and β) nuclear estrogen receptors (ERs), with different activation responses to different estrogenic compounds. In situ hybridization of sectioned embryos with mRNA probes for these genes demonstrated that both genes were expressed in the MM of E12.5–E14.5 (other ages not tested) mouse embryos, with higher levels of ER-α [122, 123]. Transcripts of both genes were also detected in the E18 MR, while only ER-α is expressed at immunohistochemically detectable levels in the fat pad precursor [111]. Extracts of E12.5, E14.5, and E16.5 male and female mouse embryos activate ER-α—though not ER-β—in vitro, indicating that estrogens do naturally circulate in embryos of both sexes [124]. It is conceivable that these estrogens may activate the ERs in the MM. Progesterone receptor expression has been detected in the E14.5 ME, but whether it is functional has not been assessed [125].

As mentioned above, embryonic MR development does respond to treatment of the mother with normal or synthetic estrogens [63, 115–120], and more recently, the xeno-estrogen and endocrine disruptor bisphenol-A, a phenol-derivative that leaks from most plastics, has been demonstrated to affect embryonic MR development in mice as well [111]. Although each of these studies describes different effects—which may be due to differences in the timing and length of exposure and chemical structures used—together they certainly underscore the sensitivity of embryonic mammary development to ER-activation. Even in the absence of a significant role for endogenous ER-signaling in normal embryonic mammary development, this sensitivity to xeno-estrogens is highly relevant for further study, as exposing pregnant female animals (e.g., in agriculture) and humans to estrogenic compounds may lead to serious malformations of the mammary gland and nipple in the embryos, and thus to functional insufficiency in postnatal life [47]. Furthermore, inappropriate ER signaling in the embryonic MR may well predispose the mammary gland to cancer in postnatal life [126–130].

6   Models and Methods to study the Molecular Regulation of Embryonic Mammary Development

6.1   In Situ Detection of Protein (Activity) and RNA Molecules

Since the 1950s, studies on mammary gland development include questions pertaining to the activity and regulatory roles of molecules. For example, Balinsky [68] and Propper [131] observed fluctuating levels of alkaline phosphatase activity and RNA content in the ME and MM of the developing MRs of sectioned embryos, but could only speculate about the implications of these molecules and their fluctuations. When techniques for protein purification, antibody production and labeling also became available, they were first used to localize for example matrix molecules such as tenascin-C, laminin, and fibronectin, as well as milk proteins in histological tissue preparations of MRs [70, 132–134], soon followed by a plethora of other proteins. More recently, techniques to assay protein expression in preparations of whole mount MRs [56] and 3D-reconstructions of stained histological or optical sections of MRs were developed [54, 57, 58]. Meanwhile, techniques were also developed to synthesize labeled RNA probes, which are used to study gene expression patterns by whole mount in situ hybridization of whole embryos up to E13/ E14, or by in situ hybridization of sectioned embryos of any age [52, 135–137].

6.2   Spontaneous Mutant Mouse Models

Almost four decades lapsed between Raynaud's discovery of hormonal control of mammary gland development [112, 113, 116–120, 138] and the identification of another molecular regulator of mammary development. This began with the observation of absent MRs in E13 embryos of the spontaneous mouse mutant *Extratoes (Xt)* [139], but it took until 1993 until this mutation was identified as a functional null allele of the transcription factor Gli3 [140]. Other spontaneous mutations leading to mammary defects are the *X-linked testicular feminization* (X^{Tfm}) encoding a dysfunctional androgen receptor [106], *Scaramanga (Ska)* representing a misregulated allele encoding the soluble factor neuregulin 3 (Nrg3) [141–143], *Tabby* encoding a functional null allele for the soluble protein ectodysplasinA1 (EdaA1) [85, 144], and *Splotch* encoding a functionally null Pax3 transcription factor [60].

6.3 Genetically Engineered Mouse models

Meanwhile, gene targeting techniques to generate genetically engineered mice (GEMs) [42] became widely used, and produced a myriad of constitutive, tissue-specific, and inducible mutant mice, in which the endogenous gene no longer produces a functional transcript of protein (knockout), or carries a domain deletion or point mutations that alters protein properties such as localization, binding affinity or enzymatic activity. In addition, transgenic mice were produced that carry exogenous DNA encoding a normal or mutant gene to increase expression levels of normal protein or produce high quantities of mutant protein, which outcompetes the normal. Most models studied for embryonic mammary gland development (Table 1) are constitutive knockouts and tissue-specific transgenic mice in which the promoter of either cytokeratin5 (Krt5) or cytokeratin 14 (Krt14) generates a functional null deletion or drives transgenic overexpression in the ectoderm/epidermis and the epithelial compartment of epidermal appendages such as teeth, hairs and mammary glands. In some cases these mutations are combined with lacZ or fluorescent (GFP) reporters that either mark the mammary line or rudiments (e.g., TOPGAL, s-Ship-GFP) or replace the expression of the endogenous gene (e.g., *Sostdc1^{LacZ}*) (Table 2).

The observation of a mammary defect in mutant embryos is usually accompanied by an analysis of the expression pattern of the normal gene in wild type (wt) embryos. This leads to an expansion of a database of suitable expression markers for the mammary tissues at various stages, as well as to hypotheses about the relevance of specific aspects of the spatiotemporal expression pattern for mammogenesis. Similarities in expression patterns of two genes in wt mice respectively in mammary defects in mutants of these genes may lead to additional hypotheses about epistatic interactions between these genes. Most of these hypotheses are tested ex vivo with explant assays, or in vivo by combining several mutations in one mouse to determine if one mutation restores or alters the mammary phenotype caused by the other mutation. During the past 25 years and especially since the beginning of this century, this has led to many insights in the molecular regulation of various stages of embryonic mammary gland development. Most of these

Table 1
Genetically engineered mice studied for embryonic mammary development

mouse genotype	description of allelic mutation											defect observed at stage: From induction to hillock stage						bud to bulb stage	sexual dimorphism	sprout stage	nipple stage	branching stage	remarks	references
	spontaneous: functional null	hypomorphic allele	ENU-induced LOF	targeted: constitutive	tissue-specific	inducible	null	hypomorphic	loss of domain	overexpression	ectopic expression / LacZ knock-in	MR1	MR2	MR3	supernumerary	MR4	MR5							
wild type												●	●	●		●	●							
p63-/-				*			*					○	○	○		○	○							[145, 146]
Krt14-Dkk1 (loss of Wnt-signalling)					*							○	○	○		○	○							[147]
Krt5-rtTA;tetO-Dkk1 (loss of Wnt signalling)					*	*				*		○	○	○		○	○							[148]
Tbx3^tm1Pa/tm1Pa				*			*					•/○	•/○	○		○	○	*					v.p.	[149,150]
Fgfr2b-/-				*			*					○	○	○		●	●	*						[52,60]
CMV-Cre;Rosa26-rtTA^flox;tetO-sFgfr2b (dom.neg.)						*			*	*		○	○	○		•/○	○							[151]
Krt14-Cre;Gata3^flox/flox	*				*		*		*	*		•/○	•/○	•/○		•/○	•/○	*		*	*	*	no nipples (no MRs??) at birth	[54, 60, 139, 153, 154]
Gli3^Xt-J/Xt-J or Gli3^Xt/Xt (extra-toes)				*			*					●	●	●		●	○						v.p.	[150]
Tbx2^tm1Pa/tm1Pa				*			*					●	○	●		●							v.p.	[150]
Tbx3^tm1Pa/+												●	●	●		•							v.p.	[141-143]
Nrg3^ska (scaramanga)		*										●	●	○		•							v.p.	[60]
Pax3^Sp/Sp (splotch)	*											●	●	•/○		●	●							[60]
Pax3^AZ/n2											*	●	●	•/○		●	●							[60]
Krt14-tTA;tetO-Wise				*	*	*				*		•	•	•		●	●	*				*	v.p.	[135,156]
Lef1-/-				*			*					•	•	•		•	•			*	*	*		[157]
Pygo2-/-				*	*	*						•	•	•		•	•			*	*	*		[157]
Krt14-Cre;Pygo2^flox/flox					*		*					•	•	•		●	●					*	MR4 develops normal	[155]
TCF-tTA;tetO-Wise												●	●	●		●	●	*						[155]
Krt14-Cre;β-catenin^flox/flox					*		*					●	●	●		●	●						MR2 and 3 fuse	[155]
Lrp4^mdig/mdig	*											●	●	•	•	•	•							[155]

Single gene mutations

Genotype	Phenotype	References
Lrp4^mitt/mitt	MR2 and 3 fuse	[155]
Lrp4^mte/mte	MR2 and 3 fuse	[155]
Hoxc6^-/-		[158]
Tbx2^tm1Pa/*		[150]
Gli1^izki/izki		[153,154]
Gli2^izki/izki		[153,154]
Gli2^zki/zki (Gli1 expressed from Gli2 promoter)		[153,154]
Shh^-/-	none	[95,96]
Eda^Ta/Ta (tabby)		[57, 85, 144]
Edar^dl/dl (downless)		[85,144]
Krt14-EdaA2	none	[144]
Krt14-EdaA1		[57, 85, 144]
Krt14-Nrg3	supernum. MGs + nipples seen postnatally	[159]
c^IkBaDN/+ (super-ikBα into β-catenin null locus)		[57]
Sostdc1^LacZ/LacZ (Wise^-/-)	MR2 and MR3 fuse	[155,160]
Lrp5^-/-		[161]
Lrp6^-/-		[162]
IGF1R^-/-		[163]
p190B Rho-GAP^-/-		[163]
PTHrP^-/-		[163]
PTHrP-R1^-/-		[67, 84, 109, 137, 164]
PTHrP^Δ1-84 (loss of NLS)		[67, 84, 137, 164]
Dermo-Cre;β-catenin^flox/flox;TOPGAL-C		[165]
Dermo-Cre;dtTomato;β-catenin^flox/flox		[166]
X^Tfm (X-linked testicular feminization; AR null)		[166]
BDNF^LacZ/LacZ		[106,167]
TrkB^-/-		[167]
TrkB^T65EA/T65EA		[167]
TrkB.T1^-/-		[167]
Avil^Cre/+;TrkB^flox/flox		[167]
Msx2^-/-		[168]
Krt14-PTHrP		[67, 164, 169]

Table 1 (continued)

Compound gene	Notes	Ref.
Krt14-PTHrP-R1		[67]
Krt14-tTA;tetO-PTHrP	hair-to-MR transformation	[67, 170]
Krt14-Cre;Smo^flox/flox (loss of Hh signaling)	none	[171]
Gli2^-/-	none	[172]
Ihh^-/-	none	[95,96]
Ptc1^-/-	none	[173]
Msx1^-/-	none	[174]
Troy^-/-	none	[175]
Egfr^-/-	none	[176]
Itga3^-/-	none	[177]
Itga6^-/-	none	[177]
Itgβ4^-/-	none	[177]
Gata3^+/NLSlacZ	none	[152]
Pea3^NLSlacZ/NLSlacZ	none	[178]
Bax^-/-	none	[167]
Fgf10^mlcx24-lcz)/-		[60]
Gli3^Xt-J/+;Gli2^lzki/lzki	MR1-3 regress before E18.5	[153]
Tbx2^m1Pa/+;Tbx3^m1Pu/+		[150]
Krt14-tTA;tetO-Wise	rescues MR2-3 fusion	[155]
Lrp4^mitt/mdlg;Krt14-Cre;β-catenin^flox/-	MR2 and 3 fuse	[155]
Lrp4^mitt/mdlg	MR2 and 3 fuse	[155]
Wise^-/-;Lrp4^mdlg/mdlg	MR2 and 3 fuse	[155]
Wise^-/-;Lrp4^mitt/mitt	MR2 and 3 fuse	[155]
Krt14-tTA;tetO-Wise-GFP;Lrp4^mdlg/mdlg	MR2 and 3 fuse	[155]
Krt14-tTA;tetO-Wise-GFP;Lrp4^mitt/mitt	MR2 and 3 fuse	[155]
Lrp4^mdlg/mdlg;Lrp5^-/-	rescues MR2-3 fusion	[155]
Lrp4^mdlg/mdlg;Lrp6^-/-	rescues MR2-3 fusion	[155]
Lrp4^mdlg/mdlg;Lrp5^-/-	rescues MR2-3 fusion	[155]
Wise^-/-;Lrp5^-/-	rescues MR2-3 fusion	[155]
Wise^-/-;Krt14-Cre;β-catenin^flox/-	rescues MR2-3 fusion	[155]
Wise^-/-;TCF-tTA;tetO-Wise		[155]
Eda^-/-;Troy^-/-		[175]

Gene mutation	Phenotype	Ref.
p53^tm1Tyj;Tbx3^tm1Pa		[150]
Arf^-/-;Tbx3^tm1Pa		[150]
IRS1^-/-;IRS2^-/-		[163]
Msx1^-/-;Msx2^-/-		[168]
PTHrP^-/-;Krt14-PTHrP	regress by E15.5	[168]
PTHrP-R1^-/-;Krt14-PTHrP	ectopic MM	[109,137]
PTHrP^-/-;CoII-II-PTHrP;K14-PTHrP		[67]
PTHrP^-/-;CoII-II-PTHrP		[137]
Krt14-PTHrP;Dermo-Cre;tdTomato;β-catenin^flox/flox	no MG/nipple postnatally	[137]
Krt14-PTHrP;Lef1^-/-	rescues nipple skin phenotype	[166]
K14-PTHrP;Msx2^-/-		[164]
Krt14-Eda;IκBα^DN	partial rescue of ectopic nipple skin phenotype	[164]
		[57]
Krt14-Cre;Pygo2^flox/-;krt14-β-catenin^ΔN	none	[157]

This table lists all gene mutations studied to date (2014) for an embryonic mammary phenotype. Phenotypes at induction and hillock stage do not do necessarily affect all MRs, as indicated by absence of specific MRs (open circles ○), hypoplastic MRs (smaller dots ● or • than those for wild types ● at the top of the table), supernumerary MRs (indicated with additional dots ●, sometimes hypoplastic ● or •, at the position where they occur) or hyperplastic MRs (indicated with larger dots ● than wt ●). Some mutations also affect (or only affect) later stages of development, in which case publications often do not mention specifically which MRs are affected. Abbreviations: MG: mammary gland; MR: mammary rudiment; nd.: not determined. NLS: nuclear localisation signal; v.p: variable penetrance. Table elaborated from [27].

Table 2
Reporter mice used in studies of embryonic mammary development

Reporter mice	Marks	References
TOPGAL-F	Wnt signalling in epithelium; ML, MR	[148, 179]
Fgf10⁻;Topgal-F		[60]
Fgfr2b⁻;Topgal-F		[60]
Gli3ˣᵗ⁻ᴶ;Topgal-F		[60, 153]
Lrp4ᵐᵈⁱᵍ;TOPGAL-F		[155]
Sostdc1ᴸᵃᶜᶻ;TOPGAL-F		[160]
Wise⁻;TOPGAL-F		[155]
Nrg3ˢᵏᵃ;TOPGAL-F		[180]
TOPGAL-C	Wnt signalling in epithelium and mesenchyme	[181]
Lef1⁻;TOPGAL-C		[156]
Krt14-PTHrP;TOPGAL-C		[166]
Dermo-Cre;β-cateninᶠˡᵒˣ/ᶠˡᵒˣ;TOPGAL-C		[166]
BATGAL	Wnt signalling in epithelium and mesenchyme	[182]
Lrp5⁻;BATGAL		[161]
Lrp6⁻;BATGAL		[162]
Pygo2⁻;BATGAL		[157]
Sostdc1ᴸᵃᶜᶻ;BATGAL		[160]
Conductinˡᶻ/⁺ (=Axin2ˡᶻ/⁺)	Wnt signalling in epithelium and mesenchyme	[155]
Axin2ᶜʳᵉᴱᴿᵀ²/⁺;R26Rˡᵃᶜᶻ/⁺		[183]
Axin2ᶜʳᵉᴱᴿᵀ²/⁺;R26Rˡᵃᶜᶻ/⁺		[183]
Gli3ˣᵗ⁻ᴶ;Conductinᴸᵃᶜᶻ		[154]
TCF/LEF:H2B-GFP	Wnt signalling, similar to TOPGAL-F	[155, 184]
Lrp4ᵐᵈⁱᵍ;TCF/LEF:H2B-GFP		[155]
Edaᴿᴱᴾ	Eda signalling	[175]
Edaᵗᵃ;Edaᶻᴿᴱᴾ		[175]
Krt14-Eda;Edaᴸᵃᶜᶻᴿᴱᴾ		[175]
Krt17-GFP	*Krt17* expression; epidermis	[160, 185]
Sostdc1ᴸᵃᶜᶻ;Krt17-GFP		[160]
s-Ship-GFP	*Ship1* expression; ML	[186] [41]
Nrg3ˢᵏᵃ;s-Ship-GFP		[180]
Krt14cre:R26-ᶠˡᵒˣˢᵗᵒᵖ⁻ᴸᵃᶜᶻ	Cre, LacZ specifically in MRs from E12 onwards	[155]
Krt14-tTA:tetO-Wise-GFP	transgenic *Wise* expression in MRs from E12 onwards	[155]
Msx1-LacZ		[174]
Msx2-LacZ	transgenic *Msx2* expression	[174]
BMP4-LacZneo	transgenic *BMP4* expression	[164, 187]
TrkBᴳᶠᴾ/⁺	neurons	[167, 188]
Lrp4-LacZ	*Lrp4* promoter activity	[155]
Wise-LacZ	*Wise* promoter activity	[155]

This table lists all reporter mice, and their combination with gene mutations causing an embryonic mammary phenotype, known to date (early 2014) insights have recently been comprehensively reviewed elsewhere [32, 33, 43, 44, 189–191]. Below, the focus lies on the experimental approaches that led to some of these insights.

7   Molecular Regulation of Patterning of the MRs in the Surface Ectoderm

From their tissue recombination experiments Propper, Kratochwil, and Cunha and Hom had concluded that the differentiation of ectoderm into mammary epithelium is induced by (then unknown) mesodermal/mesenchymal factors [49, 50, 69, 70, 76–79, 81, 82, 192]. Correspondingly, some GEMS with defective mammary induction (Table 1) carry a mutated version of a gene which in wt is among others expressed in the dermal mesenchyme at the time of ML and MR induction, e.g., the growth factor $Nrg3^{ska}$, and transcription factors $Tbx2$ and $Tbx3$ [28, 33, 149, 159, 193]. However, most GEMS with a known induction defect lack a gene that in wt is expressed in the somites, i.e., the mesodermal structures that give rise to vertebrae, ribs, muscles, and the dermal mesenchyme. These genes encode the transcription factors Gli3, Pax3, Tbx2, Tbx3, and likely Hoxc6, the growth factor FGF10, or retinoic acid receptors [32, 43, 44, 158, 189, 191, 194, 195]. This somitic expression was of particular and dual and interest, because (1) the dermal mesenchyme is a derivative of the somites, and (2) the induction of mammogenesis, characterized as a combination of cell elongation and $Wnt10b$ expression [60], first manifests itself as a line of fragments overlying the ventral (hypaxial) tips of the somites between forelimb and hindlimb on the flank, which suggests the involvement of hypaxial somitic signals in the onset of mammogenesis [61]. The relevance of the somites in the induction of mammogenesis was supported by the finding that hypaxial truncation of the somites, as in $Pax3$ null embryos, is associated with a narrower and dorsally displaced ML on the flank, and delayed formation of MR3 forms compared to wt embryos [60].

In wt embryos, this hypaxial area has the highest $Fgf10$ expression within the somites. At the time of onset of mammogenesis in wt embryos, $Fgf10$ is expressed in the somites and limb buds, while the gene encoding its main receptor $Fgfr2b$ is expressed in the surface ectoderm. $Fgfr2b^{-/-}$ and $Fgf10^{-/-}$ embryos do not form a mammary streak/line on the flank, and no MRs (except MR4). By contrast, hypomorphic $Fgf10^{-/mlcv24Lacz}$ embryos do form a ML and MRs, but not MR3. $Gli3^{Xt-J/Xt-J}$ (null) embryos resemble $Fgf10^{-/mlcv24Lacz}$ embryos with regards to ML and MR3 formation, and have reduced somitic $Fgf10$ expression levels while $Fgf10$ expression in the limbs is unchanged or elevated. Stand alone, each of these evidences for somitic involvement in the induction of mammogenesis on the flank is circumstantial. Nonetheless, the combined analysis of mammary phenotypes and gene expression patterns in these mutants makes a strong case for involvement of somitic signals, i.e., *Gli3* and *Fgf10* in the induction of mammogenesis between the limbs [60].

The expression patterns in wt and mutant embryos suggested that FGF10 acts downstream of Gli3, but are no proof of such. As FGF10 is a soluble factor, it can be added to culture assays. Implantation of a bead soaked in FGF10 in explant cultures of E11.5 *Gli3$^{Xt-J/Xt-J}$* embryonic flanks rescued the formation of MR3, indicating that *Fgf10* indeed acts downstream of somitic Gli3 and is sufficient to induce MR3 in the absence of Gli3 [60, 189].

Gli3 is a transcription factor with two family members, Gli1 and Gli2. The Gli1 protein is a transcriptional activator that is usually produced in response to Hedgehog signaling. By contrast, Gli2 and Gli3 are often co-expressed at sites with no Hedgehog signaling, which allows their cleavage and consequent functioning as transcriptional repressors. In the presence of high Hedgehog signaling, they can however remain uncleaved and act as transcriptional activators. By replacing two *Gli2* alleles by *Gli1* activator in the absence of one allele of *Gli3*, Hatsell and Cowin were able to restore the *Gli3* mammary phenotype, demonstrating that *Gli3* acts as a repressor [153] as previously predicted [171, 196]. Since the absence of *Gli3* leads to reduced somitic *Fgf10* expression [60], *Gli3* regulates *Fgf10* transcription indirectly.

But how do *Gli3* and *Fgf10* relate to the other somitic/dermal genes, e.g., *Tbx*-genes (Fig. 3)? Around E10.5, wt embryos begin to express *Tbx2* in a band of ventral dermal mesenchyme encompassing the prospective mammary streak between forelimb and hindlimb, and *Tbx3* in a similar but wider band spanning approximately the ventral half of the underlying somites. *Tbx3* is also expressed in the mammary placode epithelium once it is formed. While heterozygous nulls for either gene do not have a mammary phenotype, 20 % of compound *Tbx2/Tbx3* heterozygous nulls have no MR2 at E13.5 (earlier not investigated). This indicates that these *Tbx* genes complement each other or interact with each other via yet unknown mechanisms in early development of MR2 [150]. Wt embryos express *Bmp4* in the ventral dermal mesenchyme in the subaxillary and suprainguinal region at E11-E11.5. The somitic/dermal expression domain of *Tbx3* is narrower in *Gli3Xt/Xt* (null) mutants than in wt embryos [154]. Electroporation of wt flank explants with *Tbx3* downregulates *Bmp4* expression, and broadens the ML. Conversely, electroporation of *Bmp4* downregulates *Tbx3* expression but did not affect the breadth of the ML, while co-electroporation of *Bmp4* and *Tbx3* had the same effect as *Tbx3* alone or caused additional broadening of the ML in the ventral direction. All variables led to an increase of *Lef1* expression as a marker for ME formation. These data indicate a reciprocal negative interaction (direct or indirect) between *Tbx3* and *Bmp4* whose

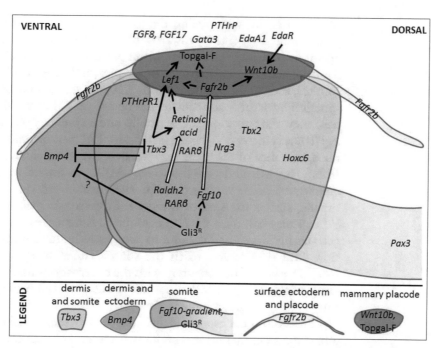

Fig. 3 Molecular players and interactions for the induction of mammary rudiment 3. The molecular cascades regulating the initiation of mammogenesis have been best studied for the mammary streak between the fore limb and hind limb (encompassing MR2, MR3, and MR4) and in particular for MR3, as this MR pair seems most susceptible to loss of gene function and is coincidentally the most accessible for manipulation by for example bead implantation or electroporation in studies with flank explant cultures. This *cartoon* shows the ventral end of somite 15 or 16, with overlying surface ectoderm and developing MR3. *Fgf10* is expressed in a gradient along the somites, with highest expression in the ventral tip, indirectly downstream of Gli3-repressor activity. FGF10 activates the ectodermal *Fgfr2b*, leading to *Wnt10b* expression and Wnt signaling (reported by TOPGAL-F and Lef1 expression). The site and level of *Fgf10* expression (co-dependent on for example the length of somites which is controlled by Pax3) as well as the reciprocal repression between BMP4 and TBX3, likely downstream of Gli3R, are determinants of the dorsoventral position of this mammary rudiment. Other molecular players at early stages are indicated in this cartoon as well, although they relationships still have to be determined. Note that other MRs require different tissue and molecular interactions for their induction. Modified from refs. [60, 71, 154, 195], with permission

interface and relative expression levels determine the dorsoventral position and width of the ML [71]. *Gli3$^{Xt/Xt}$* embryos have a slightly upregulated, dorsalized and posteriorized *Bmp4* expression and correspondingly, the *Tbx3/Bmp4* interface seems to be dorsalized, suggesting that the reciprocal inhibitory interaction between *Tbx3/Bmp4* functions downstream of somitic *Gli3*. Given that *Bmp4* has Gli binding sites, Gli3 may repress *Bmp4* directly [154]. Interactions of these genes with *Hox* genes, *Nrg3*, and retinoic acid signaling remain to be investigated.

Interestingly, it became clear that at different locations along the ML, the MRs have different requirements for or sensitivities to these mesenchymal factors (Table 1). Despite these differences in mesenchymal inducers, the cellular response in the overlying ectoderm is the same, being the loss of proliferative activity and gain of migratory activity towards the prospective ML and placode positions [46, 54]. The dorsoventral position of the streak on the flank depends on the length of the somites, and gene activity in the somites, such as *Gli3* and *Fgf10*. At E10, the somites are still small spherical structures adjacent to the neural tube, but within half a day, they start to elongate ventrally and express *Fgf10*. Somitic *Fgf10* levels increase between E10.5 and E11.5, concomitant with the appearance of the mammary streak between the limbs, and the MRs. The ML and all MRs except MR4 are absent in *Fgf10*$^{-/-}$ embryos. FGF10 is known for its chemotactic function in other organs, and as the surface ectoderm expresses its main receptor, FGF10 may exhibit a similar chemotactic function on the surface ectoderm, "dragging" it along to progressively more ventral position until the somites reach the ventral lateral plate mesoderm [60]. This would be consistent with the observations that multilayering of the ML and MRs does not result from cell proliferation, but from cell aggregation or influx [54].

Furthermore, despite the differences in mesenchymal inducers along the mammary line, the known molecular responses of the overlying ectoderm are also the same, namely de novo or increased expression of genes such as *Wnt10b*, *Wnt6* and an engagement in Wnt signaling along the entire ML [61, 148, 191], soon followed by expression of *Lef1* [52], *EdaA1* and *EdaR* [43], *Gata3* [152], *Nrg3* [193] several FGFs [62], and PTHrP [164, 166], specifically in the placode epithelium.

As mentioned, *Wnt10b* first appears as an array of fragments overlying the ventral tips of the somites [61]. However, not all *Wnt10b*$^{+ve}$ fragments use their potential to become a MR. Whether they do, depends in part on the level of somitic *Fgf10* expression, as deduced from the non-induction of MR3 in *Fgf10* hypomorphic mutants. Moreover, *Fgf10* is expressed in a bilateral gradient across somites 12–18, and MR3 is formed above the somites (#15/#16) with the highest *Fgf10* expression in wt embryos [60]. In part, it also depends on sufficient levels of canonical Wnt signaling in the ectoderm, as mutants with a complete reduction in Wnt signaling fail to form the ML and MRs [148], and mutants with a partial reduction of Wnt signaling form MRs with impaired growth and which often regress [135, 147, 155–157, 160–162]. Conversely, increased Wnt signaling by addition of for example Wnt3A to explant cultures, or creating tissue-specific knockouts for inhibitory co-receptors or transgenic overexpression of activators of the pathway, leads to enlarged MRs [148, 155]. By contrast, tissue-specific overexpression of EdaA1 or its receptor, or Nrg3 [159] leads to conversion of more *Wnt10b*$^{+ve}$ fragments into MRs [85, 144]. Nrg3 seems to regulate migration of mammary epithelial precursors towards the placode sites [180], whereas Eda/TNF signaling represses ectodermal Wnt signaling and enhances mesenchymal Wnt signaling at E13.5 [57]. It is now of interest whether the same interaction exists between Eda and Wnt signaling pathways at the induction stage. In any case, by regulating the size and number of MRs, they are important determinants of the patterning of ME in the ectoderm.

One role for Wnt-signaling may be to downregulate the proliferative activity of ectodermal cells in association with their acquisition of a ME fate. This conclusion is based on data from Ahn et al. [155] who show that MR2 and MR3 fuse in the absence of the Wnt-antagonists Lrp4 or Wise, preceded by a loss of proliferation in the interplacodal region.

Notably, the various ligands, receptors and antagonists of Wnt signaling vary widely in their expression domain, from broad expression in the dermal mesenchyme or surface ectoderm, to restricted expression in the ventral or dorsal domain or flank, ML, rudimentary ME or MM. Thus, various modes or subsets of canonical Wnt signaling may exist in the mammary region, both in the epithelium and in the mesenchyme. This is exemplified by the different expression patterns in MRs [60, 148, 153–157, 160–162, 166] as well as in other organs [197] of the reporters for canonical Wnt-signaling: Topgal-F [179], Topgal-C [181], Batgal [182], and Axin2-LacZ [198] (Table 2). It remains a challenge to identify separate roles for mesenchymal and epithelial Wnt-signaling, let alone whether subsets of Wnt signaling locally act alone or in concert with other subsets.

8 Molecular Regulators of Growth and Survival of the MRs Until E16

While a complete abolishment of Wnt-signaling through overexpression of the inhibitor Dkk1 prevents the formation of all MRs [147, 148], MRs are induced if Wnt-signaling is only partially reduced due to a null mutation for *Lef1*, *Lrp5*, or *Pygo2* [135, 156, 157, 161]. However, such MRs are small, grow poorly if at all, and may regress with variable penetrance before E15.5. Whereas *Msx1*$^{-/-}$ single knockouts have no embryonic mammary defects and MRs in *Msx2*$^{-/-}$ develop normally until sprouting stage, *Msx1*$^{-/-}$; *Msx2*$^{-/-}$ double knockouts develop MRs that fail to express *Lef1* and regress by E15.5 [168].

As Lef1 is a transcriptional target and mediator of Wnt signaling, it is tempting to speculate that the regression in *Msx1*$^{-/-}$;*Msx2*$^{-/-}$ mutants is due to reduced Wnt signaling; perhaps because ectodermal cells retain their proliferative activity and fail to acquire a mammary fate or commit to it if Wnt signaling is low. However, in wild types, *Lef1* expands its expression domain from the ME to include the MM by E15.5 [67] while it mediates the converse expansion of Topgal-C expression (a reporter for a subset of Wnt signaling) from the MM to include the ME between E13.5 and E15.5 [156]. Notably, during this time span, the epithelial compartment of the wt MR transits from growth by epidermal cell recruitment to growth by proliferation of the ME cells proper [46, 54]. It is thus possible that the absence of Wnt-signaling in the ME in E15.5 *Lef1-/-* embryos also disrupts the functional transition that ME cells need to undergo around that time, leading to lack of growth and eventual regression of the MRs.

Interestingly, loss of p190B-RhoGAP allows for MR induction, but at E14.5 the buds are small despite a slight increase in epithelial proliferation and a lack of apoptosis [163]. Given that p190RhoGAP is expressed in the ME of E12.5 embryos onwards, and that the enzyme is known for its roles in cytoskeletal remodeling to promote cell migration and inhibit mitosis, it is conceivable that the mammary phenotype in *p190B-RhoGAP-/-* embryos is caused by both impaired cell migration and sustained cell proliferation.

Contrary to *Tbx3-/-* embryos, *Tbx3+/-* embryos induce all five MR pairs, but the three thoracic pairs are often lost between E13.5 and E18.5. This defect is exacerbated in *Tbx2+/-;Tbx3+/-* double heterozygotes [150]. Both genes are well known for their role in cell cycling control through p19^{Arf}/p53 signaling. While this mechanism is intact in these mutants, it remains of interest to investigate which signaling cascade is impaired and causes the haplo-insufficiency in mammary development of these mutants [150].

9   Molecular Regulators of Sexual Dimorphism

As mentioned far above in section 5 about steroid hormones, the sexual dimorphism of mammary gland development is created by the absence or presence of androgen receptor activation in the mammary mesenchyme of female respectively male mice [49, 50, 63, 106, 107, 114]. Analysis of null mutants for the genes encoding PTHrP or its receptor PTHrP-R1 revealed their lack of sexual dimorphism in mammary gland development: Mammary glands in these mutant males and females lack androgen receptor and tenascin-C expression in the MM, and develop similar to their counterparts in wild type female embryos [84, 109]. In wild type mice, *PTHrP* is expressed in the ME from placode stage onwards, while *PTHrP-R1* becomes broadly expressed in the dermal mesenchyme [84, 109]. These expression patterns may suggest that the defect in mutants is due to an absence of PTHrP/PTHrP-R1 signaling between the ME and prospective MM. However, far prior to the onset of mammary gland development, PTHrP and PTHrP-R1 are expressed in several extra-embryonic and embryonic tissues [199]. Therefore, further testing was required to exclude the possibility that the mammary defect is a secondary effect of lack of PTHrP/PTHrP-R1 signaling earlier in embryogenesis.

Rescue experiments in which PTHrP was reintroduced in the ectodermis/epidermis of *PTHrP−/−* embryos by crossing in a Krt14-PTHrP transgene, restored androgen receptor expression. This facilitated the androgen response in male mutants. These experiments confirmed that the mammary defect is indeed caused by the absence of PTHrP signaling in the MR proper, and that no earlier PTHrP signaling is required [109]. Similarly, *PTHrP−/−* and *PTHrP-R1−/−* mutant mice lack Topgal-F expression (a marker for a subset of Wnt signaling) in the MM. Conversely, transgenic over-expression of PTHrP in the entire flank induces ectopic Topgal-F expression in the underlying dermal mesenchyme, confirming that mesenchymal Wnt signaling requires no PTHrP/PTHrP-R1 signaling prior to mammary placode formation.

Epistasis assays in which the Wnt-transducer β-catenin was removed from the mesenchyme in Krt14-PTHrP transgenic mice, showed that dermal β-catenin is required downstream of PTHrP/PTHrP-R1 signaling between the mammary placode epithelium and its contiguous dermal mesenchyme, to induce mammary mesenchymal specific markers such as Wnt signaling and expression of lef1, estrogen receptor and androgen receptor [166]. Like PTHrP/PTHrP-R, also Gli3 is required for androgen receptor and tenascin-C expression, and it now becomes interesting to study if and how PTHrP/PTHrP-R signaling, Wnt signaling and Gli3 interact to regulate expression of androgen receptor and tenascin-C as differentiation markers for MM [154]. Despite a normal testicular histology and androgen receptor expression in the MM, the MRs of some Krt14-Eda males may escape the androgen-mediated destruction and even form a nipple. The ME manages to sprout and enter the fat pad precursor, where it undergoes a modest degree of branching morphogenesis, albeit with a lack of canalization. Most likely the escape from destruction is provided by precocious proliferation of the ME, which allows penetration into the androgen receptor negative fat pad precursor [57]. Remarkably, there also exists a sexual dimorphism in sensory innervation of the mammary gland. This is due to the expression of a truncated form of TrkB, a receptor for the neurotrophic factor BDNF, downstream of androgen receptor activation. This truncated receptor prevents normal BDNF/TrkB signaling in sensory axons, which leads to a loss of innervation of the mammary gland in males [167].

10   Molecular Regulators of Nipple Formation

The nipple is a late appendage to the skin and mammary gland, both in terms of evolution and in embryonic development, as they only develop in marsupials and placentals, and as a secondary structure to the mammary gland [6]. The supernumerary MRs in *Krt14-EdaA1* transgenic mutants do form nipples, albeit it with an aberrant shape, and not all connected to a ductal network and associated with a fat pad [85, 144]. Nipples of $Eda^{Ta/Ta}$ (null) mice were abnormally flat, but nonetheless both the loss and gain of function mutants nursed their offspring normally [144]. *PTHrP*$^{-/-}$ and *PTHrP-R1*$^{-/-}$ mutants do not develop nipples, nor can their nipple development be rescued with transgenic Krt14-PTHrP [67, 169, 200]. However, the entire ventral epidermis transforms into nipple skin when transgenic Krt14-PTHrP is expressed on a wt background, ectopically in the entire ventral epidermis instead of in the ME only [67]. These analyses led to the conclusion that PTHrP/PTHrP-R1 signaling is required and instructive for nipple development. Normally, PTHrP-R1 is ubiquitously expressed in the ventral dermal mesenchyme, whereas PTHrP expression is restricted to the ME only. Thus, despite the ubiquitous expression of PTHrP-R1 in wt embryos, activation of this receptor is restricted to just a few layers of mesenchyme in close proximity to the ME. This mesenchyme differentiates into MM and signals back to the overlying epidermis, which responds locally by differentiating into nipple skin [67]. As PTHrP/PTHrP-R1 signaling activates Wnt signaling to specify the MM [166] it is perhaps not surprising that mutants lacking the Wnt co-receptor Lrp6 have smaller nipples [162].

One feature of nipple formation is the suppression of hair follicle formation. Indeed Krt14-PTHrP transgenic embryos lack hair follicles on their ventral (nipple) skin [201], in conjunction with reduced BMP signaling due to reduced transcription of the BMP receptor BMPR1A [164]. Loss of *Msx2* in *Krt14-PTHrP* embryos (*Krt14-PTHrP;Msx2*$^{-/-}$ mutants) rescues hair follicle formation. As BMP4 and PTHrP have a synergistic stimulatory effect on *Msx2* expression in cultured dermal mesenchymal cells, it was concluded that *Msx2* mediates the repressive effect of PTHrP/PTHrP-R1-augmented BMP signaling on hair follicle development in the nipple area [164]. Indeed, suppression of BMP signaling by transgenic expression of *Krt14-Noggin* allows the formation of *Shh*-expressing hair follicles in the nipple area [202]. Moreover, transgenic Noggin suppresses *PTHrP* expression, whereas addition of BMP4 to cultured cells augments PTHrP-promoter activity [202]. This points to a feed-forward loop between PTHrP and BMP signaling. In the absence of the Gli3 repressor of (sonic) Hedgehog signaling, hair follicles develop in the nipple area [154]. It is now of interest to study the relationship between PTHrP, BMP, and Hh signaling in establishing a properly differentiated nipple tissue without hairs.

Interestingly, the time frame allowing nipple development is very wide, as supernumerary nipples are formed in *Sostdc1*$^{-/-}$ mice at the end of puberty around 6 weeks postpartum [160]. Remarkably, these nipples are not connected to a ductal network, and both the normal and supernumerary nipples contain hair follicles [160].

11   Molecular Regulators of Sprouting and Branching Morphogenesis

In reduction or loss of function mutants for *PTHrP*, *PTHrP-R1*, *Msx2*, *Fgf10*, *Tbx2/Tbx3*, *Pygo2*, *Lrp6*, *Gli3*, or *Eda*, and in transgenic mice overexpressing the super-repressor of Eda/NFkB signaling, IkBα∆N, mammary buds all fail to properly elongate into sprouts or are impaired in branching morphogenesis [44, 52, 54, 57, 150, 157, 162, 168]. In wild types, all these molecules are expressed in the MM and/or fat pad precursor, with exception of Pygo2, Lrp6, the Tbx transcription factors, and PTHrP which is expressed in the ME but finds its receptor in the MM. It was therefore likely that sprouting and branching morphogenesis of the ME are regulated by molecular interactions of the ME with its surrounding mesenchymal tissues. This has been tested and validated for PTHrP and FGF10 signaling: *Fgf10*$^{-/-}$ ME was able to generate a branched tree when grafted into a cleared fat pad of a 3-week-old wt [52]. Similarly, tissue recombinants of E13.5 *PTHrP-R1*$^{-/-}$ ME with wt MM that were grafted under the kidney capsule, did show ductal outgrowths similar to wt/wt recombinations, while recombinants of wt ME with *PTHrP-R1*$^{-/-}$ MM did not grow out [84]. These data showed that FGF10 and PTHrP-R1 expression are only required in the mesenchyme for normal branching. The level and timing of PTHrP-R1 activation is important, as transient overexpression of PTHrP in the epidermis using the Krt14-driven inducible tet-off system [203] during prenatal branching morphogenesis causes branching defects during puberty [170]. PTHrP/PTHrP-R1 signaling regulates *Msx2* expression in the MM [164], and the similarity in sprouting and branching defects in null mutants for PTHrP, PTHrP-R1, and Msx2 suggests that Msx2 is a mediator of PTHrP-induced sprouting and branching. Overexpression of Eda in Krt14-Eda transgenics induces precocious branching. Microarray expression profiling of *Eda*$^{-/-}$ skin cultured in the absence or presence of recombinant Eda, showed an upregulation of amongst others *Wnt10b* and *PTHrP* in response to Eda. In accordance, higher levels of these mRNAs were detected by in situ hybridization of *Krt14-Eda* embryos. In an ex vivo explant culture setup adapted to monitor branching morphogenesis [89], recombinant Wnt3a and PTHrP accelerate branching morphogenesis in mammary. It is therefore likely that Eda promotes branching morphogenesis via its regulation of PTHrP and Wnt signaling [57].

Other evidence for an involvement of Wnt signaling in branching morphogenesis comes from the severely impaired ductal branching in constitutive and skin-specific null mutants for *Pygo2* [157] and *Lrp6-/-* mutants [162].

Tbx2 and *Tbx3* are expressed in the mesenchyme surrounding the nipple sheath, and *Tbx3* but not *Tbx2* is also expressed in the mammary epithelium at E18.5. Heterozygous *Tbx2* nulls have no mammary defects, but heterozygous *Tbx3* nulls display reduced branching in all their MRs at E18.5. Whereas double heterozygotes for both genes more often lose MR1–3 between E13.5 and E18.5, the branching defect in the rudiments that do survive is not more severe than in *Tbx3* heterozygotes [150].

12   Embryonic Mammary Gland and "Omics"

With a modification of Kratochwil's enzymatic tissue separation technique [82, 88, 204], the ME and MM of MR4 of several E12 embryos have been isolated and pooled per tissue for the subsequent extraction of mRNA and transcriptome analysis [29]. RNA was then amplified and subject to microarray analysis. By comparing the transcriptional profiles of both tissues with that of a non-treated intact MR (ME + MM), the gene pool that was activated by the enzyme treatment could be filtered out, and relevant transcriptome profiles were obtained with many new potential regulators of early mammogenesis. Interestingly, the ME profile showed many similarities with the mammary stem and progenitor cell populations of adult mammary gland #4 [29], and subsets of its profile also showed similarities to breast cancer profiles [205]. With similar tissue isolation techniques, the expression of miRNAs was also analyzed and led to the discovery of miR206 in the mammary mesenchyme [206]. Overexpression of miR206 by electroporation in flank explant cultures led to significant changes in gene expression in the MM, amongst others a reduction of estrogen receptor expression [206].

To reduce the effect of enzyme treatment on gene expression profiles and to speed up the tissue separation and processing time for increased RNA integrity, Sun et al. developed a tissue separation and harvesting technique based on the dehydrating effect of RNA-Later [86]. Analysis of these tissues have revealed that each of the five MRs has different expression profiles [ref. Sun and Veltmaat unpublished, http://www.veltmaatlab.net/research.html#sunli]. Any regulatory role of these differentially expressed genes in the identity of the MRs needs yet to be established.

13   Stem Cell Activity in the Embryonic MR

In 1979, Sakakura transplanted an E13.5 MR into the fat pad of a prepubertal mouse and demonstrated that the transplant could grow out, branch, and produce milk like an endogenous mammary gland [92]. With the exception of testing for milk production, similar outgrowth potential has been observed for intact E12.5 MRs [18]. These outcomes indicate that cells of the E12.5 ME have a pluripotent capacity and enormous proliferative potential, possibly via self-renewing stem or progenitor cells.

The intron5/6 region of the gene encoding Ship1 phosphatase contains stem-cell specific promoter activity [186]. Interestingly, this transgenic promoter construct drives GFP expression even in the ML and uniformly in the MRs at E11.5 and E12.5 ([180, 186] and cover illustration of [41]), suggesting the presence of mammary stem cells from the onset of mammogenesis onwards. In that context, it is of interest that (1) Wnt signaling is required for the induction and development of the embryonic mammary gland [191], as well as for self-renewal of mammary stem cells in the adult [207]; and that (2) Pygo2, which converges with Wnt-signaling, is enriched in adult mammary stem cells and required for proper induction and development of the embryonic MRs [157].

The phenotypic identification of mammary stem cell populations began in adult mammary glands, by fluorescence-activated cell sorting (FACS) of single cell suspensions of partial mammary glands. This technique is based on fluorescent labeling of tissue-specific cell-surface markers, which facilitates the separation of mammary epithelial cells from endothelial and stromal cells. Epithelial subpopulations can be further sorted based on fluorescent labeling of subpopulation-specific markers, and transplanted in limiting dilutions into cleared fat pads of prepubertal mice, to be scored for mammary repopulation units (MRUs) in these fat pads. Such studies identified a high MRU-potential of the CD24^{high};CD49f^{high} subpopulation, whose regenerative potential was demonstrated by their ability to generate daughter MRUs upon retransplantation to a new cleared fat pad [208, 209]. This technique has recently been used to identify subpopulations with high MRU-capacity in the E18.5 ME [17–19]. These studies demonstrated that the stem cell activity of the embryonic ME resides entirely in the CD24^{high};CD49f^{high} subpopulation, and that embryonic ME has a higher regenerative potential than adult ME [17–19].

However, when single ME cells of an embryonic MR are transplanted in a cleared fat pad, they rarely generate mammary glands. Moreover, when the donor embryo is younger than E15.5, outgrowths are only observed when the ME cells are co-transplanted with Matrigel. Perhaps this can be explained by lineage-restricted stemcellness at E12.5 as follows: Cells can be labeled in a tissue-specific manner and under temporal control by combining the *CreERT2* and *mT/mG* transgenes. The *mT/mG* transgene (encoding the fluorochromes **T**omato-Red and **G**reen Fluorescent Protein, GFP), can be inserted in for example the Rosa26 (R26R) locus for ubiquitous expression. Under normal conditions, such transgenic cells express Tomato-Red, whereas upon exposure to Cre-recombinase (from the *CreERT2* transgene), they switch to GFP expression. The *Cre-ERT2* transgene expresses Cre-recombinase upon occupation of its ERT2 binding sites by estrogenic compounds like tamoxifen. Insertion of this transgene in the locus of a tissue-specific gene and temporal control of administering tamoxifen provides temporospatial control of the color switch of *mTmG* transgenic cells, and subsequently all progeny of switched cells will express GFP. Axin2 is a mediator and target of Wnt/β-catenin signaling, and is expressed throughout the MR epithelium at E12.5 [183]. When female mice pregnant of Axin2$^{CreERT2/+}$;R26R$^{mTmG/+}$ embryos are given tamoxifen mice on the 12th, 14th or 17th day of pregnancy, the mammary glands of their offspring in adulthood will only express GFP in luminal cell, indicating that embryonic mammary cells engaged in canonical Wnt signaling are progenitors for exclusively the luminal lineage [183]. These data suggest there may already be separate stem or progenitor cell populations for the luminal, the basal, and perhaps both cell layers at that time. This lineage restriction of at least some cells in the embryonic MR may explain the low take rate of transplanted single cells of embryonic MRs. On the other hand, transplantation of FACS-sorted lineage-restricted stem cells in cleared fat pads still yields normal outgrowths with a basal and luminal compartment, strongly suggesting that lineage-restriction is a facultative state in real life, which can be converted into bipotency upon disturbance of the normal cell and tissue integrity [183].

The success rate of generating a mammary gland increases dramatically when single ME cells are used of E15.5 and E16.5 MRs, and keeps on increasing by using E17.5 and E18.5 ME. These data suggest that critical properties required for the outgrowth of a mammary gland in such experiments are required at E15.5 [18]. It is worth noting that E15.5 is also the decisive stage for MRs in certain mutants (e.g., *Lef1*$^{-/-}$,*Msx1*$^{-/-}$;*Msx2*$^{-/-}$) to either survive or revert to an epidermal fate [135, 156, 168], just prior to keratinization and impermeabilization of the epidermis. As ME cells are thus not committed to a mammary fate prior to E15.5, an alternative explanation for the low take rate of single ME cells in transplantation assays, it that the harsh circumstances during cell dissociation may change their expression pattern such that they cannot maintain their identity as mammary stem cells of any kind [41].

Microarray analysis of this subpopulation revealed that the gene expression profile of E18.5 fetal mammary stem cells (fMaSCs) cells shows overlap with, but is very different from that of adult MaSCs. By contrast, the expression profile of fetal mammary stroma (fSTR) more closely resembles that of aMaSCs. fMaSCs express markers of multiple adult mammary lineages (indicating multipotency) in addition to gene sets that are unique for embryonic ME [18]. Although the expression signatures of fMaSCs and fSTR are significantly different from those of E12.5 ME respectively MM, it is of great interest that some breast cancer subtypes are enriched for any of these profiles [18, 29]. However, it must be noted that the entire experimental procedure prior to the gene-profiling step may have altered the expression pattern, given the observed differences in potential (bipotent versus lineage-restricted) observed for the same cell population in lineage-tracing experiments versus FACS + transplantation assays [183].

14 Experimental Design and Pitfalls in Interpretation of Own and Published Data

Studies on the embryonic mammary gland rely partly on distinct techniques, some of which differ from those in the adult mammary gland, such as explant culture, tissue separation and recombination, grafting, whole mount in situ hybridization, and immunodetection. The development and applications of those techniques has been described in this review. Figure 4 illustrates how these field-specific techniques can be combined in parallel or sequentially with generic molecular and biochemistry techniques, as well as with the most recent stem cell and "omics" techniques, to address most questions related to embryonic mammary gland development.

Until about the 1970s, only few experimental interventions were possible, and consequently most studies were based on histology and microscopy solely. Such studies revealed differences in histological appearance and organ morphology between different species and developmental stages. However, some researchers would speculate or draw conclusions about possible mechanisms that would cause these appearances and changes, without having the proper experimental basis for such conclusions. Some of these conclusions were wrongfully propagated in the literature and extrapolated to other species, and almost became dogmatic to the field. For example mammary gland development was published to start with the formation of a continuous ectodermal band/line/ridge from and on which the MRs develop [45]. However, recent studies with molecular techniques and genetically engineered mice with more than the usual five pairs of MRs, contradict this: First many individual sites of possible MR development are formed, which then temporarily fuse into a continuous line (one line on each flank),

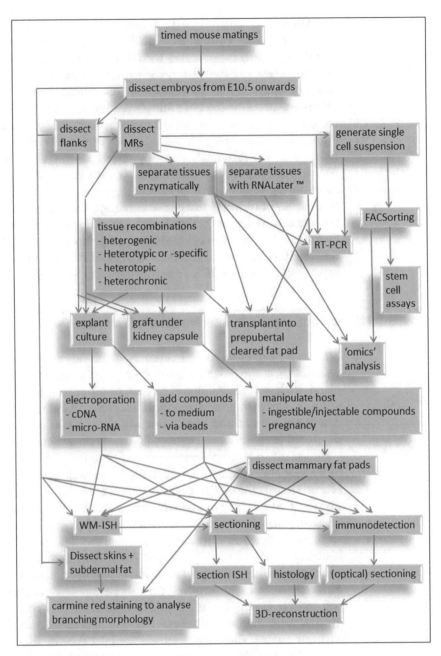

Fig. 4 Flowchart for experimental setup for studies of embryonic mammary development. These studies start with the husbandry of (genetically engineered) mice, and may include explant culture, and a variety of molecular and histological analyses

after which MR development continues only at a subset of the initial sites [41, 61]. The histological observation that the ML and MRs were already multilayered before the surface ectoderm, led to a similar unfounded conclusion that this precocious multilayering was due to locally enhanced cell proliferation [45]. Decades later, a study with tritiated thymidine incorporation demonstrated the near absence of proliferative activity in MRs between E12.5 and E13.5 [13]. In subsequent literature, these two conclusions were combined and propagated as the misconception that MRs would undergo 24 h of proliferative arrest - after supposedly initial high proliferative activity - between E12.5 and E13.5, even though Balinsky had already contested the assumption that initial multilayering was due to cell proliferation [46]. These examples underscore two often-made mistakes: drawing mechanistic conclusions from static data without experimental variables, and the wrongful combination and rephrasing of published conclusions.

Nowadays, gene and protein expression data are often similarly misinterpreted. For example, whole mount in situ hybridization patterns are often judged without sectioning. However, due to the transparency of the embryo, hybridization signals of deeper tissues can be seen through the embryonic skin, but not attributed to a particular organ or tissue. It may be tempting to interpret a stacked array of dorsoventral hybridization stripes on the flank as somitic gene expression, while it also possible that the signal is generated by the somite-derived dermis or overlying ectoderm. Only sectioning of the embryo can reveal which (combination) of these tissues generates the hybridization signal.

In addition, the absence of a hybridization signal is often interpreted as the absence of a structure, e.g., the absence of *Wnt10b* or *Lef1* expression as markers for MRs, is often interpreted as an absence of MRs. This implies that the researcher assumes that these markers are required for the formation of MRs. This assumption is understandable, since Wnt signaling is known to be required for placode formation [148]. Nonetheless, the assumption is incorrect. Whereas *Wnt10b* is a very suitable marker for the ML and MRs of C57BL/6J mice [61], some albino mouse strains do not express this marker yet develop functional mammary glands (J. Veltmaat, unpublished observations) and *Wnt10b* null mice have no reported mammary gland defect [210]. This illustrates that mere gene expression should not be confused with (or misinterpreted as) gene function, and an absence of gene expression may not be interpreted as an absence of a structure. Similarly, *Lef1* is a marker for and mediator of canonical Wnt-signaling. Whereas MRs of *Lef1* nulls show severe hypoplasia at E12.5 [156] and arrest in bud stage or disappear by E15.5 [135], all MRs are induced at E11.5 [pers. comm Kratochwil in ref. [53]; and personal observations]. Therefore, an absence of *Lef1* expression should not be interpreted as an absence of MR formation. TOPGAL-F is a suitable reporter for only a subset of all Wnt-signaling. Even though it also nicely marks the ML and MRs, its absence of expression does not necessarily indicate an absence of all Wnt signaling or MR formation. The absence of marker expression should always be accompanied by histological analysis to warrant a conclusion that structures are indeed absent.

Conversely, the presence of gene-expression may indicate that a structure is there, but does not necessarily mean the structure is normal. In some mutant mouse strains on a TOPGAL background, the MRs may appear as narrower or wider dots, which is often interpreted as smaller or larger MRs. However, the size but not morphology may still be normal, as the MRs may have a relatively elongated respectively flattened shape compared to wild type littermates. In conclusion, it is always advisable to combine gene expression analysis with histological analysis.

If new mutant mice are generated and published "with no mammary defects" or "to nurse their offspring normally," this does not exclude possible anomalies in the number, morphology, or full functionality of mammary glands, especially if the publishing lab has no interest in mammary development per se.

Only since the beginning of this century has the notion grown that all pairs of MRs in mouse embryos are different with respect to the timing of their appearance [52, 53, 62], their molecular requirements and morphogenetic program [27, 52–54]. When reading older literature, but even when reading recent literature, one should keep in mind that findings and models may be published as if valid for all MRs, while perhaps only one, two, or three pairs of MRs were used for the study without specific mentioning. MR3 is especially easily accessible for experimentation; whereas MR1 and MR5 are hidden behind the limbs and hard to view or retrieve, and consequently are often not taken along in the analysis. Thus, if a publication states that for example embryonic mammary glands of embryonic lethal mutants develop with—or without—abnormalities upon transplantation in a cleared fat pad, this may not hold true for all MRs. On the same note, it is advisable to design future studies such, that all MRs are examined separately in each experiment, and reported as separate entities in the literature as well.

Modern techniques are becoming increasingly sensitive, allowing even stem cell assays and transcriptome analysis to be performed with embryonic mammary rudiments.

A few technical territories remain unexplored, such as proteomics and biochemical assays such as immunoprecipitations or pull-down, due to their requirement for greater quantities of sample material. But a true technical challenge seems to be live imaging of cell behavior during embryonic mammary gland development, due to the continual shift of the plane of interest during growth ex vivo. The establishment of good live imaging protocols would be extremely helpful in establishing the area and direction of cell migration in the establishment of the ML and MRs, or the behavior of cells within the developing MRs.

15 Conclusion

This review describes how, with perhaps the exception of some live imaging and biochemical techniques that require large amounts of protein as input, all techniques that are used to study the postnatal mammary gland can also be used to study the embryonic mammary gland. But the embryonic mammary gland has other advantages: It can be easily dissected, and optionally its tissues can be separated and recombined in various combinations, for growth ex vivo or as a transplant, which facilitates the study the role of tissue interactions in morphogenesis and function. Such studies are more difficult to carry out with adult mammary glands, due to their greater tissue complexity. Moreover, in cases where the role of a gene or its mutation in the postnatal gland cannot be studied due to perinatal lethality of constitutive mutants, and tissue-specific mutants are not available, mutant embryonic MRs can be transplanted into a wild type prepubertal mammary gland for further study.

Studies on the embryonic mammary gland are certainly relevant to postnatal mammary gland development, function, and pathology, because the embryonic mammary gland displays many features of the postnatal mammary gland: It already contains stem cells [17–19], commits to a mammary fate by producing milk when stimulated by pregnancy hormones [92], and undergoes a series of morphogenetic changes that are reiterated during puberty and pregnancy. There is a high degree of similarity in tissue interactions and molecular controls of these changes during embryonic and postnatal life in the mouse [13, 16]. Moreover, such molecular similarities have also begun to be discovered between mice and human, even extending between murine embryonic mammary development and postnatal mammary tumorigenesis in mouse and human [211]. Another resemblance lies in the influence of the mammary stroma on the functional differentiation and homeostasis of the mammary epithelium during embryonic mammary gland development, and postnatal formation of hyperplasia and neoplastic lesions [30, 90, 212]. Such parallels make studies of the embryonic mammary gland important even beyond the questions concerning the embryonic phase per se [38].

In conclusion, given the relative lack of tissue complexity of the embryonic mammary glands and the ease with which they can be accessed and manipulated for study, the embryonic mammary glands are a very suitable starting point or alternative or additional model to study a wide range of questions pertaining to normal and pathological postnatal breast development as well.

Acknowledgements

The author is grateful for the financial support of the European Commission Fifth Framework, Institut Curie (Paris, France), California Breast Cancer Research Program, and Children's Hospital Los Angeles (CA, USA), as well as the A*STAR Institute of Molecular and Cell Biology and National University of Singapore (Singapore) that enabled her so far to explore and contribute to this fascinating field of research. She would also like to thank her colleagues through the years both in the lab and in the field for sharing the passion to further this field of research. She regrets if any contribution to this field is not cited in this review.

References

1. Aristotle (approx. 350 B.C.) On the parts of animals – in four books (trans: Ogle W). Kegan Paul, Trench & Co, London
2. Evarts JL, Rasweiler JJ, Behringer RR, Hennighausen L, Robinson GW (2004) A morphological and immunohistochemical comparison of mammary tissues from the short-tailed fruit bat (Carollia perspicillata) and the mouse. Biol Reprod 70(6):1573–1579. doi:10.1095/biolreprod.103.022988
3. Francis CM, Anthoy EP, Brunton JA, Kunz TH (1994) Lactation in male fruit bats. Nature 367:691–692
4. Linnaeus C (1758–1759) Systema naturae per regna tria natura, secundum classes, ordines, genera, species, cum characteribus, differentiis synonymis, locis (trans: Turton W), 10 edn. Laurentius Salvius, Stockholm
5. Oftedal OT, Dhouailly D (2013) Evo-devo of the mammary gland. J Mammary Gland Biol Neoplasia 18(2):105–120. doi:10.1007/s10911-013-9290-8
6. Koyama S, Wu HJ, Easwaran T, Thopady S, Foley J (2013) The nipple: a simple intersection of mammary gland and integument, but focal point of organ function. J Mammary Gland Biol Neoplasia 18(2):121–131. doi:10.1007/s10911-013-9289-1
7. Oftedal OT (2002) The mammary gland and its origin during synapsid evolution. J Mammary Gland Biol Neoplasia 7(3):225–252
8. Bresslau E (1920) The mammary apparatus of the Mammalia: in the light of ontogenesis and phylogenesis. Methuen & Co, Ltd, London
9. Diamond JM (1987) Evolutionary adaptations. Aristotle's theory of mammalian teat number is confirmed. Nature 325(6101):200. doi:10.1038/325200a0
10. Koo W, Tank S, Martin S, Shi R (2014) Human milk and neurodevelopment in children with very low birth weight: a systematic review. Nutr J 13:94. doi:10.1186/1475-2891-13-94
11. Goldman AS (2012) Evolution of immune functions of the mammary gland and protection of the infant. Breastfeed Med 7(3):132–142. doi:10.1089/bfm.2012.0025
12. Butler JE (1979) Immunologic aspects of breast feeding, antiinfectious activity of breast milk. Semin Perinatol 3(3):255–270
13. Robinson GW, Karpf AB, Kratochwil K (1999) Regulation of mammary gland development by tissue interaction. J Mammary Gland Biol Neoplasia 4(1):9–19
14. Robinson GW (2007) Cooperation of signalling pathways in embryonic mammary gland development. Nat Rev Genet 8(12):963–972. doi:10.1038/nrg2227, nrg2227 [pii]
15. Robinson GW (2004) Identification of signaling pathways in early mammary gland development by mouse genetics. Breast Cancer Res 6(3):105–108. doi:10.1186/bcr776

16. Watson CJ, Khaled WT (2008) Mammary development in the embryo and adult: a journey of morphogenesis and commitment. Development 135(6):995–1003. doi:10.1242/dev.005439

17. Makarem M, Spike BT, Dravis C, Kannan N, Wahl GM, Eaves CJ (2013) Stem cells and the developing mammary gland. J Mammary Gland Biol Neoplasia 18(2):209–219. doi:10.1007/s10911-013-9284-6

18. Spike BT, Engle DD, Lin JC, Cheung SK, La J, Wahl GM (2012) A mammary stem cell population identified and characterized in late embryogenesis reveals similarities to human breast cancer. Cell Stem Cell 10(2):183–197. doi:10.1016/j.stem.2011.12.018

19. Makarem M, Kannan N, Nguyen LV, Knapp DJ, Balani S, Prater MD, Stingl J, Raouf A, Nemirovsky O, Eirew P, Eaves CJ (2013) Developmental changes in the in vitro activated regenerative activity of primitive mammary epithelial cells. PLoS Biol 11(8), e1001630. doi:10.1371/journal.pbio.1001630

20. Bell A (1899) Bell on the development by selection of supernumerary mammae in sheep. Science 9(227):637–639

21. Hsu MJ, Moore J, Lin JF, Agoramoorthy G (2000) High incidence of supernumerary nipples and twins in formosan macaques (Macaca cyclopis) at Mt. Longevity, Taiwan. Am J Primatol 52(4):199–205. doi:10.1002/1098-2345(200012)52:4<199::AID-AJP4>3.0.CO;2-2

22. Cellini A, Offidani A (1992) Familial supernumerary nipples and breasts. Dermatology 185(1):56–58

23. Schmidt H (1998) Supernumerary nipples: prevalence, size, sex and side predilection – a prospective clinical study. Eur J Pediatr 157(10):821–823

24. Duijvesteijn N, Veltmaat JM, Knol EF, Harlizius B (2014) High-resolution association mapping of number of teats in pigs reveals regions controlling vertebral development. BMC Genomics 15:542. doi:10.1186/1471-2164-15-542

25. Drickamer LC, Rosenthal TL, Arthur RD (1999) Factors affecting the number of teats in pigs. J Reprod Fertil 115(1):97–100

26. Merks JW, Mathur PK, Knol EF (2012) New phenotypes for new breeding goals in pigs. Animal 6(4):535–543. doi:10.1017/S1751731111002266

27. Veltmaat JM, Ramsdell AF, Sterneck E (2013) Positional variations in mammary gland development and cancer. J Mammary Gland Biol Neoplasia 18(2):179–188. doi:10.1007/s10911-013-9287-3

28. Howard B, Ashworth A (2006) Signalling pathways implicated in early mammary gland morphogenesis and breast cancer. PLoS Genet 2(8), e112. doi:10.1371/journal.pgen.0020112, 06-PLGE-RV-0134R2 [pii]

29. Wansbury O, Mackay A, Kogata N, Mitsopoulos C, Kendrick H, Davidson K, Ruhrberg C, Reis-Filho JS, Smalley MJ, Zvelebil M, Howard BA (2011) Transcriptome analysis of embryonic mammary cells reveals insights into mammary lineage establishment. Breast Cancer Res 13(4):R79. doi:10.1186/bcr2928

30. Sakakura T, Suzuki Y, Shiurba R (2013) Mammary stroma in development and carcinogenesis. J Mammary Gland Biol Neoplasia 18(2):189–197. doi:10.1007/s10911-013-9281-9

31. Takebe N, Warren RQ, Ivy SP (2011) Breast cancer growth and metastasis: interplay between cancer stem cells, embryonic signaling pathways and epithelial-to-mesenchymal transition. Breast Cancer Res 13(3):211. doi:10.1186/bcr2876

32. Kim EJ, Jung HS, Lu P (2013) Pleiotropic functions of fibroblast growth factor signaling in embryonic mammary gland development. J Mammary Gland Biol Neoplasia 18(2):139–142. doi:10.1007/s10911-013-9278-4

33. Douglas NC, Papaioannou VE (2013) The T-box transcription factors TBX2 and TBX3 in mammary gland development and breast cancer. J Mammary Gland Biol Neoplasia 18(2):143–147. doi:10.1007/s10911-013-9282-8

34. American Cancer Society (2014) Breast cancer key statistics. http://www.cancer.org/cancer/breastcancer/detailedguide/breast-cancer-key-statistics. Accessed 17 Nov 2014

35. Cancer Research UK (2014) Breast cancer key facts. http://publications.cancerresearchuk.org/downloads/Product/CS_KF_BREAST.pdf. Accessed 17 Nov 2014

36. Cancer Research UK (2014) Breast cancer risk factors. http://www.cancerresearchuk.org/cancer-info/cancerstats/types/breast/riskfactors/breast-cancer-risk-factors#Family. Accessed 17 Nov 2014

37. American Cancer Society (2014) What are the risk factors for breast cancer? http://www.cancer.org/cancer/breastcancer/detailedguide/breast-cancer-risk-factors. Accessed 17 Nov 2014

38. Howard BA, Veltmaat JM (2013) Embryonic mammary gland development; a domain of

fundamental research with high relevance for breast cancer research. Preface. J Mammary Gland Biol Neoplasia 18(2):89–91. doi:10.1007/s10911-013-9296-2

39. Propper AY (1973) Développement Embryonnaire de la Gland Mammaire Chez le Lapin (Oryctolagus Cuniculus L.). Ph.D., Université de Besançon, Besançon

40. Cardiff RD, Wellings SR (1999) The comparative pathology of human and mouse mammary glands. J Mammary Gland Biol Neoplasia 4(1):105–122

41. Propper AY, Howard BA, Veltmaat JM (2013) Prenatal morphogenesis of mammary glands in mouse and rabbit. J Mammary Gland Biol Neoplasia 18(2):93–104. doi:10.1007/s10911-013-9298-0

42. Capecchi MR (1989) The new mouse genetics: altering the genome by gene targeting. Trends Genet 5(3):70–76

43. Lindfors PH, Voutilainen M, Mikkola ML (2013) Ectodysplasin/NF-kappaB signaling in embryonic mammary gland development. J Mammary Gland Biol Neoplasia 18(2):165–169. doi:10.1007/s10911-013-9277-5

44. Hiremath M, Wysolmerski J (2013) Parathyroid hormone-related protein specifies the mammary mesenchyme and regulates embryonic mammary development. J Mammary Gland Biol Neoplasia 18(2):171–177. doi:10.1007/s10911-013-9283-7

45. Turner CW, Gomez ET (1933) The normal development of the mammary gland of the male and female albino mouse. I Intrauterine. Missouri Agric Exp Stat Res Bull 182:3–20

46. Balinsky BI (1950) On the prenatal growth of the mammary gland rudiment in the mouse. J Anat 84(3):227–235

47. Raynaud A (1961) Morphogenesis of the mammary gland. In: Kon S, Cowie A (eds) Milk: The mammary gland and its secretion vol 1. Academic, New York, NY, pp 3–46

48. Dalton AJ (1945) Histogenesis of the mammary gland of the mouse. The Science Press Printing Company, Lancaster, PA

49. Durnberger H, Kratochwil K (1980) Specificity of tissue interaction and origin of mesenchymal cells in the androgen response of the embryonic mammary gland. Cell 19(2):465–471, doi:0092-8674(80)90521-8 [pii]

50. Durnberger H, Heuberger B, Schwartz P, Wasner G, Kratochwil K (1978) Mesenchyme-mediated effect of testosterone on embryonic mammary epithelium. Cancer Res 38(11 Pt 2):4066–4070

51. Propper AY (1978) Wandering epithelial cells in the rabbit embryo milk line. A preliminary scanning electron microscope study. Dev Biol 67(1):225–231, doi:0012-1606(78)90311-1 [pii]

52. Mailleux AA, Spencer-Dene B, Dillon C, Ndiaye D, Savona-Baron C, Itoh N, Kato S, Dickson C, Thiery JP, Bellusci S (2002) Role of FGF10/FGFR2b signaling during mammary gland development in the mouse embryo. Development 129(1):53–60

53. Veltmaat JM, Mailleux AA, Thiery JP, Bellusci S (2003) Mouse embryonic mammogenesis as a model for the molecular regulation of pattern formation. Differentiation 71(1):1–17, doi:S0301-4681(09)60262-0 [pii] 10.1046/j.1432-0436.2003.700601.x

54. Lee MY, Racine V, Jagadpramana P, Sun L, Yu W, Du T, Spencer-Dene B, Rubin N, Le L, Ndiaye D, Bellusci S, Kratochwil K, Veltmaat JM (2011) Ectodermal influx and cell hypertrophy provide early growth for all murine mammary rudiments, and are differentially regulated among them by Gli3. PLoS One 6(10), e26242. doi:10.1371/journal.pone.0026242

55. Rasmussen SB, Young LJT, Smith GH (2000) Preparing mammary gland whole mounts from mice. In: Ip MM, Asch BB (eds) Methods in mammary gland biology and breast cancer research. Kluwer, New York, NY, pp 75–85

56. Panchal H, Wansbury O, Howard BA (2010) Embryonic mammary anlagen analysis using immunolabelling of whole mounts. Methods Mol Biol 585:261–270. doi:10.1007/978-1-60761-380-0_18

57. Voutilainen M, Lindfors PH, Lefebvre S, Ahtiainen L, Fliniaux I, Rysti E, Murtoniemi M, Schneider P, Schmidt-Ullrich R, Mikkola ML (2012) Ectodysplasin regulates hormone-independent mammary ductal morphogenesis via NF-kappaB. Proc Natl Acad Sci U S A 109(15):5744–5749. doi:10.1073/pnas.1110627109

58. Kogata N, Howard BA (2013) A whole-mount immunofluorescence protocol for three-dimensional imaging of the embryonic mammary primordium. J Mammary Gland Biol Neoplasia 18(2):227–231. doi:10.1007/s10911-013-9285-5

59. Sakakura T, Kusano I, Kusakabe M, Inaguma Y, Nishizuka Y (1987) Biology of mammary fat pad in fetal mouse: capacity to support development of various fetal epithelia in vivo. Development 100(3):421–430

60. Veltmaat JM, Relaix F, Le LT, Kratochwil K, Sala FG, van Veelen W, Rice R, Spencer-Dene B, Mailleux AA, Rice DP, Thiery JP, Bellusci

S (2006) Gli3-mediated somitic Fgf10 expression gradients are required for the induction and patterning of mammary epithelium along the embryonic axes. Development 133(12):2325–2335, doi:133/12/2325 [pii] 10.1242/dev.02394

61. Veltmaat JM, Van Veelen W, Thiery JP, Bellusci S (2004) Identification of the mammary line in mouse by Wnt10b expression. Dev Dyn 229(2):349–356. doi:10.1002/dvdy.10441

62. Eblaghie MC, Song SJ, Kim JY, Akita K, Tickle C, Jung HS (2004) Interactions between FGF and Wnt signals and Tbx3 gene expression in mammary gland initiation in mouse embryos. J Anat 205(1):1–13. doi:10.1111/j.0021-8782.2004.00309.x

63. Kratochwil K (1971) In vitro analysis of the hormonal basis for the sexual dimorphism in the embryonic development of the mouse mammary gland. J Embryol Exp Morphol 25(1):141–153

64. Sakakura T, Sakagami Y, Nishizuka Y (1982) Dual origin of mesenchymal tissues participating in mouse mammary gland embryogenesis. Dev Biol 91(1):202–207, doi:0012-1606(82)90024-0 [pii]

65. Sakakura T (1987) Mammary embryogenesis. In: Neville MC, Daniel CW (eds) The mammary gland: development, regulation and function. Plenum Press, Michigan, pp 37–66

66. Hogg NA, Harrison CJ, Tickle C (1983) Lumen formation in the developing mouse mammary gland. J Embryol Exp Morphol 73:39–57

67. Foley J, Dann P, Hong J, Cosgrove J, Dreyer B, Rimm D, Dunbar M, Philbrick W, Wysolmerski J (2001) Parathyroid hormone-related protein maintains mammary epithelial fate and triggers nipple skin differentiation during embryonic breast development. Development 128(4):513–525

68. Balinsky BI (1952) On the developmental processes in mammary glands and other epidermal structures. Trans R Soc Edinb 62:1–31

69. Propper AY (1976) Modalités et déterminisme du développement embryonnaire de la glande mammaire. Senelogia 1(4):19–26

70. Cunha GR, Young P, Christov K, Guzman R, Nandi S, Talamantes F, Thordarson G (1995) Mammary phenotypic expression induced in epidermal cells by embryonic mammary mesenchyme. Acta Anat (Basel) 152(3):195–204

71. Cho KW, Kim JY, Song SJ, Farrell E, Eblaghie MC, Kim HJ, Tickle C, Jung HS (2006) Molecular interactions between Tbx3 and Bmp4 and a model for dorsoventral positioning of mammary gland development. Proc Natl Acad Sci U S A 103(45):16788–16793. doi:10.1073/pnas.0604645103, 0604645103 [pii]

72. Fell HB, Robison R (1929) The growth, development and phosphatase activity of embryonic avian femora and limb-buds cultivated in vitro. Biochem J 23(4):767–784

73. Hardy MH (1950) The development in vitro of the mammary glands of the mouse. J Anat 84(4):388–393

74. Lasfargues EY, Murray MR (1959) Hormonal influences on the differentiation and growth of embryonic mouse mammary glands in organ culture. Dev Biol 1(4):413–435. doi:10.1016/0012-1606(59)90037-5

75. David D, Propper A (1964) Sur la culture organotypique de la glande mammaire embryonnaire du lapin. C R Soc Seances Soc Biol Fil 158:2315–2317

76. Propper A, Gomot L (1967) Interactions tissulaires au cours de l'organogen'ese de la glande mammaire de l'embryon de lapin. C R Acad Sci Hebd Seances Acad Sci D 264(22):2573–2575

77. Propper A (1968) Relations épidermo-mésodermiques dans la différenciation de l'ébauche mammaire d'embryon de lapin. Ann Embriol Morphog 1(2):151–160

78. Propper A (1972) Rôle du mésenchyme dans la différenciation de la glande mammaire chez l'embryon de lapin. Bull Soc Zool France 97(3):505–512

79. Propper A (1969) Compétence de l'épiderme embryonnaire d'oiseau vis-à-vis de l'inducteur mammaire mésenchymateux. C R Acad Sci Hebd Seances Acad Sci D 268(10):1423–1426

80. Grobstein C (1956) Trans-filter induction of tubules in mouse metanephrogenic mesenchyme. Exp Cell Res 10(2):424–440

81. Kratochwil K (1969) Organ specificity in mesenchymal induction demonstrated in the embryonic development of the mammary gland of the mouse. Dev Biol 20(1):46–71, doi:0012-1606(69)90004-9 [pii]

82. Kratochwil K (1986) Tissue combination and organ culture studies in the development of the embryonic mammary gland. Dev Biol (New York, NY: 1985) 4:315–333

83. Lee JM, Kim EJ, Jung HS (2013) A method for electroporation to study gene function in mammary gland development. J Mammary Gland Biol Neoplasia 18(2):233–237. doi:10.1007/s10911-013-9292-6

84. Dunbar ME, Young P, Zhang JP, McCaughern-Carucci J, Lanske B, Orloff JJ,

Karaplis A, Cunha G, Wysolmerski JJ (1998) Stromal cells are critical targets in the regulation of mammary ductal morphogenesis by parathyroid hormone-related protein. Dev Biol 203(1):75–89

85. Mustonen T, Ilmonen M, Pummila M, Kangas AT, Laurikkala J, Jaatinen R, Pispa J, Gaide O, Schneider P, Thesleff I, Mikkola ML (2004) Ectodysplasin A1 promotes placodal cell fate during early morphogenesis of ectodermal appendages. Development 131(20): 4907–4919. doi:10.1242/dev.01377, dev. 01377 [pii]

86. Sun L, Lee MY, Veltmaat JM (2011) A non-enzymatic microsurgical dissection technique of mouse embryonic tissues for gene expression profiling applications. Int J Dev Biol 55:969–974. doi:10.1387/ijdb.113424ls

87. Veltmaat JM (2013) Investigating molecular mechanisms of embryonic mammary gland development by bead-implantation in embryonic flank explant cultures - a protocol. J Mammary Gland Biol Neoplasia 18(2):247–252. doi:10.1007/s10911-013-9297-1

88. Cunha GR (2013) Tissue recombination techniques for mouse embryonic mammary glands. J Mammary Gland Biol Neoplasia 18(2): 221–225. doi:10.1007/s10911-013-9295-3

89. Voutilainen M, Lindfors PH, Mikkola ML (2013) Protocol: ex vivo culture of mouse embryonic mammary buds. J Mammary Gland Biol Neoplasia 18(2):239–245. doi:10.1007/s10911-013-9288-2

90. Sakakura T, Nishizuka Y, Dawe CJ (1976) Mesenchyme-dependent morphogenesis and epithelium-specific cytodifferentiation in mouse mammary gland. Science 194(4272): 1439–1441

91. Deome KB, Faulkin LJ Jr, Bern HA, Blair PB (1959) Development of mammary tumors from hyperplastic alveolar nodules transplanted into gland-free mammary fat pads of female C3H mice. Cancer Res 19(5): 515–520

92. Sakakura T, Nishizuka Y, Dawe CJ (1979) Capacity of mammary fat pads of adult C3H/HeMs mice to interact morphogenetically with fetal mammary epithelium. J Natl Cancer Inst 63(3):733–736

93. Medina D, Vaage J, Sedlacek R (1973) Mammary noduligenesis and tumorigenesis in pathogen-free C3Hf mice. J Natl Cancer Inst 51(3):961–965

94. Robinson GW, Accili D, Hennighausen L (2000) Rescue of mammary epithelium of early lethal phenotypes by embryonic mammary gland transplantation as exemplified with insulin receptor null mice. In: Ip MM, Asch BB (eds) Methods in mammary gland biology and breast cancer research. Kluwer, New York, NY, pp 307–316

95. Gallego MI, Beachy PA, Hennighausen L, Robinson GW (2002) Differential requirements for shh in mammary tissue and hair follicle morphogenesis. Dev Biol 249(1):131–139, doi:S001216060290761X [pii]

96. Michno K, Boras-Granic K, Mill P, Hui CC, Hamel PA (2003) Shh expression is required for embryonic hair follicle but not mammary gland development. Dev Biol 264(1):153–165, doi:S0012160603004019 [pii]

97. Joshi PA, Chang H, Hamel PA (2006) Loss of Alx4, a stromally-restricted homeodomain protein, impairs mammary epithelial morphogenesis. Dev Biol 297(1):284–294. doi:10.1016/j.ydbio.2006.05.032

98. Raynaud A, Pieau C (1974) Inhibitory capacity of testosterone towards sexual and mammary embryonic primordia. C R Soc Seances Soc Biol Fil 168(2-3):207–210

99. Raynaud A, Raynaud J (1953) Process of destruction of the second inguinal pair of mammary buds in male mouse fetus. C R Soc Seances Soc Biol Fil 147(23-24): 1962–1967

100. Richardson FL, Cloudman AM (1947) The mammary gland development in male mice at nine weeks of age. Anat Rec 97(2): 223–237

101. Hoshino K (1965) Development and function of mammary glands of mice prenatally exposed to testosterone propionate. Endocrinology 76:789–794. doi:10.1210/endo-76-4-789

102. Elger W, Neumann F (1966) The role of androgens in differentiation of the mammary gland in male mouse fetuses. Proc Soc Exp Biol Med 123(3):637–640

103. Lyon MF, Hawkes SG (1970) X-linked gene for testicular feminization in the mouse. Nature 227(5264):1217–1219

104. Attardi B, Ohno S (1978) Physical properties of androgen receptors in brain cytosol from normal and testicular feminized (Tfm/y hermaphrodite) mice. Endocrinology 103(3): 760–770. doi:10.1210/endo-103-3-760

105. Gehring U, Tomkins GM, Ohno S (1971) Effect of the androgen-insensitivity mutation on a cytoplasmic receptor for dihydrotestosterone. Nature 232(30):106–107

106. Kratochwil K, Schwartz P (1976) Tissue interaction in androgen response of embryonic mammary rudiment of mouse: identification of target tissue for testosterone. Proc Natl Acad Sci U S A 73(11):4041–4044

107. Heuberger B, Fitzka I, Wasner G, Kratochwil K (1982) Induction of androgen receptor formation by epithelium-mesenchyme interaction in embryonic mouse mammary gland. Proc Natl Acad Sci U S A 79(9):2957–2961

108. Wasner G, Hennermann I, Kratochwil K (1983) Ontogeny of mesenchymal androgen receptors in the embryonic mouse mammary gland. Endocrinology 113(5):1771–1780. doi:10.1210/endo-113-5-1771

109. Dunbar ME, Dann PR, Robinson GW, Hennighausen L, Zhang JP, Wysolmerski JJ (1999) Parathyroid hormone-related protein signaling is necessary for sexual dimorphism during embryonic mammary development. Development 126(16):3485–3493

110. vom Saal FS, Quadagno DM, Even MD, Keisler LW, Keisler DH, Khan S (1990) Paradoxical effects of maternal stress on fetal steroids and postnatal reproductive traits in female mice from different intrauterine positions. Biol Reprod 43(5):751–761

111. Vandenberg LN, Maffini MV, Wadia PR, Sonnenschein C, Rubin BS, Soto AM (2007) Exposure to environmentally relevant doses of the xenoestrogen bisphenol-A alters development of the fetal mouse mammary gland. Endocrinology 148(1):116–127. doi:10.1210/en.2006-0561, en.2006-0561 [pii]

112. Raynaud A (1955) Frequency and distribution of mammary deformities in the mouse fetus following estrogen injection. C R Soc Seances Soc Biol Fil 149(11-12):1229–1233

113. Raynaud A, Raynaud J (1956) Production of experimental mammary malformations of the mammary gland in mouse fetus by sex hormones. Ann Inst Pasteur (Paris) 90(1):39–91

114. Kratochwil K (1977) Development and loss of androgen responsiveness in the embryonic rudiment of the mouse mammary gland. Dev Biol 61(2):358–365

115. Raynaud A, Chaulin-Serviniere J (1954) Effect of estrogenic hormone on transformation of the mammary anlage of the single primary cord in the mouse fetus in the enlarge with multiple nodes or two primary mammary cords. C R Hebd Seances Acad Sci 239(2):191–193

116. Raynaud A, Raynaud J (1955) Frequency of mammary deformities in female mouse fetus from mice receiving estradiol dipropionate injection during gestation. C R Soc Seances Soc Biol Fil 149(11-12):1233–1236

117. Raynaud A, Raynaud J (1955) Formation of epidermic cone supporting the mammary cord in the fetus of mice treated with estrogens. C R Hebd Seances Acad Sci 240(7):810–812

118. Raynaud A, Raynaud J (1954) Various mammary malformations produced in mouse fetus by sex hormones. C R Soc Seances Soc Biol Fil 148(11-12):963–968

119. Raynaud A, Raynaud J (1954) Effects of injection of less than a milligram of estradiol in pregnant mice on primordial cells of mammary glands of female mouse fetuses. C R Soc Seances Soc Biol Fil 148(9-10):848–853

120. Raynaud A, Raynaud JC (1956) Successive fetal stages of inhibition of the development of the mammary rudiments of the mouse, under the effect of an estrogenic hormone. C R Hebd Seances Acad Sci 243(4):424–427

121. Narbaitz R, Stumpf WE, Sar M (1980) Estrogen receptors in mammary gland primordia of fetal mouse. Anat Embryol 158(2):161–166

122. Lemmen JG, Broekhof JL, Kuiper GG, Gustafsson JA, van der Saag PT, van der Burg B (1999) Expression of estrogen receptor alpha and beta during mouse embryogenesis. Mech Dev 81(1-2):163–167

123. van der Burg B, Sonneveld E, Lemmen JG, van der Saag PT (1999) Morphogenetic action of retinoids and estrogens. Int J Dev Biol 43(7):735–743

124. Lemmen JG, van den Brink CE, Legler J, van der Saag PT, van der Burg B (2002) Detection of oestrogenic activity of steroids present during mammalian gestation using oestrogen receptor alpha- and oestrogen receptor beta-specific in vitro assays. J Endocrinol 174(3):435–446

125. Ismail PM, Li J, DeMayo FJ, O'Malley BW, Lydon JP (2002) A novel LacZ reporter mouse reveals complex regulation of the progesterone receptor promoter during mammary gland development. Mol Endocrinol 16(11):2475–2489. doi:10.1210/me.2002-0169

126. Acevedo N, Davis B, Schaeberle CM, Sonnenschein C, Soto AM (2013) Perinatally administered bisphenol a as a potential mammary gland carcinogen in rats. Environ Health Perspect 121(9):1040–1046. doi:10.1289/ehp.1306734

127. Cabaton NJ, Canlet C, Wadia PR, Tremblay-Franco M, Gautier R, Molina J, Sonnenschein C, Cravedi JP, Rubin BS, Soto AM, Zalko D (2013) Effects of low doses of bisphenol A on the metabolome of perinatally exposed CD-1 mice. Environ Health Perspect 121(5):586–593. doi:10.1289/ehp.1205588

128. Soto AM, Brisken C, Schaeberle C, Sonnenschein C (2013) Does cancer start in the womb? altered mammary gland development and predisposition to breast cancer due to in utero exposure to endocrine disruptors. J Mammary Gland Biol Neoplasia 18(2):199–208. doi:10.1007/s10911-013-9293-5

129. Vandenberg LN, Schaeberle CM, Rubin BS, Sonnenschein C, Soto AM (2013) The male mammary gland: a target for the xenoestrogen bisphenol A. Reprod Toxicol (Elmsford, NY) 37:15–23. doi:10.1016/j.reprotox.2013.01.002

130. Wadia PR, Cabaton NJ, Borrero MD, Rubin BS, Sonnenschein C, Shioda T, Soto AM (2013) Low-dose BPA exposure alters the mesenchymal and epithelial transcriptomes of the mouse fetal mammary gland. PLoS One 8(5), e63902. doi:10.1371/journal.pone.0063902

131. Gomot L, Propper A (1965) Localisation de la phosphatase alcaline et des acides nucleaiques dans la glande mammaire de l'embryon de lapin. Ann Facult Sci Univ Clermont 26(2):47–56

132. Kimata K, Sakakura T, Inaguma Y, Kato M, Nishizuka Y (1985) Participation of two different mesenchymes in the developing mouse mammary gland: synthesis of basement membrane components by fat pad precursor cells. J Embryol Exp Morphol 89:243–257

133. Sakakura T, Ishihara A, Yatani R (1991) Tenascin in mammary gland development: from embryogenesis to carcinogenesis. Cancer Treat Res 53:383–400

134. Inaguma Y, Kusakabe M, Mackie EJ, Pearson CA, Chiquet-Ehrismann R, Sakakura T (1988) Epithelial induction of stromal tenascin in the mouse mammary gland: from embryogenesis to carcinogenesis. Dev Biol 128(2):245–255

135. van Genderen C, Okamura RM, Farinas I, Quo RG, Parslow TG, Bruhn L, Grosschedl R (1994) Development of several organs that require inductive epithelial-mesenchymal interactions is impaired in LEF-1-deficient mice. Genes Dev 8(22):2691–2703

136. Kalembeyi I, Yoshida T, Iriyama K, Sakakura T (1997) Analysis of tenascin mRNA expression in the murine mammary gland from embryogenesis to carcinogenesis: an in situ hybridization study. Int J Dev Biol 41(4):569–573

137. Wysolmerski JJ, Philbrick WM, Dunbar ME, Lanske B, Kronenberg H, Broadus AE (1998) Rescue of the parathyroid hormone-related protein knockout mouse demonstrates that parathyroid hormone-related protein is essential for mammary gland development. Development 125(7):1285–1294

138. Raynaud A, Raynaud J (1956) Experimental production of mammary malformations in rat fetuses by sex hormones. Ann Inst Pasteur (Paris) 90(2):187–220

139. Johnson DR (1967) Extra-toes: a new mutant gene causing multiple abnormalities in the mouse. J Embryol Exp Morphol 17(3):543–581

140. Hui CC, Joyner AL (1993) A mouse model of Greig cephalopolysyndactyly syndrome: the extra-toesJ mutation contains an intragenic deletion of the Gli3 gene. Nat Genet 3(3):241–246. doi:10.1038/ng0393-241

141. Howard B, Panchal H, McCarthy A, Ashworth A (2005) Identification of the scaramanga gene implicates Neuregulin3 in mammary gland specification. Genes Dev 19(17):2078–2090, doi:19/17/2078 [pii] 10.1101/gad.338505

142. Howard BA, Gusterson BA (2000) The characterization of a mouse mutant that displays abnormal mammary gland development. Mamm Genome 11(3):234–237

143. Howard BA, Gusterson BA (2000) Mammary gland patterning in the AXB/BXA recombinant inbred strains of mouse. Mech Dev 91(1-2):305–309, doi:S0925-4773(99)00268-3 [pii]

144. Mustonen T, Pispa J, Mikkola ML, Pummila M, Kangas AT, Pakkasjarvi L, Jaatinen R, Thesleff I (2003) Stimulation of ectodermal organ development by ectodysplasin-A1. Dev Biol 259(1):123–136, doi:S001216060300157X [pii]

145. Yang A, Schweitzer R, Sun D, Kaghad M, Walker N, Bronson RT, Tabin C, Sharpe A, Caput D, Crum C, McKeon F (1999) p63 is essential for regenerative proliferation in limb, craniofacial and epithelial development. Nature 398(6729):714–718. doi:10.1038/19539

146. Mills AA, Zheng B, Wang XJ, Vogel H, Roop DR, Bradley A (1999) p63 is a p53 homologue required for limb and epidermal morphogenesis. Nature 398(6729):708–713. doi:10.1038/19531

147. Andl T, Reddy ST, Gaddapara T, Millar SE (2002) WNT signals are required for the initiation of hair follicle development. Dev Cell 2(5):643–653

148. Chu EY, Hens J, Andl T, Kairo A, Yamaguchi TP, Brisken C, Glick A, Wysolmerski JJ, Millar SE (2004) Canonical WNT signaling promotes mammary placode development and is essential for initiation of mammary gland morphogenesis. Development 131(19):4819–4829. doi:10.1242/dev.01347, dev.01347 [pii]

149. Davenport TG, Jerome-Majewska LA, Papaioannou VE (2003) Mammary gland,

limb and yolk sac defects in mice lacking Tbx3, the gene mutated in human ulnar mammary syndrome. Development 130(10):2263–2273

150. Jerome-Majewska LA, Jenkins GP, Ernstoff E, Zindy F, Sherr CJ, Papaioannou VE (2005) Tbx3, the ulnar-mammary syndrome gene, and Tbx2 interact in mammary gland development through a p19Arf/p53-independent pathway. Dev Dyn 234(4):922–933. doi:10.1002/dvdy.20575

151. Parsa S, Ramasamy SK, De Langhe S, Gupte VV, Haigh JJ, Medina D, Bellusci S (2008) Terminal end bud maintenance in mammary gland is dependent upon FGFR2b signaling. Dev Biol 317(1):121–131, doi:S0012-1606(08)00110-3 [pii] 10.1016/j.ydbio.2008.02.014

152. Asselin-Labat ML, Sutherland KD, Barker H, Thomas R, Shackleton M, Forrest NC, Hartley L, Robb L, Grosveld FG, van der Wees J, Lindeman GJ, Visvader JE (2007) Gata-3 is an essential regulator of mammary-gland morphogenesis and luminal-cell differentiation. Nat Cell Biol 9(2):201–209. doi:10.1038/ncb1530, ncb1530 [pii]

153. Hatsell SJ, Cowin P (2006) Gli3-mediated repression of Hedgehog targets is required for normal mammary development. Development 133(18):3661–3670, doi:dev.02542 [pii] 10.1242/dev.02542

154. Chandramouli A, Hatsell SJ, Pinderhughes A, Koetz L, Cowin P (2013) Gli activity is critical at multiple stages of embryonic mammary and nipple development. PLoS One 8(11), e79845. doi:10.1371/journal.pone.0079845

155. Ahn Y, Sims C, Logue JM, Weatherbee SD, Krumlauf R (2013) Lrp4 and Wise interplay controls the formation and patterning of mammary and other skin appendage placodes by modulating Wnt signaling. Development 140(3):583–593. doi:10.1242/dev.085118

156. Boras-Granic K, Chang H, Grosschedl R, Hamel PA (2006) Lef1 is required for the transition of Wnt signaling from mesenchymal to epithelial cells in the mouse embryonic mammary gland. Dev Biol 295(1):219–231. doi:10.1016/j.ydbio.2006.03.030, S0012-1606(06)00232-6 [pii]

157. Gu B, Sun P, Yuan Y, Moraes RC, Li A, Teng A, Agrawal A, Rheaume C, Bilanchone V, Veltmaat JM, Takemaru K, Millar S, Lee EY, Lewis MT, Li B, Dai X (2009) Pygo2 expands mammary progenitor cells by facilitating histone H3 K4 methylation. J Cell Biol 185(5):811–826. doi:10.1083/jcb.200810133, jcb.200810133 [pii]

158. Garcia-Gasca A, Spyropoulos DD (2000) Differential mammary morphogenesis along the anteroposterior axis in Hoxc6 gene targeted mice. Dev Dyn 219(2):261–276. doi:10.1002/1097-0177(2000)9999:9999<::AID-DVDY1048>3.0.CO;2-3 10.1002/1097-0177(2000)9999:9999<::AID-DVDY1048>3.0.CO;2-3 [pii]

159. Panchal H, Wansbury O, Parry S, Ashworth A, Howard B (2007) Neuregulin3 alters cell fate in the epidermis and mammary gland. BMC Dev Biol 7:105, doi:1471-213X-7-105 [pii] 10.1186/1471-213X-7-105

160. Närhi K, Tummers M, Ahtiainen L, Itoh N, Thesleff I, Mikkola ML (2012) Sostdc1 defines the size and number of skin appendage placodes. Dev Biol 364(2):149–161

161. Lindvall C, Evans NC, Zylstra CR, Li Y, Alexander CM, Williams BO (2006) The Wnt signaling receptor Lrp5 is required for mammary ductal stem cell activity and Wnt1-induced tumorigenesis. J Biol Chem 281(46):35081–35087, doi:M607571200 [pii] 10.1074/jbc.M607571200

162. Lindvall C, Zylstra CR, Evans N, West RA, Dykema K, Furge KA, Williams BO (2009) The Wnt co-receptor Lrp6 is required for normal mouse mammary gland development. PLoS One 4(6), e5813. doi:10.1371/journal.pone.0005813

163. Heckman BM, Chakravarty G, Vargo-Gogola T, Gonzales-Rimbau M, Hadsell DL, Lee AV, Settleman J, Rosen JM (2007) Crosstalk between the p190-B RhoGAP and IGF signaling pathways is required for embryonic mammary bud development. Dev Biol 309(1):137–149. doi:10.1016/j.ydbio.2007.07.002, S0012-1606(07)01163-3 [pii]

164. Hens JR, Dann P, Zhang JP, Harris S, Robinson GW, Wysolmerski J (2007) BMP4 and PTHrP interact to stimulate ductal outgrowth during embryonic mammary development and to inhibit hair follicle induction. Development 134(6):1221–1230. doi:10.1242/dev.000182, dev.000182 [pii]

165. Boras-Granic K, Dann P, Vanhouten J, Karaplis A, Wysolmerski J (2014) Deletion of the nuclear localization sequences and C-terminus of PTHrP impairs embryonic mammary development but also inhibits PTHrP production. PLoS One 9(5), e90418. doi:10.1371/journal.pone.0090418

166. Hiremath M, Dann P, Fischer J, Butterworth D, Boras-Granic K, Hens J, Van Houten J, Shi W, Wysolmerski J (2012) Parathyroid hormone-related protein activates Wnt signaling to specify the embryonic mammary

mesenchyme. Development 139(22):4239–4249. doi:10.1242/dev.080671

167. Liu Y, Rutlin M, Huang S, Barrick CA, Wang F, Jones KR, Tessarollo L, Ginty DD (2012) Sexually dimorphic BDNF signaling directs sensory innervation of the mammary gland. Science 338(6112):1357–1360. doi:10.1126/science.1228258

168. Satokata I, Ma L, Ohshima H, Bei M, Woo I, Nishizawa K, Maeda T, Takano Y, Uchiyama M, Heaney S, Peters H, Tang Z, Maxson R, Maas R (2000) Msx2 deficiency in mice causes pleiotropic defects in bone growth and ectodermal organ formation. Nat Genet 24(4):391–395. doi:10.1038/74231

169. Abdalkhani A, Sellers R, Gent J, Wulitich H, Childress S, Stein B, Boissy RE, Wysolmerski JJ, Foley J (2002) Nipple connective tissue and its development: insights from the K14-PTHrP mouse. Mech Dev 115(1-2):63–77, doi:S0925477302000928 [pii]

170. Dunbar ME, Dann P, Brown CW, Van Houton J, Dreyer B, Philbrick WP, Wysolmerski JJ (2001) Temporally regulated overexpression of parathyroid hormone-related protein in the mammary gland reveals distinct fetal and pubertal phenotypes. J Endocrinol 171(3):403–416, doi:JOE04411 [pii]

171. Gritli-Linde A, Hallberg K, Harfe BD, Reyahi A, Kannius-Janson M, Nilsson J, Cobourne MT, Sharpe PT, McMahon AP, Linde A (2007) Abnormal hair development and apparent follicular transformation to mammary gland in the absence of hedgehog signaling. Dev Cell 12(1):99–112, doi:S1534-5807(06)00569-7 [pii] 10.1016/j.devcel.2006.12.006

172. Lewis MT, Ross S, Strickland PA, Sugnet CW, Jimenez E, Hui C, Daniel CW (2001) The Gli2 transcription factor is required for normal mouse mammary gland development. Dev Biol 238(1):133–144. doi:10.1006/dbio.2001.0410

173. Lewis MT, Ross S, Strickland PA, Sugnet CW, Jimenez E, Scott MP, Daniel CW (1999) Defects in mouse mammary gland development caused by conditional haploinsufficiency of Patched-1. Development 126(22):5181–5193

174. Phippard DJ, Weber-Hall SJ, Sharpe PT, Naylor MS, Jayatalake H, Maas R, Woo I, Roberts-Clark D, Francis-West PH, Liu YH, Maxson R, Hill RE, Dale TC (1996) Regulation of Msx-1, Msx-2, Bmp-2 and Bmp-4 during foetal and postnatal mammary gland development. Development 122(9): 2729–2737

175. Pispa J, Pummila M, Barker PA, Thesleff I, Mikkola ML (2008) Edar and Troy signalling pathways act redundantly to regulate initiation of hair follicle development. Hum Mol Genet 17(21):3380–3391. doi:10.1093/hmg/ddn232

176. Wiesen JF, Young P, Werb Z, Cunha GR (1999) Signaling through the stromal epidermal growth factor receptor is necessary for mammary ductal development. Development 126(2):335–344

177. Klinowska TC, Alexander CM, Georges-Labouesse E, Van der Neut R, Kreidberg JA, Jones CJ, Sonnenberg A, Streuli CH (2001) Epithelial development and differentiation in the mammary gland is not dependent on alpha 3 or alpha 6 integrin subunits. Dev Biol 233(2):449–467. doi:10.1006/dbio.2001.0204

178. Kurpios NA, MacNeil L, Shepherd TG, Gludish DW, Giacomelli AO, Hassell JA (2009) The Pea3 Ets transcription factor regulates differentiation of multipotent progenitor cells during mammary gland development. Dev Biol 325(1):106–121. doi:10.1016/j.ydbio.2008.09.033

179. DasGupta R, Fuchs E (1999) Multiple roles for activated LEF/TCF transcription complexes during hair follicle development and differentiation. Development 126(20): 4557–4568

180. Kogata N, Oliemuller E, Wansbury O, Howard BA (2014) Neuregulin-3 regulates epithelial progenitor cell positioning and specifies mammary phenotype. Stem Cells Dev 23(22):2758–2770. doi:10.1089/scd.2014.0082

181. Cheon SS, Cheah AY, Turley S, Nadesan P, Poon R, Clevers H, Alman BA (2002) beta-Catenin stabilization dysregulates mesenchymal cell proliferation, motility, and invasiveness and causes aggressive fibromatosis and hyperplastic cutaneous wounds. Proc Natl Acad Sci U S A 99(10):6973–6978. doi:10.1073/pnas.102657399

182. Maretto S, Cordenonsi M, Dupont S, Braghetta P, Broccoli V, Hassan AB, Volpin D, Bressan GM, Piccolo S (2003) Mapping Wnt/beta-catenin signaling during mouse development and in colorectal tumors. Proc Natl Acad Sci U S A 100(6):3299–3304. doi:10.1073/pnas.0434590100

183. van Amerongen R, Bowman AN, Nusse R (2012) Developmental stage and time dictate the fate of Wnt/beta-catenin-responsive stem cells in the mammary gland. Cell Stem Cell 11(3):387–400. doi:10.1016/j.stem.2012.05.023

184. Ferrer-Vaquer A, Piliszek A, Tian G, Aho RJ, Dufort D, Hadjantonakis AK (2010) A sensi-

tive and bright single-cell resolution live imaging reporter of Wnt/ss-catenin signaling in the mouse. BMC Dev Biol 10:121. doi:10.1186/1471-213X-10-121

185. Bianchi N, Depianto D, McGowan K, Gu C, Coulombe PA (2005) Exploiting the keratin 17 gene promoter to visualize live cells in epithelial appendages of mice. Mol Cell Biol 25(16):7249–7259. doi:10.1128/mcb.25.16.7249-7259.2005

186. Rohrschneider LR, Custodio JM, Anderson TA, Miller CP, Gu H (2005) The intron 5/6 promoter region of the ship1 gene regulates expression in stem/progenitor cells of the mouse embryo. Dev Biol 283(2):503–521. doi:10.1016/j.ydbio.2005.04.032

187. Zhang J, Tan X, Contag CH, Lu Y, Guo D, Harris SE, Feng JQ (2002) Dissection of promoter control modules that direct Bmp4 expression in the epithelium-derived components of hair follicles. Biochem Biophys Res Commun 293(5):1412–1419. doi:10.1016/s0006-291x(02)00416-3

188. Li L, Rutlin M, Abraira VE, Cassidy C, Kus L, Gong S, Jankowski MP, Luo W, Heintz N, Koerber HR, Woodbury CJ, Ginty DD (2011) The functional organization of cutaneous low-threshold mechanosensory neurons. Cell 147(7):1615–1627. doi:10.1016/j.cell.2011.11.027

189. Lee MY, Sun L, Veltmaat JM (2013) Hedgehog and Gli signaling in embryonic mammary gland development. J Mammary Gland Biol Neoplasia 18(2):133–138. doi:10.1007/s10911-013-9291-7

190. Kogata N, Zvelebil M, Howard BA (2013) Neuregulin 3 and erbb signalling networks in embryonic mammary gland development. J Mammary Gland Biol Neoplasia 18(2):149–154. doi:10.1007/s10911-013-9286-4

191. Boras-Granic K, Hamel PA (2013) Wnt-signalling in the embryonic mammary gland. J Mammary Gland Biol Neoplasia 18(2):155–163. doi:10.1007/s10911-013-9280-x

192. Cunha GR, Hom YK (1996) Role of mesenchymal-epithelial interactions in mammary gland development. J Mammary Gland Biol Neoplasia 1(1):21–35

193. Wansbury O, Panchal H, James M, Parry S, Ashworth A, Howard B (2008) Dynamic expression of Erbb pathway members during early mammary gland morphogenesis. J Invest Dermatol 128(4):1009–1021. doi:10.1038/sj.jid.5701118

194. Dillon C, Spencer-Dene B, Dickson C (2004) A crucial role for fibroblast growth factor signaling in embryonic mammary gland development. J Mammary Gland Biol Neoplasia 9(2):207–215. doi:10.1023/B:JOMG.0000037163.56461.1e

195. Cho KW, Kwon HJ, Shin JO, Lee JM, Cho SW, Tickle C, Jung HS (2012) Retinoic acid signaling and the initiation of mammary gland development. Dev Biol 365(1):259–266. doi:10.1016/j.ydbio.2012.02.020

196. Lewis MT, Veltmaat JM (2004) Next stop, the twilight zone: hedgehog network regulation of mammary gland development. J Mammary Gland Biol Neoplasia9(2):165–181, doi:10.1023/B:JOMG.0000037160.24731.35.490064 [pii]

197. Al Alam D, Green M, Tabatabai Irani R, Parsa S, Danopoulos S, Sala FG, Branch J, El Agha E, Tiozzo C, Voswinckel R, Jesudason EC, Warburton D, Bellusci S (2011) Contrasting expression of canonical Wnt signaling reporters TOPGAL, BATGAL and Axin2(LacZ) during murine lung development and repair. PLoS One 6(8), e23139. doi:10.1371/journal.pone.0023139

198. Lustig B, Jerchow B, Sachs M, Weiler S, Pietsch T, Karsten U, van de Wetering M, Clevers H, Schlag PM, Birchmeier W, Behrens J (2002) Negative feedback loop of Wnt signaling through upregulation of conductin/axin2 in colorectal and liver tumors. Mol Cell Biol 22(4):1184–1193

199. van de Stolpe A, Karperien M, Lowik CW, Juppner H, Segre GV, Abou-Samra AB, de Laat SW, Defize LH (1993) Parathyroid hormone-related peptide as an endogenous inducer of parietal endoderm differentiation. J Cell Biol 120(1):235–243

200. Kobayashi T, Kronenberg HM, Foley J (2005) Reduced expression of the PTH/PTHrP receptor during development of the mammary gland influences the function of the nipple during lactation. Dev Dyn 233(3):794–803. doi:10.1002/dvdy.20406

201. Wysolmerski JJ, Broadus AE, Zhou J, Fuchs E, Milstone LM, Philbrick WM (1994) Overexpression of parathyroid hormone-related protein in the skin of transgenic mice interferes with hair follicle development. Proc Natl Acad Sci U S A 91(3):1133–1137

202. Mayer JA, Foley J, De La Cruz D, Chuong CM, Widelitz R (2008) Conversion of the nipple to hair-bearing epithelia by lowering bone morphogenetic protein pathway activity at the dermal-epidermal interface. Am J Pathol 173(5):1339–1348. doi:10.2353/ajpath.2008.070920

203. Gossen M, Bujard H (1992) Tight control of gene expression in mammalian cells by tetracycline-responsive promoters. Proc Natl Acad Sci U S A 89(12):5547–5551

204. Sahlberg C, Mustonen T, Thesleff I (2002) Explant cultures of embryonic epithelium. Analysis of mesenchymal signals. Methods Mol Biol 188:373–382. doi:10.1385/1-59259-185-x:373

205. Zvelebil M, Oliemuller E, Gao Q, Wansbury O, Mackay A, Kendrick H, Smalley MJ, Reis-Filho JS, Howard BA (2013) Embryonic mammary signature subsets are activated in Brca1−/− and basal-like breast cancers. Breast Cancer Res 15(2):R25. doi:10.1186/bcr3403

206. Lee MJ, Yoon KS, Cho KW, Kim KS, Jung HS (2013) Expression of miR-206 during the initiation of mammary gland development. Cell Tissue Res 353(3):425–433. doi:10.1007/s00441-013-1653-3

207. Zeng YA, Nusse R (2010) Wnt proteins are self-renewal factors for mammary stem cells and promote their long-term expansion in culture. Cell Stem Cell 6(6):568–577. doi:10.1016/j.stem.2010.03.020

208. Shackleton M, Vaillant F, Simpson KJ, Stingl J, Smyth GK, Asselin-Labat ML, Wu L, Lindeman GJ, Visvader JE (2006) Generation of a functional mammary gland from a single stem cell. Nature 439(7072):84–88, doi:nature04372 [pii] 10.1038/nature04372

209. Stingl J, Eirew P, Ricketson I, Shackleton M, Vaillant F, Choi D, Li HI, Eaves CJ (2006) Purification and unique properties of mammary epithelial stem cells. Nature 439(7079):993–997. doi:10.1038/nature04496, nature04496 [pii]

210. Stevens JR, Miranda-Carboni GA, Singer MA, Brugger SM, Lyons KM, Lane TF (2010) Wnt10b deficiency results in age-dependent loss of bone mass and progressive reduction of mesenchymal progenitor cells. J Bone Miner Res 25(10):2138–2147. doi:10.1002/jbmr.118

211. Special issue on embryonic mammary gland development (2013) J Mamm Gland Biol Neoplasia 18(2):89-252. doi:10.1007/s10911-013-9292-6

212. Howard BA, Lu P (2014) Stromal regulation of embryonic and postnatal mammary epithelial development and differentiation. Semin Cell Dev Biol 25–26:43–51. doi:10.1016/j.semcdb.2014.01.004

Chapter 3

Pubertal Mammary Gland Development: Elucidation of In Vivo Morphogenesis Using Murine Models

Jean McBryan and Jillian Howlin

Abstract

During the past 25 years, the combination of increasingly sophisticated gene targeting technology with transplantation techniques has allowed researchers to address a wide array of questions about postnatal mammary gland development. These in turn have significantly contributed to our knowledge of other branched epithelial structures. This review chapter highlights a selection of the mouse models exhibiting a pubertal mammary gland phenotype with a focus on how they have contributed to our overall understanding of in vivo mammary morphogenesis. We discuss mouse models that have enabled us to assign functions to particular genes and proteins and, more importantly, have determined when and where these factors are required for completion of ductal outgrowth and branch patterning. The reason for the success of the mouse mammary gland model is undoubtedly the suitability of the postnatal mammary gland to experimental manipulation. The gland itself is very amenable to investigation and the combination of genetic modification with accessibility to the tissue has allowed an impressive number of studies to inform biology. Excision of the rudimentary epithelial structure postnatally allows genetically modified tissue to be readily transplanted into wild type stroma or vice versa, and has thus defined the contribution of each compartment to particular phenotypes. Similarly, whole gland transplantation has been used to definitively discern local effects from indirect systemic effects of various growth factors and hormones. While appreciative of the power of these tools and techniques, we are also cognizant of some of their limitations, and we discuss some shortcomings and future strategies that can overcome them.

Key words Mammary gland, Breast, Development, In vivo morphogenesis, Pubertal mammary gland development, Mouse models, Knockout, Transgenic, Mammary gland transplantation, Epithelial branching, Branch patterning, Terminal endbud, Mammary stem cells, Cell polarity, Cell–cell adhesion, Extracellular matrix

1 Introduction

Pubertal development of the mammary gland is a fascinating demonstration of rapid, highly organized epithelial ductal outgrowth and branching. Unlike other branched epithelial organs whose development occurs predominantly in the embryo, the pubertal time period makes the mammary gland relatively easy to study. At birth, the mammary gland consists of a simple, rudimentary ductal

Finian Martin et al. (eds.), *Mammary Gland Development: Methods and Protocols*, Methods in Molecular Biology, vol. 1501, DOI 10.1007/978-1-4939-6475-8_3, © Springer Science+Business Media New York 2017

network occupying only a fraction of the mammary fat pad. The gland grows isometrically with the body until puberty when ovarian hormones initiate allometric mammary growth. At the onset of puberty, terminal end bud (TEB) structures form at the tip of growing epithelial ducts and drive ductal elongation, invasion, and branching of the epithelial tree. The TEB structure consists of a single outer layer of cap cells and multiple inner layers of body cells. The cap cells will differentiate into myoepithelial cells while body cells will form a single layer of luminal epithelial cells along the established duct. The remaining body cells will undergo apoptosis to create the hollow ductal lumen. Branching occurs as a result of TEB bifurcation and also due to lateral sprouting of the epithelium to create secondary and tertiary side branches. TEBs are unique to puberty, a time period that lasts only 4–5 weeks in the mouse, and TEBs regress when the epithelial ducts reach the edges of the mammary fat pad.

At the molecular level, much of our understanding of the players involved in regulating this phase of development has arisen from the study of genetically modified mouse models. By manipulating the expression of specific genes we have gained a detailed understanding of the myriad factors required to orchestrate this rapid phase of development. Mouse models displaying mammary phenotypes such as reduced ductal outgrowth, ductal hyperplasia, distended TEBs and abnormal branch patterning, have highlighted the complexity and degree of control required to successfully complete normal pubertal mammary gland development. Along the way, various techniques have been established to overcome issues such as embryonic lethality which would otherwise prevent pubertal phenotypes from being identified. These include the use of inducible systems such as Cre-loxP recombination, tetracycline-inducible Tet-On/Off systems, as well as the use of mammary-specific promoters such as WAP (whey acidic protein) and MMTV (mouse mammary tumor virus). Understandably, due to the large number of mouse models that have been generated since the early 1990s, not all can be discussed in detail here but a comprehensive list is provided in Table 1. We extend our apologies to authors whose works are not cited due to practical constraints.

2 Systemic Hormonal Control

2.1 Ovarian Steroids: Estrogen and Progesterone

A rise in the levels of gonadotrophins defines the onset of puberty and leads to ovarian secretion of the hormones, estrogen and progesterone. Estrogen receptor knockout (ERKO) mammary glands have both epithelial and stromal components as well as a rudimentary mammary ductal tree implying that estrogen signaling via its receptor is dispensable for embryonic development of the mammary gland [1]. Additionally, as embryonic mammary development

Table 1
Animal models exhibiting a pubertal mammary gland phenotype

Model	Phenotype	Reference
Hormones and hormone receptors		
ERα KO	Absence of ductal outgrowth	[1, 4]
ERα KO (MMTV-Cre)	Impaired ductal elongation and side branching	[5]
Aromatase	Accelerated ductal outgrowth	[171]
PRKO	Impaired ductal outgrowth, distended TEBs, defects in ductal side branching	[11, 12]
PR-A overexpression	Persistence of TEBs with extensive lateral branching (hyperplasia)	[15]
GHR KO	Impaired ductal outgrowth and side branching	[172]
GR^{dim}	Impaired ductal outgrowth	[23]
GR KO	Distended lumina, multiple epithelial cell layers, increased periductal stroma	[22]
VDR KO	Accelerated growth, increased branching and number of TEBs	[29]
PTHrP (K14)/Tet-PTHrP (K14-tTA) or PTH (K14)	Impaired ductal outgrowth and branching morphogenesis	[27, 28]
Growth factors and growth factor receptors		
AREG KO	Impaired ductal outgrowth	[44]
EGFR1 KO	Impaired ductal outgrowth	[46]
EGFR2	Impaired ductal outgrowth	[48]
EGFR3 KO	Impaired ductal outgrowth, smaller TEBs, aberrant ductal spacing	[53]
NRG1 (NDF/Heregulin)	Persistence of TEBs	[54]
IGF1 KO	Impaired ductal outgrowth, reduced TEB number reduced duct number	[17]
IGF1 (K5)	Increased ductal proliferation	[59]
IGF1R KO	Impaired ductal outgrowth TEB formation	[60]
IGF1R (MMTV)	Delayed ductal elongation but increased side branching and hyperplastic lesions	[61]
constitutively active IGF1R (MMTV)	Delayed ductal outgrowth and reduced numbers of TEBs with apparent adenocarcinomas	[62]
FGF3 (MMTV)	Impaired ductal outgrowth and reduced side branching	[66]
Mosaic FGFR2 (MMTV-Cre)	Reduced ductal invasion and proliferation	[67]

(continued)

Table 1
(continued)

Model	Phenotype	Reference
HGF (transplanted retroviral induced expression in PMECs	Increased ductal branching, increased size and number of TEBs	[68]
c-Met (MMTV-Cre/lox)	Reduced secondary and side branching of the ductal tree	[70]
(Pellets) TGFβ1	Suppression of TEB formation and ductal outgrowth	[126]
Constitutively active TGFβ1$^{S223/225}$ (MMTV)	Hypoplasia, impaired ductal outgrowth	[127]
TGFβ1 KO	Accelerated ductal development	[128]
WNT5A KO	Accelerated ductal development	[130]
Endocrine disrupters		
Bisphenol-A	Reduced ductal outgrowth, increased numbers of TEB/area (in offspring of exposed mothers)	[34]
Genistein	Increased TEB density (in offspring exposed in utero)	[35]
Cadmium	Increased numbers of TEB (in offspring exposed in utero)	[36]
Transcription factors and coregulators		
SRC-1 KO	Impaired ductal outgrowth and reduced branching	[7, 173]
CITED1 KO	Reduced ductal outgrowth and dilated ducts	[8]
ATBF1 (MMTV-Cre)	Increased ductal elongation and branching	[9]
Cytokines		
CSF-1 KO	Impaired ductal outgrowth, branching and TEB formation	[145]
Eotaxin KO	Reduced branching and TEB formation, reduced TEB number	[145]
(Overexpression in T-lymphocytes) IL-5	Retarded ductal outgrowth and impaired TEB formation	[146]
MSC and progenitor lineage determinants		
GATA3	Failure to form TEBs, ducts consisting of myoepithelial cells	[82, 174]
P18^{INK4C}	Hyperproliferative luminal cells, spontaneous DCIS	[84]
Bmi-1 KO	Impaired ductal outgrowth and premature lobuloalveologenesis	[85]
Pea3 KO	Increase in TEB number, persistence of the TEB structures	[86]

(continued)

Table 1
(continued)

Model	Phenotype	Reference
MED1 KO	Retarded ductal elongation and decreased lobuloalveolar development	[87]
MED1 (LxxLL motif-mutant knockin)	Impaired mammary ductal growth and branch morphogenesis, insensitivity to E2	[88]
MED1/MED24 double heterozygous KO	Impaired mammary ductal growth and branch morphogenesis, insensitivity to E2	[89]
STAT5 KO	Defective lateral branching	[92]
STAT5 (transplanted following lentiviral transduction)	Epithelial hyperproliferation and precocious alveolar development	[94, 95]
(Conditional mutation) BRCA1	Impaired ductal outgrowth	[99]
ECM and cell adhesion		
ADAM17 KO	Impaired ductal outgrowth	[45]
Est1 and Hs2st (MMTV-Cre/lox)	Reduced primary side branching; reduction in secondary and side branching	[70]
Syndecan KO	Reduced primary secondary branching	[73]
Heparinase (CMV)	Increased ductal branching with precocious alveolar development	[75]
Heparinase KO	Increased ductal branching with precocious alveolar development	[175]
(Antibody releasing pellets) E-Cadherin	Disruption of the cap cell layer in TEBs	[108]
(Antibody releasing pellets) P-Cadherin	Disruption of the body cells in TEBs	[108]
P-Cadherin KO	Hyperplasia and precocious alveolar development	[109]
Cytoplasmic domain E-Cadherin (MMTV)	Precocious alveolar development	[110]
Netrin-1 KO	Dissociation of cap cells and breaks in basal lamina surrounding TEBs	[124]
Neogenin KO	Dissociation of cap cells and breaks in basal lamina surrounding TEBs	[124]
EPHB4 (MMTV)	Impaired ductal outgrowth and reduced branching	[125]
DDR1	Delayed ductal invasion with enlarged TEBs, hyperproliferation and increased branching	[131]
MMP3 (WAP)	Increased ductal branching	[136, 137]
MMP3 KO	Reduced lateral branching	[135]

(continued)

Table 1
(continued)

Model	Phenotype	Reference
MMP2 KO	Delayed ductal invasion but increased lateral branching, apoptotic TEBs	[135]
TIMP1 (Beta-Actin); TIMP1 (pellets)	Impaired ductal outgrowth	[135, 138]
TIMP1 KO	Enlarged TEBs	[135]
β1-Integrin (K5-Cre)	Abnormal ductal branching	[139]
(Antibody releasing pellet) β1-integrin	Impaired ductal outgrowth and reduced number of TEBs	[140]
(Antibody releasing pellet) Laminin	Impaired ductal outgrowth and reduced number of TEBs	[140]
FAK (MMTV-Cre)	Impaired ductal elongation	[141]
FAK KO	Dilated ducts	[142]
DDPI KO	Impaired ductal outgrowth	[147]
β1,4-galactosyltransferase (metallothionein)	Reduced size of TEBs and retarded ductal elongation	[176]
β1, 4-galactosyltransferase KO	Increased branching morphogenesis	[177]
Gelsolin KO	Impaired ductal outgrowth	[178]
Cell signaling (various)		
Cdc42 (MMTV-Tet)	Hyperbudding of TEBs and hyperbranching but a reduced ductal tree area, collagen deposition around TEBs	[121]
P190BRhoGAP (MMTV-Tet)	Aberrant TEBs, hyperbranching, and alterations in the adjacent stroma	[122]
Scribble (MMTV-Cre)	Increased numbers of TEB and excessive ductal branching	[123]
Stabilized form of β-Catenin (MMTV)	Delayed outgrowth with precocious alveolar development	[113]
APC[mut] (BGL-Cre)	Delayed ductal outgrowth	[112]
NKCC1 KO	Retarded ductal outgrowth and increased number of branches	[179]
Truncated Patched1 knockin	Block in ductal elongation	[180]
PxmP2 KO	Retarded ductal outgrowth	[181]
(Conditional deletion) EZH2	Impaired terminal end bud formation and ductal elongation	[182]
KRCT (MMTV-Tet)	Supernumerary TEBs with increased periductal stroma	[183]

The specific promoter is indicated in brackets where relevant, "-Cre" indicates where a Cre-loxP excision strategy was employed and "-Tet," where a tetracycline-inducible expression system was used. "KO" indicates various methods of gene deletion/ablation

is normal following X-irradiation of the ovaries, we can conclude that although the embryonic gland is *responsive* to estrogen, it is not *required* until the onset of puberty [2]. By puberty, ductal outgrowth in the ERKO mouse is completely stunted, no further mammary development occurs, and the mammary glands resemble those of a new-born wild type mouse [1, 3]. The ERKO mice referred to above are those that lack the ERα. By contrast, mice lacking ERβ exhibit no pubertal phenotype indicating that ERβ, although expressed during puberty, is dispensable for this period of mammary development [1]. Transplant experiments placing ERKO epithelium into cleared fat pads of wild type mice and vice versa have demonstrated that ERα is only essential in the epithelial cells [4, 5]. Initial transgenic models had proposed a role for ER in the stroma but it was later established that these mice produced a truncated ER protein, which likely masked the true requirements for ER. The second generation of ERKO mouse models using Cre-loxP based conditional knockout mice have more clearly demonstrated the need for epithelial ERα [5]. Interestingly, ERKO epithelial cells can persist in a mammary tree when transplanted together with wild type epithelial cells supporting the hypothesis that ERα signaling occurs in a paracrine manner [4]. Estrogen receptors require the recruitment of additional coregulators in order to function successfully as transcription factors [6]. The steroid receptor coactivator, SRC-1, is one such cofactor that has been implicated in pubertal mammary gland development. SRC-1 KO mice have an underdeveloped ductal network which is consistent with reduced ERα signaling [7]. Similarly the CITED1 coregulator, the mRNA of which displays increased expression during puberty has also been implicated in pubertal development since CITED1 KO mice exhibit delayed ductal outgrowth at puberty [8]. These phenotypes of mice lacking ER coregulators are highly consistent with, but slightly less severe, than the phenotype of ERKO mice. This highlights the functional redundancy of coregulators and the ability of some ERα signaling to persist despite their absence. CITED1 KO mice, in addition to delayed ductal outgrowth, also exhibit dilated ductal structures with an apparent lack of spatial restriction. This abnormal ductal patterning is not necessarily attributed solely to CITED1's role as an ERα coactivator as coregulators can contribute to a variety of signaling pathways. CITED1 is also known to function as a coregulator for SMAD4, downstream of TGFβ that is a more likely player in ductal patterning [8].

The transcription factor ATBF1 is also upregulated during puberty but acts as a negative regulator or corepressor of ERα via direct interaction. It also functions within an autoregulatory feedback loop whereby ERα induces its expression but ATBF1 is degraded by the estrogen responsive ubiquitin ligase, EFP. Deletion of ATBF1 results in increased ductal elongation and branching in the pubertal gland although recovery of this stunted outgrowth phenotype is evident by the end of puberty. Interestingly, the increased outgrowth appears to be due to increased cell proliferation of the ER-positive cell population that do not normally proliferate themselves but rather provide paracrine stimulation to ER-negative cells [9].

The progesterone receptor (PR) is an established target of estrogen-ERα signaling, mediating pubertal and later lobuloalveolar development in the adult. It appears that ATBF1, is also a transcriptional target of progesterone-PR signaling, which may explain its increased expression in the lactating gland. Although no lobuloalveolar mouse phenotype has been described, it has been demonstrated that ATBF1 is required for some function of PR in vivo [9, 10]. At puberty, progesterone receptor knockout (PRKO) mice display impaired ductal growth and a lack of side branching [11, 12]. Transplant experiments have demonstrated that PR is predominantly required in the epithelium although a possible indirect role for stromal PR, signaling via secondary growth factor signals, has also been suggested [12]. Two isoforms of the progesterone receptor exist: PR-A and PR-B with PR-A being approximately twice as prevalent in the pubertal gland. Despite this imbalance, selective PR-A or PR-B knockout mice indicate that the PR-B isoform is the one required for pubertal development with PR-A KO mice failing to exhibit any pubertal mammary phenotype [13, 14]. Overexpression of PR-A, however, leads to ductal hyperplasia and persistence of TEBs [15]. Disruption in the normal 2:1 ratio of isoforms PR-A and PR-B is hypothesized to contribute to the observed developmental abnormalities.

2.2 Pituitary Hormone Signaling: Growth Hormone and Prolactin

Ovarian steroid hormones, although necessary for initiation of pubertal development, are not sufficient. Classical studies with pituitary gland ablation revealed that several pituitary hormones such as growth hormone (GH) and prolactin (Prl) are also required [16]. Growth hormone receptor knockout (GHRKO) mice exhibit retarded ductal outgrowth with limited side branching, and transplantation experiments demonstrated that GHR is required in the stroma and is sufficient to support growth of GHRKO epithelium. Unlike the nuclear steroid hormone receptors, GHR is a transmembrane protein that activates internal signaling cascades to produce its effects. Growth hormone is known to signal via insulin-like growth factor 1 (IGF-1) and GHRKO mice also show decreased serum IGF-1 levels. Not surprisingly, therefore, IGF-1 KO mice also show a mammary phenotype similar to that of GHRKO [17]. The requirement for prolactin in the mammary gland was demonstrated by prolactin receptor (PrlR) null mice, where reduced ductal outgrowth at puberty and the persistence of TEBs were more apparent in the homozygous null relative to the heterozygote [18, 19]. This was despite the fact that prolactin gene knockout mice had previously been reported to have normal ductal outgrowth up to puberty but defective lobuloalveolar development. The discrepancy between the ligand and receptor knockout strains at puberty could potentially be explained by maternal supply of prolactin to the offspring in utero via the maternal circulation and subsequently via nursing [20]. Alternatively and more likely, the pubertal effects seen in the receptor-deficient mice were also suggested to be attributable to secondary loss of ovarian steroid hormones. To test whether abnormalities in ductal development were really due to lack of PrlR in the mammary epithelium or secondary to systemic effects due to loss of PrlR, Brisken et al. performed transplantation experiments where PrlR null mammary epithelium was transplanted into wild type fat pads. This led to complete rescue of the pubertal phenotype and clearly demonstrated that the pubertal requirement for PrlR was not local but restricted to other cell types or organs. Notably, the lobuloalveolar defect could not be rescued by transplantation of PrlR null epithelium into wild type fat pads demonstrating the requirement of epithelial PrlR for this later phase of mammary development [21].

2.3 Glucocorticoids, Parathyroid Hormone, and Vitamin D

The transplantation of glucocorticoid receptor (GR) knockout epithelium into wild type stroma was necessary to rescue an otherwise perinatal lethal phenotype. Subsequently, these mice exhibited abnormal ductal morphogenesis characterized by distended lumina and dilated ducts with an atypical branching pattern. Some of the ducts contained multiple epithelial cell layers and an increase in the surrounding stroma [22]. Another model aimed to distinguish the DNA-binding dependent functions of GR to those dependent on GR protein-protein interactions. Mice carrying a DNA-binding defective GR (GR^{dim}) displayed reduced ductal outgrowth, reduced side-branching and persistence of TEB structures in the adult gland [23]. In both models it appeared, somewhat surprisingly perhaps, that GR was dispensable for lobuloalveolar development given the absence of any overt phenotype during pregnancy, lactation or involution. A more thorough investigation used conditional deletion (Cre-loxP) restricted to the lobuloalveolar epithelium. Although this revealed evidence of a slight delay due to reduced proliferation, it was ultimately compensated for [24]. The effects of GR are, therefore, in all practicality, limited to pubertal development.

Parathyroid related hormone (PTHrP) is a homologue of parathyroid hormone (PTH) and both molecules utilize a common receptor, PTHR1. Unlike PTH, which is a classic peptide hormone produced by the parathyroid glands, PTHrP is produced locally by cells of the mammary gland and other tissues but does not circulate. Deletion of PTHrP results in neonatal death; however, rescue experiments by targeted overexpression of PTHrP to overcome lethality, demonstrated the requirement for PTHrP and its receptor PTHR1 in determining the fate of the mammary mesenchyme. PTHrP or PTHR1 null mice accordingly arrest development of the mammary gland at the embryonic bud stage [25, 26]. Conversely, overexpression of PTHrP or PTH in the mammary epithelium, driven by an epithelial keratin 14 (K14) promoter leads to reduced ductal outgrowth, reduced branching, and impairment of subsequent lobuloalveolar development [27]. The latter defect was demonstrated to be as a consequence of impairment at puberty, rather than intrinsic, since temporal overexpression using a Tet-On/Off system driven by a K14 promoter in the mature gland demonstrated no adverse effects on alveologenesis [28].

The 1,25-(OH)2D3 (vitamin D_3) receptor (VDR) has been implicated in mammary development since VDRKO mice display abnormal ductal morphogenesis characterized by accelerated ductal outgrowth, increased secondary branching and an increased number of TEBs at puberty [29]. In agreement, branching in vivo using whole organ culture was inhibited by exposure to vitamin D_3. VDR deficiency predisposes mice to tumorigenesis and has led to the hypothesis that even dietary deficiency of vitamin D_3 could impact breast cancer susceptibility [30].

2.4 Endocrine Disruptors

The endocrine disruptor hypothesis was proposed in the 1990s in response to several key observations. Of these, perhaps the most cited, is the effect of the widespread diethylstilbestrol (DES) administration to pregnant women, the subsequent predisposition of their female children (exposed in utero) to vaginal clear-cell carcinoma, and their increased incidence of breast cancer [31, 32]. Rodent studies have played a vital role in understanding this phenomenon and substantiating the claim that even environmentally derived chemical, pharmaceutical, or dietary agents can affect mammary gland development and susceptibility to later neoplastic transformation. These models have perhaps unsurprisingly highlighted the necessity of adequate control of hormone exposure during sensitive developmental stages of many endocrine target organs. A full description of endocrine disrupting agents and controversies in the field are outside the scope of this review but we mention here some examples of models that exhibit pubertal mammary gland phenotypes. Bisphenol-A (BPA) a chemical capable of binding both ERα and ERβ and formally widely used in the manufacturing of plastics, can leach from consumer end products resulting in inadvertent human exposure and ingestion [33]. Despite the fact that the mammary gland develops independently of estrogen until puberty, perinatal exposure to BPA has been shown to accelerate the onset of puberty and disrupt estrogenic cycling. In addition, offspring of BPA treated mothers exhibit reduced ductal invasion coupled with an increase in the number of TEBs relative to the ductal area occupied (although total number was not significantly different). These animals also had an increased number and clustering of PR-positive epithelial cells with subsequent evidence of increased lateral branching of the ducts. The effects are proposed to be due to altered sensitivity of the gland to estrogen, and indeed this is supported by the increase in TEB number and size in response to exogenous estrogen treatment. However, some have also attributed the mammary phenotype to a potentially defective hypothalamic–pituitary–ovarian axis of BPA-exposed offspring as evidenced by ovarian changes and reduced serum luteinizing hormone [34]. Similar "estrogenic" effects on the mammary gland in F1 generation of animals exposed during pregnancy have been reported for the phytoestrogen, genistein and the metal, cadmium [35, 36]. The increased susceptibility to mammary tumorigenesis observed for the offspring of mothers treated with estrogen or estrogen modulators has naturally sparked interest in the effects of human exposure to environmental endocrine disrupting compounds (EDCs) [37]. However, when it comes to environmental chemical exposure the old adage that "the dose makes the poison" is very relevant. In this regard it should be noted that some studies that attempt to use compound concentrations more in line with real-world human exposure levels fail to find such dramatic effects in vivo [38]. Additionally, the more publicized concern that pharmaceuticals produced as oral contraceptives for birth control in humans were a significant source of environmentally derived EDCs has been largely put aside given the fact that the contribution of such medicines to the level of estrogen detectable in the water supply is minimal, and in fact, is dwarfed by the contribution of naturally occurring estrogen derived from pregnancy events in a population [39, 40].

3   Local Growth Control

3.1   Amphiregulin and the EGFR Family

Downstream of the ovarian and pituitary hormones, locally acting growth factors translate signals into local paracrine messages. Amphiregulin (AREG), an epidermal growth factor receptor (EGFR) ligand produced in the mammary epithelium, is the prototypical example of a downstream local mediator of hormonal signals [41] (Fig. 1). During puberty, AREG is the key paracrine mediator of estrogen-ERα signaling and accordingly. AREG KO mice have severely impaired ductal outgrowth at puberty; a similar phenotype to that of ERKO mice. In agreement with AREG's proposed role as a paracrine mediator, ERKO mice could be rescued by exogenous AREG administration [42]. The AREG gene is thought to be a direct transcriptional target of ERα and the identification of an ERα occupied AREG-associated ERE has been reported. Additionally, the stunted ductal outgrowth seen in the CITED1 KO mice could be explained in part by diminished expression of AREG due to lack of availability of the ERα

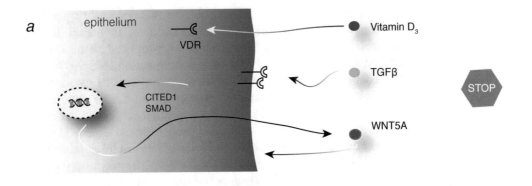

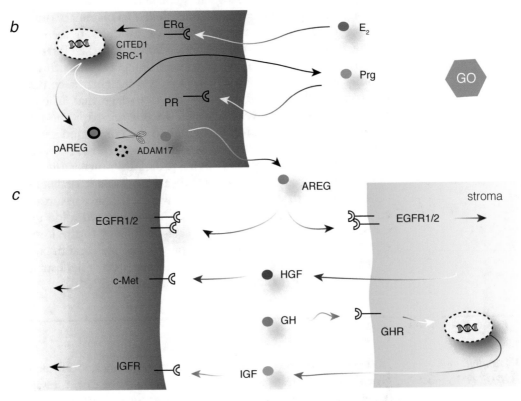

Fig. 1 Systemic and local growth control. A depiction of some of the hormones and growth factors necessary for pubertal ductal outgrowth as elucidated using mouse models (**a**) The inhibitory morphogens proposed to date include vitamin D₃, TGFβ, and WNT5A. Vitamin D₃ binds to its steroid hormone receptor (VDR) in the cytoplasm and leads to downstream activation of its target genes. TGFβ binds TGFβ receptors on the cell surface and affects target gene expression via SMADs and other coregulators such as CITED1. One of the target genes mediating the effects of TGFβ on growth inhibition is WNT5A. These pathways are ultimately believed to effectively oppose estrogen driven proliferation in the mammary epithelium. (**b**) Positive growth regulation is mediated by the primary mammogens, estrogen (E₂) and progesterone (Prg), via their epithelial steroid hormone receptors (ERα, PR) and subsequent transcription of their target genes dependent on interaction with coregulators such as CITED1 and SRC1. Production of the estrogen responsive gene amphiregulin (AREG) and its subsequent activation by ADAM17 mediated cleavage, leads to further downstream activation of the EGFR pathway in potentially both stromal and epithelial compartments. (**c**) Signaling pathways comprising both stromal and

coregulator [8, 43]. AREG is found in the luminal epithelium and TEBs at puberty and exists as a membrane-anchored precursor that requires cleavage for activity [44]. The enzyme responsible for the cleavage of AREG is a member of the disintegrin and metallopro-tease family (ADAM) family, ADAM17. It is also known as tumor necrosis alpha cleaving enzyme (TACE) since it can cleave and activate more than one EGFR ligand. Unsurprisingly, therefore, the ADAM17 KO mouse phenocopies the AREG KO mammary gland, and that of the EGFR KO where pubertal ductal outgrowth is also impaired. Transplantation experiments demonstrated that epithelium derived AREG was necessary and sufficient for ductal outgrowth regardless of the genotype of the stroma [45]. Conversely, stromal EFGR is a prerequisite for ductal outgrowth and epithelial EGFR is dispensable. Despite the numerous poten-tial ligands for EGFR the reason the pubertal EGFR KO pheno-copies the AREG KO is due to the fact that AREG is by far the dominant ligand expressed at this developmental stage, and in agreement, the pubertal phenotypes of EGF or TGFα KOs are less severe. However, there appears to be some division of labor among the EGFR ligands with TGFα and EGF instead being required for later lobuloalveolar development [44]. Interestingly, transplanta-tion experiments performed by Wiesen et al. demonstrated that it was possible to complete normal lobuloalveolar development in EGFR KO mice when prolactin was secreted from a co-trans-planted pituitary extract, suggesting that in fact EGFR was only required for pubertal ductal outgrowth and not later development of the gland [46].

The EGF receptor family also includes the ErbB2 (Her2/neu), ErbB3, and ErbB4 receptors and it is thought that EGFR/ErbB2 heterodimers primarily drive puberty given that these are expressed and colocalized in the virgin gland, while ErbB3 and ErbB4 are rather expressed at higher levels during pregnancy and lactation [47, 48]. Consistent with this hypothesis, estrogen can stimulate tyrosine phosphorylation and activation of EGFR (ErbB1) and ErbB2 null transplanted glands exhibit delayed pubertal ductal elongation but no defect in lobuloalveolar development [48, 49]. ErbB4 null mice have no defects in ductal morphogenesis; how-ever, the same is not true for the ErbB3-deficient mammary epi-thelium [50]. ErbB4 appears to be necessary for terminal differentiation of the gland and has been shown to be essential for lobuloalveolar development and lactation via its activation of STAT5a [51, 52]. In contrast, ErbB3 null mammary epithelial

---

Fig. 1 (continued) epithelial compartments include the growth hormone (GH) mediated transcription of insu-lin-like growth factor (IGF) via its activation of stromal growth hormone receptor (GHR). IGF binds and activates the epithelial insulin-like growth factor receptor (IGFR). Similarly, stromal derived HGF (from fibroblasts) acti-vates the c-Met receptor in the epithelial compartment

trees only partly fill the fat pad of wild type mice and this phenotype is maintained through adulthood, although lobuloalveolar development proceeds normally. The ErbB3 null pubertal gland also exhibits decreased TEB size and abnormal branch spacing [53]. Mice overexpressing heregulin under control of the MMTV promoter further support a role for ErbB3 in pubertal mammary development. NRG1 (NDF/heregulin) is a secreted growth factor and a ligand for the EGFR family. In heregulin overexpressing mice only the ErbB3 receptor was phosphorylated and these mice also displayed persistent TEBs beyond the normal pubertal period [54].

3.2 Insulin-Like Growth Factor Family

As mentioned, IGF mediates the effects of GH and accordingly IGF-1 KO mice display a phenotype similar to that of GHR KO mice. IGF-1 is expressed both in the stroma and in the TEBs of pubertal glands. IGF-1 KO mice have been generated by several mechanisms and consistently display stunted pubertal development with reduced ductal outgrowth [17, 55–57]. IGF-1 is believed to be important specifically as an initiator of pubertal mammary development. Culturing of prepubertal glands with IGF-1 led to extensive ductal outgrowth [58]. Similarly, overexpression of IGF-1 under control of the bovine keratin 5 (K5) promoter also resulted in increased ductal proliferation in prepubertal mice [59]. Consistent with these phenotypes, transplanted IGF1R KO mammary glands (used to overcome embryonic lethality in IGF1R KO mice) also fail to support ductal outgrowth. Defects in TEB formation and growth were due to a lack of cell proliferation rather than altered apoptosis [60]. Overexpression of the IGF1R using either a constitutively active receptor or an inducible doxycycline-MMTV driven mechanism also led to aberrant mammary gland development. Notably, the IGF1R overexpression phenotype is not simply accelerated ductal outgrowth but rather it includes impaired ductal elongation as well as formation of mammary tumors from as young as 8 weeks of age [61, 62]. The IGF binding proteins (IGFBPs) are also likely to play a role in ductal morphogenesis as IGFBP3 and IGFBP5 localize to pubertal epithelial cells while IGFBP2 and IGFBP4 are primarily detected in the stroma [63].

Interestingly, cross talk between the GH-IGF-1 and ERα-AREG signaling pathways has also been proposed. IGFs can modulate the action of ERα, and ERα action can regulate expression of IGF ligands, receptors and binding proteins. Studies with IGF-1 KO mice demonstrated that their defect could be rescued by administration of IGF-1 together with estrogen [17]. IGF-1 alone did promote development but this was further enhanced by the addition of estrogen. Of note, GH in combination with estrogen was insufficient to rescue the IGF-1 KO defect, confirming that IGF-1 acts downstream of the GHR [17].

3.3 Fibroblast Growth Factor

Members of the fibroblast growth factor (FGF) family have been implicated in pubertal mammary gland development, initially due to their expression patterns although mouse models have since confirmed a number of roles. FGFs 1, 2, 7, and 10 are all expressed in the mammary gland during ductal outgrowth. Somewhat unexpectedly, FGF7 null mice display no mammary gland phenotype [64]. FGF10 KO mice die at birth due to disrupted pulmonary branching morphogenesis so although a mammary phenotype has not yet been identified, FGF10 is still hypothesized to play a role in other forms of branching morphogenesis [65]. Transgenic overexpression of FGF3 however, under control of the MMTV promoter, results in impaired ductal outgrowth and reduced side branching [66]. An elegant study using genetic mosaic analysis to inactivate the receptor, FGFR2, in specific cells demonstrated a function for FGFR2 in proliferating and invading TEBs but not in mature ducts. The mosaic gland initially consisted of both homozygous null and heterozygous FGFR2 cells, the latter of which was seen to out compete the former during ductal expansion owing, at least in part, to the differential effect of FGFR2 on proliferative capacity [67].

3.4 Hepatocyte Growth Factor

The hepatocyte growth factor (HGF) or scatter factor, which binds to the c-Met tyrosine kinase receptor, has also been implicated in pubertal mammary gland development. Retrovirally transduced HGF expression in mammary epithelial cells transplanted into wild type mice results in increased ductal branching and TEB defects including larger, more numerous TEBs [68]. In agreement, interference with HGF signaling in mammary gland cultures inhibits in vitro branching morphogenesis [69]. Additionally, subsequent to pubertal development it could be observed that animals with a conditional deletion of the HGF receptor, c-Met, had significantly reduced secondary side branching of the ductal tree [70]. Fibroblasts, which are present in the mammary stroma along with several other migratory cells, are the cells responsible for the secretion of HGF as well as a number of other growth factors [71].

3.5 Heparin Sulfate Proteoglycans

It should be noted that the function of many growth factors, including FGF and HGF is mediated by binding to heparin sulfate proteoglycans (HSPGs) located on the cell surface of epithelial cells or in the pericellular matrix. They have been proposed to regulate growth factor activity in a variety of ways, including protection against degradation and facilitating assembly of signaling complexes [72]. Accordingly, mice deficient in the heparin sulfate associated enzymes Est1 and Hs2st have been shown to have reduced primary side branching, and a reduction in secondary and tertiary branching respectively [70]. A similar but less severe phenotype is observed in transmembrane HSPG, syndecan-1-deficient mice while either overexpression or deficiency in heparinase (responsible for cleavage of HS chains) alters mammary gland morphology and results in increased ductal branching with precocious alveolar development [73–75].

4   Branching Morphogenesis

A typical cross section of a terminal end bud and subtending duct illustrates the epithelial component by exposing the cap cell outer layer at the tip, which gradually merges into the differentiated myo-epithelial or luminal epithelial cells of the duct. This is followed by the layers of body cells, which eventually disappear to reveal a hollow lumen surrounded by a continuous luminal epithelial layer and con-tractile smooth muscle cell layer that will become the channel for milk expulsion in the mature gland. TEB invasion of the surround-ing stroma ultimately forms the tree-like branching network within the fat pad of the growing mammary gland. This descriptive and static picture, however, does nothing to reflect the dynamic mor-phology of the branching process that is defined by intimate cell–cell and cell–matrix communication, local immune cell infiltration, rapid proliferation, and massive programmed cell death necessary for active ductal morphogenesis. Mouse models have contributed an enormous amount to our current understanding of this process not least because working in vivo allows an appreciation of the myriad of cell types involved in a microenvironment that is difficult to accu-rately reproduce in vitro. While an in depth view of our current understanding of the mechanism of mammary epithelial branching tubulogenesis can be found in [76], here we highlight some of the recent studies that exemplify informative in vivo approaches.

4.1   Stem Cells

At puberty the mammary gland comprises two epithelial cell types: ductal luminal and myoepithelial cells. The cap cell of the pubertal TEBs has been proposed to comprise mammary stem (MSC) or progenitor cells within a temporary niche, as TEBs ultimately disap-pear when the ductal tree is complete. Although it has long been acknowledged that a rare single isolated and transplanted murine mammary stem cell is capable of recapitulating the entire function-ing epithelial ductal tree, the exact nature of the epithelial MSC hierarchy has been intensely debated with the assertion that lineage restricted or unipotent progenitor cells rather than multipotent cells such as described in the hematopoietic system, are required for tissue maintenance, development, and homeostasis [77, 78] (Fig. 2). The confusion arises as the transplantation process itself may allow a formerly lineage committed progenitor to undergo cell state transition and access former stem cell programs [79]. The advance of lineage tracing techniques, claimed to circumvent the potential pit falls of the classical transplantation methods, led to the hypothesis that long-lived unipotent MSCs existed to drive devel-opmental epithelial expansion [80]. Other specific transgenic mod-els have helped to unravel the complexity of epithelial stem and progenitor cell hierarchy. The expression of the transcription factor, GATA3, is limited to the luminal lineage, both ductal and alveolar,

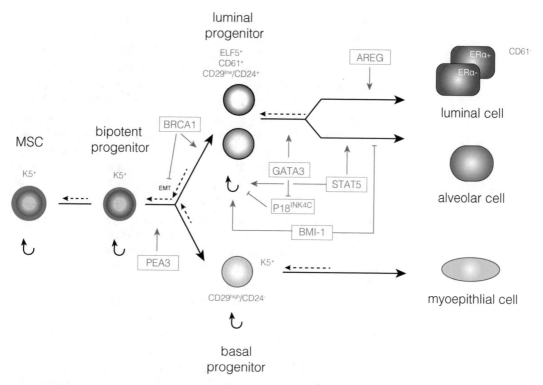

luminal
progenitor

ELF5+
CD61+
CD29low/CD24+

AREG

ERα+   CD61-

ERα-

luminal cell

BRCA1

MSC

bipotent
progenitor

K5+

K5+

EMT

GATA3

STAT5

P18INK4C

BMI-1

alveolar cell

PEA3

K5+

CD29high/CD24-

myoepithlial cell

basal
progenitor

Fig. 2 Mammary stem cell hierarchy and ductal morphogenesis. An overview of the current model of stem cell hierarchy in the mammary gland indicating key molecules affecting pubertal ductal morphogenesis via their action in promoting or inhibiting self renewal and cell fate determination as delineated from in vivo models described herein

and can be detected in the body cells and cap cells of TEBs at puberty. Conditional deletion of GATA3 has been reported to cause failure of TEB formation and lead to ducts that comprise mainly myoepithelial and ER-negative cells [81, 82]. The accumulation of $CD61^+$ cells in these mice suggested the observed defects were due to inhibition of differentiation of the luminal progenitor cell population supporting the role of GATA3 as a tumor suppressor [82, 83]. In agreement, mice deficient for the CDK inhibitor $P18^{INK4C}$ which is normally repressed by GATA3, develop spontaneous ER-positive tumors at a high rate, believed to be due to the expanded luminal progenitor population [84]. The fact that GATA3 promotes luminal progenitor differentiation and $P18^{INK4C}$ inhibits undifferentiated luminal progenitor expansion explains the observation that low $P18^{INK4C}$ expression and high GATA3 expression are indicative of luminal A-type breast tumors [84].

The polycomb group protein Bmi-1, thought to be required for stem cell maintenance, is in contrast to GATA3, expressed in all epithelial compartments of the mammary gland. The Bmi-1 deficient ductal epithelium fails to undergo ductal expansion into the fat pad although unlike in GATA3 deficiency, normal ductal architecture prevails. Although Bmi-1 deficient cells are capable of reconstituting a mammary gland they are less efficient, reflecting the reduced stem cell frequency observed. Additionally, the Bmi-1 deficient glands show signs of premature lobuloalveolar development suggesting that, at least in the mammary gland, Bmi-1, functions to maintain stem cell activity and drive proliferation while at the same time repressing terminal differentiation of the luminal alveolar lineage [85]. Another mouse model also implicated Pea3, an ets family transcription factor, in luminal and alveolar lineage-specific differentiation. Pea3 null mice, although capable of nursing, display up to a twofold increase in TEB number, persistence of the TEB structures beyond puberty, as well as retarded alveolar development. Transplantation studies confirmed that the effect was intrinsic to the epithelium and differential colony forming assays that preferentially select for unipotent or multipotent progenitors revealed that Pea3 null colonies were enriched in epithelial cells with markers of both luminal and myoepithelial lineages. Given the known limitations of in vitro assays to determine lineage potency this was also confirmed in vivo where double-labeled and thus bipotent cells, were found at a significantly lower frequency in the ducts and TEBs of wild type glands [86].

Conditional knockout of the MED1 subunit of the mediator/TRAP complex results in retarded ductal elongation and decreased lobuloalveolar development [87]. The effects could be easily attributed to the fact that MED1 functions as a coactivator for ERα; however, ablating just the specific interaction with ERα by mutation of the LxxLL domain demonstrated that in fact the contribution of this function was restricted to pubertal development and specifically, luminal differentiation. MED1 is highly expressed in luminal epithelial cells and in experiments examining the distribution of cells with specific progenitor cell markers, it could be seen that both the CD29(β1-Integrin)^{low}/CD24$^+$ and CD24$^+$/Sca1$^-$ population were enriched in MED1 LxxLL mutant epithelial cells, indicating that loss of the MED1 LxxLL domain caused an increase in the proportion of progenitors with a luminal cell fate [88]. A later study determined that there was cooperation between MED1 and MED24 subunits as mice haplo-insufficient for both subunits, but not either one alone, phenocopied the MED1 LxxLL mutant gland [89].

What has become clear is that rather than simply providing the means to sustain the gland by self-renewal and generation of iteratively more restricted progenitors, MSCs have an active role to play in remodeling events seen at specific phases of development [90]. Indeed, several previously established mediators of ductal morphogenesis where mouse models exhibited a mammary pubertal phenotype have, with newer techniques, been implicated in controlling cell fate or lineage determination. One such example is AREG. The failure of AREG KO transplanted explants to grow already suggested a role for AREG in maintenance of a mammary progenitor cell and, in line with this theory, AREG was demonstrated to be necessary for the expansion of the duct-limited subtype of mammary epithelial cells using in vitro mammosphere generation with the mammary progenitor cell line CDβGeo [42, 91]. However, to date, no in vivo evidence has been forthcoming. Another example is STAT5, reported to have a role in lobuloalveolar development and lactation downstream of EGFR4. STAT5 deficient mice also have some mild defects in nulliparous lateral branching but no defects in ductal elongation or TEB formation [92]. Consistent with this observation, STAT5a/b deficient stem cells are able to reconstitute a mammary gland suggesting no loss of multipotent or bipotent lineages [93]. Thus, as initial ductal elongation is unaffected it is thought to be the loss selectively of the lineage restricted CD61+ luminal progenitor population. In agreement with the observations of Santos et al., Vafaizadeh et al. also demonstrated defects in ductal side branching in transplants following ex vivo manipulation of MSCs to downregulate STAT5, as well as failure of alveolar differentiation [92]. Conversely, activation of STAT5 leads to epithelial hyperproliferation and precocious alveolar development [94, 95].

In an effort to address discrepancies in the various models of MSC hierarchy proposed to date, Rois et al. used in vivo clonal cell fate mapping coupled with 3D imaging to detect MSCs in situ. This work proposed that there is a division between puberty and maintenance of the adult gland such that stem and progenitor cells are required for pubertal morphogenesis while only bipotent MSCs are responsible for ductal homeostasis and contribute to later remodeling of the adult organ [96].

4.1.1 Epithelial–Mesenchymal Transition

The process of epithelial-mesenchymal transition (EMT) appears to be intimately related to stemness and cells that have undergone EMT express the same markers and phenotypically resemble stem cells isolated from the mammary gland [97]. BRCA1, the infamous tumor suppressor, which along with BRCA2, is responsible for the vast majority of hereditary breast cancers has recently been proposed to function in part via suppression of EMT in mammary epithelial cells [98]. Conditional mutation of BRCA1 results in stunted ductal outgrowth at puberty and defects in lobuloalveolar development, suggestive of a role in luminal progenitor derivation and differentiation [99]. Interestingly BRCA1, like GATA3, is also a negative regulator of $P18^{INK4C}$ but mutation and functional loss of BRCA1 in $P18^{INK4C}$ null mice blocks the expansion of the luminal progenitor population and results in tumors with basal like (rather than luminal) features [84, 100]. It was recently shown that this reversion of tumor type is due to activation of EMT resulting in effective de-differentiation of luminal cells. This supports the notion that the functional role of BRCA1 is to suppress EMT and drive stem cell differentiation in the mammary gland. It would also adequately explain the retarded ductal phenotype seen at puberty with BRCA1 deficient epithelium as EMT has also been suggested to be a mechanism employed to effect branching morphogenesis during development of many epithelial organs. In addition, BRCA1 is known to act as a transcriptional repressor of ERα target genes including AREG [101].

The collective migration and invasion of ductal epithelium, which is required for pubertal organogenesis to proceed, is believed to be dependent on reversible induction of local plasticity at the leading edges of invading branches and TEBs. Hallmarks of EMT include the loss of apical–basal polarity, expression of mesenchymal markers, reduction of E-cadherin levels and increased invasiveness. In support of this hypothesis, the expression of mesenchymal markers such as vimentin is enriched at the tips of growing branches, SNAIL1/SNAIL2 and E47 at branch initiation points, and TWIST1, TWIST2, and SNAIL in microdissected TEBs [102–104]. Furthermore, Lee et al. used the elegant in vitro tubular assay developed in the Bissell lab to demonstrate that ectopic expression of SNAIL1/SNAIL2 and E47 can induce branching, and to show that these tubules respond to growth factor stimulation by induction of mesenchymal markers and downregulation of E-cadherin [102, 105].

4.2 Cell Polarity and Cell–Cell Adhesion

The changes in epithelial polarity that characterize EMT are known to be essential for the process of collective migration proposed to be employed in ductal invasion [106]. Adherence junctions play a vital role in the establishment of apical–basolateral polarization and organization of the mammary epithelium by maintaining appropriate cell–cell contacts. Disruption of normal epithelial integrity can therefore be observed in several mouse models as a result of interference with adhesion molecule signaling. Type I cadherins, which comprise P, E, and N are all expressed in the developing mammary gland and their roles have been extensively investigated [107]. During murine ductal outgrowth, P-cadherin and E-cadherin demarcate the cap cells and body cells in the TEB, respectively. In fact, P-cadherin is restricted to the cap cells and the myoepithelial cells of the mature duct, while E-cadherin is present in body cells and luminal cells, but absent from myoepithelial cells [108, 109]. Early experiments administering anti-E-cadherin or anti-P-cadherin monoclonal antibodies in the pubertal gland caused disruption of the cap cell and body cell layers and resulted in detached cells floating in the ductal lumen [108]. Later it was shown that P-cadherin knockout mice display precocious virgin alveolar development and hyperplasia in the adult [109]. Similarly, overexpression of a truncated E-cadherin comprising only the cytoplasmic portion results in precocious alveologenesis and while normal pubertal development can occur in E-cadherin KO mice, they fail to lactate [110, 111].

One the most extensively studied aspects of E-cadherin intracellular signaling in the mammary gland is its interaction with the canonical Wnt/β-catenin pathway. Numerous mammary gland phenotypes are observed by manipulation of central players in this key signaling node. They range in severity from the LEF1 null mice that completely lack mammary glands to delayed ductal morphogenesis seen by inactivation of the negative regulator APC, and delayed outgrowth with precocious alveolar development in mice overexpressing a stabilized form of β-catenin [112–115].

Both P-cadherin and E-cadherin are capable of forming transmembrane complexes with the catenins: α, β, and p120; but the E-cadherin/β-catenin complex is the archetypal adherence junction responsible for stabilization of cell–cell contacts and polarized epithelial cell shape via connection to the actin cytoskeleton. During EMT, a cadherin-switching phenotype has been described whereby loss of epithelial E-cadherin is concomitant with an upregulation of N-cadherin expression that is more typical of motile, less polarized mesenchymal cells. Control of cadherin switching is known to be mediated by transcriptional repressors of E-cadherin such as SNAIL, SLUG, ZEB1, and ZEB2 [116]. Switching from E- to N-cadherin also has differential effects on growth factor receptor pathways. N-cadherin potentiates FGFR signaling while E-cadherin can interact with the EGF receptor and inhibit EGF-dependent signaling by restricting the mobility of the receptor [116, 117]. In accordance, it could be seen on an ErbB2 activated (MMTV-ErbB/Neu) background, N-cadherin expression leads to unfettered FGF signaling resulting in upregulation of SNAIL and SLUG, EMT and stem like properties in primary cells and tumor cell lines derived from the resulting MMTV-Neu-N-Cad mice [118].

In addition to the cadherins and catenins several other molecules are responsible for the regulation of cell junction formation including members of the Rho family GTPases and polarity complexes of Par, Crumbs and Scribble [119, 120]. Accordingly, mammary gland phenotypes have recently been described for a number of these. Overexpression of the Rho GTPase Cdc42 in mammary epithelium leads to hyperbudding of TEBs and hyperbranching but a reduced ductal tree area [121]. The P190BRhoGAP overexpressing mice display a strikingly similar phenotype and both have alterations in the adjacent stromal compartment evidenced by increased collagen deposition around TEBs [121, 122] (Fig. 3). Scribble deficient pubertal mammary glands have increased numbers of TEBs and excessive ductal branching. Closer examination revealed aberrant colocalization of apical and lateral markers and random distribution of membrane E-cadherin/β-catenin complexes suggesting a failure in establishing proper epithelial apicobasal polarity [123]. Supporting the link between polarity, EMT and stemness, limiting dilution assays and colony formation assays demonstrated that Scribble functions to inhibit expansion of a bipotent progenitor population while facilitating maturation of the ductal luminal population responsible for ductal outgrowth [123].

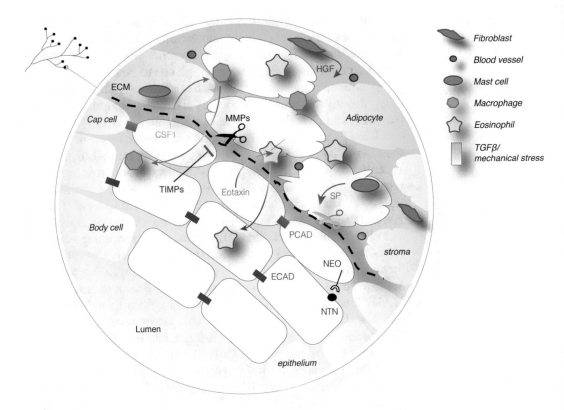

Fig. 3 TEB invasion of the mammary stroma. The TEB and its adjacent stromal environment are home to a multitude of cell types that collaborate to facilitate invasion of the fat pad. E-cadherin (ECAD) and P-cadherin (PCAD) maintain cell–cell contacts within the body and cap cells of the TEB, respectively. Netrin (NTN) and neogenin1 (NEO) form cell–cell adhesion contacts between the outer cap cells and inner body cells. CSF1 secretion recruits infiltrating macrophages while eotaxin recruits eosinophils. Matrix metalloproteases (MMPs), and serine proteases (SP) released from mast cells, are responsible for the breakdown of the ECM to allow invasion to proceed, and this is off-set by the tissue inhibitors of metalloproteases (TIMPs). Hepatocyte growth factor (HGF) and other growth factors are secreted from fibroblasts. The morphogenetic gradient of TGFβ is inversely correlated with gradient of mechanical stress exerted at the point of invasion and is depicted here by a color gradient

Originally believed to play a role only in neuronal guidance in the nervous system, mouse models have identified a number of axonal guidance proteins including Netrin-1 (Ntn1), Neogenin-1 (Neo1), and EphrinB2 (EFBN2) that are believed to perform an adhesive rather than guidance function in the mammary gland. Ntn1 is a glycoprotein ligand found on the surface of body cells and is secreted into the subcapsular space, while its receptor Neo1 is located on the cap cell surface. Both Ntn1 and Neo1 KO mice display a similar mammary phenotype with abnormal TEB formation as evidenced by an extended subcapsular space, breaks in basal lamina and dissociation of cap cells [124]. In a similar fashion, members of the Ephrin family of ligands and receptors exhibit reciprocal expression patterns during embryogenesis although postnatally they have been best described in the nervous system. EFNB2 is a ligand found in luminal epithelial cells while its interacting partner EPHB4, a receptor tyrosine kinase, is located on both myoepithelial and epithelial cells. EFNB2 and EPHB4 are both estrogen dependent and differentially expressed depending on the developmental stage of mammary development. Overexpression of EPHB4 in the mammary epithelium under the control of an MMTV promoter leads to delayed pubertal development of the mammary gland with only rudimentary epithelial trees forming [125].

4.3 The Role of TGFβ in Branch Patterning

TGFβ was considered the primary candidate inhibitory morphogen in the developing gland since Silberstein and Daniel demonstrated in 1987 that delivering it locally via implanted pellets, could inhibit the formation and growth of ducts and TEBs in vivo [126]. Subsequently, it was demonstrated that overexpression of constitutively active TGFβ1 led to marked suppression of ductal outgrowth and resulted in extremely underdeveloped glands that were nevertheless capable of alveologenesis [127]. Conversely, TGFβ1 deficient glands display accelerated ductal development and transplant experiments confirmed the effect was intrinsic to the epithelial compartment [128]. Endogenous TGFβ is believed to exert much of its sculpting control via the formation of a morphogenic gradient. Epithelial derived TGFβ is deposited in the ECM and differential concentrations can be observed depending on proximity to the TEB, bud point, and subtending ducts, such that lower levels appear permissive for outgrowth and branching while higher levels function to restrain extension of the ducts [103, 129]. Indications as to signaling molecules downstream of TGFβ that are responsible for the growth control came from experiments demonstrating that TGFβ directly regulates expression of WNT5A in the mammary gland, that the canonical WNT5A null mice phenocopied TGFβ1 deficient glands, and that WNT5A was required for TGFβ mediated ductal inhibition [130]. This suggested that modulation of cellular adhesion could be in part responsible given the role of WNT5A in epithelial cell migration and adhesion, such as its activation of the collagen binding protein DDR1. In agreement, in the dominant negative TGFβ receptor (DIIR) mutant gland, in addition to loss of WNT5A, there is reduced phosphorylation of DDR1 [130]. The DDR1 null mammary gland in turn displays a complex phenotype, with delayed pubertal ductal invasion and enlarged TEBs, but hyperproliferation and increased branching in the mature virgin. While not a strict phenocopy, the evidence would suggest DDR1 may at least in part mediate the TGFβ1/WNT5A effects in the developing gland [131]. TGFβ1/WNT5A signaling also acts to antagonize the canonical Wnt/β-catenin pathway and loss of either results in stabilization of nuclear β-catenin and expression of its target genes [132].

Interestingly, loss of the transcriptional coregulator CITED1, led to upregulation of WNT5A expression in the pubertal mammary gland, concomitant with a phenotype comprising reduced ductal outgrowth and disturbed spatial patterning. This is consistent with the role of CITED1 as potentially both a SMAD coregulator downstream of TGFβ, and a coregulator of ERα [8]. The observation that CITED1 may function in two key signaling nodes responsible for pubertal ductal outgrowth is supported by the proposal of Ewan et al. that there is a regulatory circuit where TGFβ directly limits the proliferative response to hormonal signaling. In agreement, active TGFβ could be colocalized to a subset of nonproliferating luminal epithelial cells that were ERα positive [128]. One could therefore speculate that direct competition for a transcriptional coregulator mediates the balance between the opposing actions of estrogen and TGFβ signals during pubertal outgrowth.

5   The Stromal and Extracellular Environment

The mammary stroma is composed of multiple cells types; primarily, the adipocytes of the fat pad but also fibroblasts, endothelial cells of the vasculature, infiltrating leukocytes and cells derived from nerve innervation [133]. Extracellular factors such as proteoglygcans and collagens contribute to formation of the specialized extracellular matrix, which comprises the basement membrane at the boundary of where the ductal epithelium meets the stroma [133, 134]. Although branch patterning is not yet completely understood, the extent of cell–matrix interactions suggests that the matrix is at least as much of a contributor as the epithelium in deciding the patterning of the epithelial tree.

5.1 Matrix Stability and Integrity

The importance of regulation of matrix deposition and its organization is evidenced by a number of mouse models such as that of P190BRhoGAP overexpression described above. The P190BRhoGAP overexpression phenotype was shown to be due to activation of stromal fibroblasts leading to increased extracellular collagen, fibronectin, and laminin deposition that characterized alterations seen in the periductal stroma [95]. Enzymatic degradation of the ECM, reciprocally coordinated by proteases such as the MMPs and their inhibitors, TIMPs, is also required to mediate ductal invasion. MMP2 expression is temporally regulated and specifically reduced at sites of lateral branching. MMP2 null mice display delayed ductal invasion during early puberty but increased lateral branching during late puberty [135]. The TEBs of MMP2 KO mice display increased apoptosis with enhanced activity of caspase-3, suggesting that MMP2 functions to promote cell survival within the TEB. Consistent with this, MMP14, an activator of MMP2, has elevated expression in and around TEBs while conversely, the expression of TIMP3, an inhibitor of MMP14, is reduced. MMP3, in contrast, has a more consistent expression pattern throughout the mammary stroma. MMP3 KO mice exhibit reduced lateral branching proposed to be mediated via local degradation of collagen and laminins, while mice overexpressing MMP3 under control of the WAP promoter, display increased ductal branching [135–137]. TIMP1 expression is specifically upregulated in the area around TEBs and although TIMP1 KO mice exhibit normal ductal elongation, they have significantly larger TEBs [135]. As TIMP1 deficiency, induced by the use of TIMP1 antisense RNA under control of the MMTV promoter, led to breakdown of the basement membrane around epithelial ducts, it was hypothesized that TIMP1 functions to maintain structural integrity of the ECM [138]. In agreement, in animals overexpressing TIMP1, ductal invasion was inhibited [135]. Similarly, implanting TIMP1-secreting pellets led to locally reduced ductal expansion [138]. Thus, loss of TIMP1 results in reduced ECM integrity leading to larger, less restricted TEBs, while a gain of TIMP1 leads to an ECM which is harder to break down resulting in reduced ductal outgrowth.

5.2 Integrins, Laminin and Mechanical Stress

Integrins are transmembrane receptors with a key role in maintaining communication between cells and the ECM. This communication plays a central role in cell adhesion, organization of the stroma, establishment of polarity and, potentially, maintenance of the mammary stem cell niche. Deletion of the β1-integrin selectively from basal cells by use of a K5-Cre-loxP promoter construct results in a pubertal phenotype of disorganized ductal branching and reduced numbers of side branches, with subsequent aberrant morphogenesis during pregnancy [139]. This is in agreement with earlier studies using an anti-β1-integrin antibody which resulted in impaired ductal outgrowth [140]. Interestingly, the specific loss of β1-integrin from basal cells dramatically inhibited the regenerative capacity of secondary grafts but not the ability of the primary transplants to undergo lobuloalveologenesis. This suggests depletion of a specific β1-integrin-positive basal progenitor stem cell population in these mice that is required for early gland establishment and pubertal outgrowth, but dispensable for later alveolar development [139]. A similar antibody approach to deplete laminin demonstrated that its loss results in impaired ductal outgrowth and a reduced number of TEBs, further supporting the requirement for adequate ECM organization during morphogenesis [140]. Focal adhesion kinase (FAK) is a cytoplasmic tyrosine kinase and major mediator of integrin signaling. A conditional KO of epithelial cell specific FAK led to mice with retarded ductal elongation during puberty. Although the pubertal delay did later catch up, subsequent defects in development produced mothers who were unable to feed their young [141]. An additional mouse model with loss of FAK in both mammary cell lineages resulted in dilated ducts and disruption in the normally separated myoepithelial and luminal cell layers. These mice also displayed abnormal morphogenesis during pregnancy [142].

Integrins and FAK are also thought to contribute to epithelial tissue morphogenesis by sensing and responding to biophysical cues. Mechanical stress can activate FAK directly by phosphorylation and phosphorylated FAK can be identified at the tips of tubules in a microfabricated model of epithelial morphogenesis, similar to that used to investigate TGFβ inhibition of branching. It is thought that gradients in mechanical stress, akin to morphogenic gradients, may distinguish branch points from linear ducts in vivo and in fact both biochemical and biophysical cues are required to work in tandem. Induction of EMT in mammary epithelial cells by a range of stimuli is well established and it is proposed that a transient EMT is partially responsible for the invasion of the fat pad. Mechanical stress can also induce patterned EMT in epithelial monolayers [143, 144]. Similar mechanotransduction in vivo is a likely candidate for induction of transient EMT at branch points since although TGFβ is a well-established inducer of EMT, unlike epithelial tumor cells, the normal mammary epithelium tends not to respond to TGFβ in this manner. Moreover, in the mammary epithelium, the gradient of TGFβ is inversely correlated to that of the mesenchyme markers [103, 144].

5.3 Cells of the Immune System

Eosinophils, together with mast cells and macrophages play important roles in the developing mammary gland distinct from their classical immune-related functions. Eosinophils are located in the region around TEBs and are recruited by eotaxin. As expected, eotaxin null mice display an absence of eosinophils, which is accompanied by a mammary phenotype characterized by reduced branching with reduced TEB formation and development [145]. Interestingly, the extent of ductal elongation was not altered in these mice. Although eotaxin has been demonstrated to be the primary chemokine recruiting eosinophils to the mammary gland, the cytokine IL-5 can induce eosinophil recruitment and activation. As a result, mice with constitutive overexpression of IL-5 display eosinophilia but have only a transient pubertal phenotype with decreased numbers of TEBs and retarded ductal outgrowth [146]. These phenotypes confirm that both an excess and deficiency in eosinophils affects branching and TEB formation although the exact mechanism has yet to be elucidated. Mast cells are present surrounding TEBs and the ductal epithelium of the pubertal gland, and mice lacking these exhibit reduced branching with fewer TEBs. The observed phenotype was not due to secondary loss of eosinophils or macrophages despite the fact that mast cells can function to recruit leukocytes. Changes in collagen deposition were also not evident; however, it could be shown that de-granulation of mast cells was necessary for the effect on mammary gland development essentially implicating any number of mast-cell derived factors, although proteases are arguably the most likely candidates

[147]. Despeptidyl peptidase I, DPPI (cathepsin C) is responsible for the activation of mast cell granule serine proteases and in agreement, DPPI null mice display impaired mammary ductal development. However, given that the role of DPPI is not limited to activation of mast cell derived proteases and that it can function to degrade other ECM components, it is not certain that the phenotype affects only perturbation of mast cell activity in the mammary gland [147].

Colony stimulating factor (CSF-1) is responsible for recruiting macrophages to the region around the neck of TEBs. CSF-1 null mice, as expected, display a reduced population of macrophages and also a pubertal mammary phenotype. They exhibit impaired TEB formation with reduced ductal outgrowth and branching [145]. The local reliance on CSF-1 was confirmed by rescue experiments where overexpression of CSF-1 under control of an MMTV promoter could rescue the CSF-1 KO gland, thus confirming that systemic CSF-1 was not responsible for the mammary phenotype [148]. The normal role of macrophages in this region around the neck of TEBs is thought to be in assisting branching morphogenesis by recruiting growth factors and matrix remodeling proteins. Owing to their established role in inducing epithelial cell death during involution, it has been suggested that macrophages may play a role inducing apoptosis or in phagocytosis of apoptotic body cells to enable lumen formation in the subtending duct as the TEB invades the stroma [145, 149]. In agreement, apoptotic bodies were observed in the cytoplasm of macrophages localized in TEBs [145]. Further, macrophages are required for maintenance of MSCs as demonstrated by the reduced regenerative capacity of cells from macrophage-depleted glands and even of wild type MECs transplanted into macrophage depleted stroma [150]. This illustrates yet again, the intrinsic connection between mediators of morphogenesis and maintenance of the mammary stem cell niche.

6   Caveats: Mouse Models and Pubertal Phenotypes

A number of caveats need be considered when inferring function from mouse models and we mention some of the limitations relevant for the study of mammary gland biology here. A comprehensive account of developing transgenic models and mammary transplantation techniques is outside the scope of this review but they are found adequately described within this book and elsewhere [151–154].

Given that we have included reports dating as far back as 1987, one should first caution that not all studies can be compared directly owing to, principally, differences in analysis methods to describe and quantify the phenotypes. In fact, a standardized system has never been formally established and as a result many features of some of the mice may have been overlooked. Not all investigators for example, count secondary and tertiary branching points, or quantify TEB number or size, and many subtler phenotypes that recover by maturation of the gland, may have been missed altogether. Other studies may have focussed on the more apparent later phenotypes of pregnancy and lactation. A case in point is that of cyclinD1 null mice where no specific overt pubertal phenotype has been described but rather investigators reported on the impaired lobuloalveolar development [155, 156]. However, the fact that cyclinD1 has been found to act as an ERα-coregulator required for estrogen-mediated gene expression and is recruited to the AREG promoter in vitro would suggest it has a role mediating growth factor signals utilized prior to pregnancy. Indeed in vivo, cyclinD1 null epithelium failed to transcribe AREG or EGFR, to the same extent as wild type epithelium, in response to estrogen released by locally implanted pellets. However, notably again, there was no description of morphogenesis in the cyclinD1 null gland during pubertal development [157]. In some studies the actual effect on pubertal outgrowth may be indirect and simply due to a defect in systemic hormonal signaling. For example, the SRC3 null pubertal phenotype was due to reduced systemic estrogen, while the SMAD3 KO phenotype was attributable to secondary ovarian insufficiency, but oftentimes this possibility has simply not been investigated [158, 159]. The use of organ specific promoters, e.g.: MMTV, WAP, especially prior to the advent of conditional models, meant that frequently genes were being switched on or deleted during developmental periods and in cell types they may not naturally be expressed in, complicating our understanding of the biology rather than helping elucidate function. The more recent use of native endogenous gene promoters, identification of cell type and lineage specific promoters and advances in accurate gene editing techniques offers reason for optimism. Nevertheless, we have come to rely heavily on the conclusions made previously in vivo often even putting greater trust in them than perhaps more elegantly designed and informative in vitro experiments. Transplantation studies have been the mainstay when faced with an embryonic lethal phenotype or in order to ascertain the requirement for either the stromal or epithelial compartment. However, investigation of stem and progenitor cell fate has been hampered by such experimental approaches given that it is now appreciated that the process of transplanting progenitors itself impacted the ultimate behavior of the cells. In some cases this may have led to the wrong conclusion regarding the characteristics of the progenitor population or renewal capacity in vivo. Newer techniques such as that described in [96] are attempting to address this issue.

Finally, reproducibility initiatives need to be embraced owing to the fact that when dealing with any complex biological organism, various phenotypic effects can be inadvertently misattributed. This appears to have been the case in the study where deletion of microRNA-212/132 family led to a pubertal phenotype comprising impaired ductal outgrowth that was attributed in part to the resulting upregulation of MMP3 and a decrease in the deposition of collagen around ducts [160]. Subsequently, however, two independent groups were unable to replicate the findings, albeit using a slightly different gene ablation strategy, and this prompted Kayo et al. to examine the differences in the mouse models further [161, 162]. The original study authors responded in kind and ultimately conceded that the mammary phenotype may in fact be independent of miRNA-212/132 [163]. Kayo et al. proposed that the phenotype was due to reduced expression of a gene adjacent to the targeted locus derived from the mouse strain of the ES cells used to generate the mice, whereas Ucar et al. disagreed, proposing instead that it was more likely due to deletion of an adjacent genomic region with an as yet unknown function [162, 163]. Although to date neither theory has been proven, the case in general highlights a very specific unanticipated consequence of deriving biological function from only a single mouse model.

While unfortunately this example is quite unlikely to be the sole one in the literature these too in time will ultimately serve to better our overall understanding of the biological mechanisms controlling mammary morphogenesis. Fortunately, the next generation of genomic manipulation in the form of CRISPR/Cas9 technology promises to greatly improve our editing ability to the point that imprecise gene ablation models will be confined to the past. Mouse models based on CRISPR/Cas9 techniques have the advantage in terms of both speed and accuracy for germ line manipulation, and have even proven successful for the generation of precision somatic models [164]. Germ line models can be created at the embryo stage by direct injection of Cas9/sgRNA, bypassing the need for any manipulation of ES cells while for the generation of somatic alterations the main challenge that remains is designing an adequate delivery strategy. To date, primarily the liver and lung have been successfully targeted somatically. The liver readily takes exogenous introduced DNA via hydrodynamic tail-vein injection while lung tissue for example has been successfully targeted using both lentivirus and adenovirus [165, 166]. Lenti-viral based delivery of CRISPR/Cas9 and Cre-recombinase to knockout various potential tumor suppressors in previously established Cre-loxP models of cancer such as the Cre activated Kras^{G12D} lung cancer model, has the advantage of reducing the time and effort that would otherwise be required to generate these complex changes by interbreeding [167].

The generation of a germ line Cre-dependent Cas9 knockin mouse allows for the combination of the traditional Cre-LoxP inducible system while importantly overcoming more difficult somatic delivery of the rather large Cas9 construct. Starting with Cas9 already in the germ line greatly facilitates the generation of subsequent constitutive and inducible somatic edits by targeted delivery of sgRNA. The main advantage of somatic models apart from overcoming deleterious embryonic or postnatal lethality without the need for organ transplantation, is the ability to more accurately mirror complex biology. This is particularly relevant in regard to cancers, which are overwhelming due to accumulation of various somatic alterations [168].

Apart from the exciting possibility of utilizing CRISPR/Cas9 for future mammalian corrective gene therapy [169], this technology has undoubtedly changed the game for biological mouse models. We now have, in our arsenal, tools that provide us with the ability to recreate a wide variety of genomic alterations with relative ease. While complex models including chromosomal rearrangements, and targeting of multiple loci simultaneously have been generated successfully, an important example from mammary gland biology is the elegant study recently published by the Hennighausen group. This work highlights the utility of CRISPR/Cas9 method to elucidate mammary gland specific regulation of STAT5. By identification and subsequent targeting of a mammary specific intergenic enhancer region they were for the first time able to explain how the unique lineage-specific transcriptional autoregulatory control of STAT5 is achieved and maintained via positive feedback [170].

In conclusion, despite its limitations, some of which we have only appreciated recently but most of which we can overcome, the unique niche of the mouse mammary gland as an elegant and highly successful model system is indisputably evident in the field of mammalian biology.

References

1. Couse JF, Korach KS (1999) Estrogen receptor null mice: what have we learned and where will they lead us? Endocr Rev 20:358–417. doi:10.1210/edrv.20.3.0370

2. Hovey RC, Trott JF, Vonderhaar BK (2002) Establishing a framework for the functional mammary gland: from endocrinology to morphology. J Mammary Gland Biol Neoplasia 7:17–38

3. Bocchinfuso WP, Korach KS (1997) Mammary gland development and tumorigenesis in estrogen receptor knockout mice. J Mammary Gland Biol Neoplasia 2:323–334

4. Mallepell S, Krust A, Chambon P, Brisken C (2006) Paracrine signaling through the epithelial estrogen receptor alpha is required for proliferation and morphogenesis in the mammary gland. Proc Natl Acad Sci U S A 103:2196–2201. doi:10.1073/pnas.0510974103

5. Feng Y, Manka D, Wagner K-U, Khan SA (2007) Estrogen receptor-alpha expression in the mammary epithelium is required for ductal and alveolar morphogenesis in mice. Proc Natl Acad Sci U S A 104:14718–14723. doi:10.1073/pnas.0706933104

6. Manavathi B, Samanthapudi VSK, Gajulapalli VNR (2014) Estrogen receptor coregulators and pioneer factors: the orchestrators of mammary gland cell fate and development. Front Cell Dev Biol 2:34. doi:10.3389/fcell.2014.00034

7. Han SJ, DeMayo FJ, Xu J et al (2006) Steroid receptor coactivator (SRC)-1 and SRC-3 differentially modulate tissue-specific activation functions of the progesterone receptor. Mol Endocrinol 20:45–55. doi:10.1210/me.2005-0310

8. Howlin J, McBryan J, Napoletano S et al (2006) CITED1 homozygous null mice display aberrant pubertal mammary ductal morphogenesis. Oncogene 25:1532–1542. doi:10.1038/sj.onc.1209183

9. Li M, Fu X, Ma G et al (2012) Atbf1 regulates pubertal mammary gland development likely by inhibiting the pro-proliferative function of estrogen-ER signaling. PLoS One 7, e51283. doi:10.1371/journal.pone.0051283

10. Li M, Zhao D, Ma G et al (2013) Upregulation of ATBF1 by progesterone-PR signaling and its functional implication in mammary epithelial cells. Cancer Cell 430:358–363. doi:10.1016/j.bbrc.2012.11.009

11. Brisken C, Park S, Vass T et al (1998) A paracrine role for the epithelial progesterone receptor in mammary gland development. Proc Natl Acad Sci U S A 95:5076–5081

12. Humphreys RC, Lydon JP, O'Malley BW, Rosen JM (1997) Use of PRKO mice to study the role of progesterone in mammary gland development. J Mammary Gland Biol Neoplasia 2:343–354

13. Conneely OM, Mulac-Jericevic B, Lydon JP, De Mayo FJ (2001) Reproductive functions of the progesterone receptor isoforms: lessons from knock-out mice. Mol Cell Endocrinol 179:97–103

14. Mulac-Jericevic B, Mullinax RA, DeMayo FJ et al (2000) Subgroup of reproductive functions of progesterone mediated by progesterone receptor-B isoform. Science 289:1751–1754

15. Shyamala G, Yang X, Silberstein G et al (1998) Transgenic mice carrying an imbalance in the native ratio of A to B forms of progesterone receptor exhibit developmental abnormalities in mammary glands. Proc Natl Acad Sci U S A 95:696–701

16. Kleinberg DL (1997) Early mammary development: growth hormone and IGF-1. J Mammary Gland Biol Neoplasia 2:49–57

17. Ruan W, Kleinberg DL (1999) Insulin-like growth factor I is essential for terminal end bud formation and ductal morphogenesis during mammary development. Endocrinology 140:5075–5081. doi:10.1210/endo.140.11.7095

18. Ormandy CJ, Camus A, Barra J et al (1997) Null mutation of the prolactin receptor gene produces multiple reproductive defects in the mouse. Genes Dev 11:167–178

19. Ormandy CJ, Binart N, Kelly PA (1997) Mammary gland development in prolactin receptor knockout mice. J Mammary Gland Biol Neoplasia 2:355–364

20. Horseman ND, Zhao W, Montecino-Rodriguez E et al (1997) Defective mammopoiesis, but normal hematopoiesis, in mice with a targeted disruption of the prolactin gene. EMBO J 16:6926–6935. doi:10.1093/emboj/16.23.6926

21. Brisken C, Kaur S, Chavarria TE et al (1999) Prolactin controls mammary gland development via direct and indirect mechanisms. Dev Biol 210:96–106. doi:10.1006/dbio.1999.9271

22. Kingsley-Kallesen M, Mukhopadhyay SS, Wyszomierski SL et al (2002) The mineralocorticoid receptor may compensate for the loss of the glucocorticoid receptor at specific stages of mammary gland development. Mol Endocrinol 16(9):2008–2018. doi:10.1210/me.2002-0103

23. Reichardt HM, Horsch K, Grone HJ et al (2001) Mammary gland development and lactation are controlled by different glucocorticoid receptor activities. Eur J Endocrinol 145(4):519–527

24. Wintermantel TM, Bock D, Fleig V et al (2005) The Epithelial glucocorticoid receptor is required for the normal timing of cell proliferation during mammary lobuloalveolar development but is dispensable for milk production. Mol Endocrinol 19(2):340–349. doi:10.1210/me.2004-0068

25. Dunbar ME, Dann PR, Robinson GW et al (1999) Parathyroid hormone-related protein signaling is necessary for sexual dimorphism during embryonic mammary development. Development 126(16):3485–3493

26. Hiremath M, Wysolmerski J (2013) Parathyroid hormone-related protein specifies the mammary mesenchyme and regulates embryonic mammary development. J Mammary Gland Biol Neoplasia 18:171–177. doi:10.1007/s10911-013-9283-7

27. Wysolmerski JJ, McCaughern-Carucci JF, Daifotis AG et al (1995) Overexpression of parathyroid hormone-related protein or parathyroid hormone in transgenic mice impairs branching morphogenesis during mammary gland development. Development 121(11):3539–3547

28. Dunbar ME, Dann P, Brown CW et al (2001) Temporally regulated overexpression of parathyroid hormone-related protein in the mammary gland reveals distinct fetal and pubertal phenotypes. J Endocrinol 171(3):403–416

29. Zinser G, Packman K, Welsh J (2002) Vitamin D3 receptor ablation alters mammary gland morphogenesis. Development 129(13):3067–3076

30. Welsh J (2004) Vitamin D and breast cancer: insights from animal models. Am J Clin Nutr 80(6 Suppl):1721–1724

31. Soto AM, Sonnenschein C (2010) Environmental causes of cancer: endocrine disruptors as carcinogens. Oncogene 6:363–370. doi:10.1038/nrendo.2010.87

32. Hilakivi-Clarke L (2014) Maternal exposure to diethylstilbestrol during pregnancy and increased breast cancer risk in daughters. Breast Cancer Res 16:208. doi:10.1093/ije/dyl106

33. Geens T, Aerts D, Berthot C et al (2012) A review of dietary and non-dietary exposure to bisphenol-A. Food Chem Toxicol 50:3725–3740. doi:10.1016/j.fct.2012.07.059

34. Muñoz-de-Toro M, Markey CM, Wadia PR et al (2005) Perinatal exposure to bisphenol-A alters peripubertal mammary gland development in mice. Endocrinology 146:4138–4147. doi:10.1210/en.2005-0340

35. Hilakivi-Clarke L, Cho E, Clarke R (1998) Maternal genistein exposure mimics the effects of estrogen on mammary gland development in female mouse offspring. Oncol Rep 5:609–625. doi:10.3892/or.5.3.609

36. Johnson MD, Kenney N, Stoica A et al (2003) Cadmium mimics the in vivo effects of estrogen in the uterus and mammary gland. Nat Med 9:1081–1084. doi:10.1038/nm902

37. Hilakivi-Clarke L, Cho E, Raygada M, Kenney N (1997) Alterations in mammary gland development following neonatal exposure to estradiol, transforming growth factor α, and estrogen receptor antagonist ICI 182,780. J Cell Physiol 170:279–289

38. Fielden M (2002) Normal mammary gland morphology in pubertal female mice following in utero and lactational exposure to genistein at levels comparable to human dietary exposure. Toxicol Lett 133:181–191. doi:10.1016/S0378-4274(02)00154-6

39. Wise A, O'Brien K, Woodruff T (2011) Are oral contraceptives a significant contributor to the estrogenicity of drinking water? Environ Sci Technol 45:51–60. doi:10.1021/es1014482

40. Khan U, Nicell JA (2014) Contraceptive options and their associated estrogenic environmental loads: relationships and trade-offs. PLoS One 9, e92630. doi:10.1371/journal.pone.0092630

41. McBryan J, Howlin J, Napoletano S, Martin F (2008) Amphiregulin: role in mammary gland development and breast cancer. J Mammary Gland Biol Neoplasia 13:159–169. doi:10.1007/s10911-008-9075-7

42. Ciarloni L, Mallepell S, Brisken C (2007) Amphiregulin is an essential mediator of estrogen receptor function in mammary gland development. Proc Natl Acad Sci U S A 104:5455–5460. doi:10.1073/pnas.0611647104

43. McBryan J, Howlin J, Kenny PA et al (2007) ERalpha-CITED1 co-regulated genes expressed during pubertal mammary gland development: implications for breast cancer prognosis. Oncogene 26:6406–6419. doi:10.1038/sj.onc.1210468

44. Luetteke NC, Qiu TH, Fenton SE et al (1999) Targeted inactivation of the EGF and amphiregulin genes reveals distinct roles for EGF receptor ligands in mouse mammary gland development. Development 126:2739–2750

45. Sternlicht MD, Sunnarborg SW, Kouros-Mehr H et al (2005) Mammary ductal morphogenesis requires paracrine activation of stromal EGFR via ADAM17-dependent shedding of epithelial amphiregulin. Development 132:3923–3933. doi:10.1242/dev.01966

46. Wiesen JF, Young P, Werb Z, Cunha GR (1999) Signaling through the stromal epidermal growth factor receptor is necessary for mammary ductal development. Development 126:335–344

47. Schroeder JA, Lee DC (1998) Dynamic expression and activation of ERBB receptors in the developing mouse mammary gland. Cell Growth Differ 9:451–464

48. Stern D (2003) ErbBs in mammary development. Exp Cell Res 284:89–98. doi:10.1016/S0014-4827(02)00103-9

49. Sebastian J, Richards RG, Walker MP et al (1998) Activation and function of the epidermal growth factor receptor and erbB-2 during mammary gland morphogenesis. Cell Growth Differ 9:777–785

50. Stern DF (2008) ERBB3/HER3 and ERBB2/HER2 duet in mammary development and breast cancer. J Mammary Gland Biol Neoplasia 13:215–223. doi:10.1007/s10911-008-9083-7

51. Jones FE, Welte T, Fu XY, Stern DF (1999) ErbB4 signaling in the mammary gland is required for lobuloalveolar development and Stat5 activation during lactation. J Cell Biol 147:77–88

52. Long W, Wagner K-U, Lloyd KCK et al (2003) Impaired differentiation and lactational failure of Erbb4-deficient mammary glands identify ERBB4 as an obligate mediator of STAT5. Development 130:5257–5268. doi:10.1242/dev.00715

53. Jackson-Fisher AJ, Bellinger G, Breindel JL et al (2008) ErbB3 is required for ductal morphogenesis in the mouse mammary gland. Breast Cancer Res 10(6):96. doi:10.1186/bcr2198

54. Krane IM, Leder P (1996) NDF/heregulin induces persistence of terminal end buds and

adenocarcinomas in the mammary glands of transgenic mice. Oncogene 12:1781–1788

55. Liu JL, LeRoith D (1999) Insulin-like growth factor I is essential for postnatal growth in response to growth hormone. Endocrinology 140:5178–5184. doi:10.1210/endo.140.11.7151

56. Richards RG, Klotz DM, Walker MP, Diaugustine RP (2004) Mammary gland branching morphogenesis is diminished in mice with a deficiency of insulin-like growth factor-I (IGF-I), but not in mice with a liver-specific deletion of IGF-I. Endocrinology 145:3106–3110. doi:10.1210/en.2003-1112

57. Loladze AV, Stull MA, Rowzee AM et al (2006) Epithelial-specific and stage-specific functions of insulin-like growth factor-I during post-natal mammary development. Endocrinology 147:5412–5423. doi:10.1210/en.2006-0427

58. Richert MM, Wood TL (1999) The insulin-like growth factors (IGF) and IGF type I receptor during postnatal growth of the murine mammary gland: sites of messenger ribonucleic acid expression and potential functions. Endocrinology 140:454–461. doi:10.1210/endo.140.1.6413

59. de Ostrovich KK, Lambertz I, Colby JKL et al (2008) Paracrine overexpression of insulin-like growth factor-1 enhances mammary tumorigenesis in vivo. Am J Pathol 173:824–834. doi:10.2353/ajpath.2008.071005

60. Bonnette SG, Hadsell DL (2001) Targeted disruption of the IGF-I receptor gene decreases cellular proliferation in mammary terminal end buds. Endocrinology 142(11):4937–4945

61. Jones RA, Campbell CI, Gunther EJ et al (2006) Transgenic overexpression of IGF-IR disrupts mammary ductal morphogenesis and induces tumor formation. Oncogene 26:1636–1644. doi:10.1038/sj.onc.1209955

62. Carboni JM (2005) Tumor development by transgenic expression of a constitutively active insulin-like growth factor I receptor. Cancer Res 65:3781–3787. doi:10.1158/0008-5472.CAN-04-4602

63. Flint DJ, Tonner E, Beattie J, Allan GJ (2008) Role of insulin-like growth factor binding proteins in mammary gland development. Int J Dev Biol 13:443–453. doi:10.1007/s10911-008-9095-3

64. Guo L, Degenstein L, Fuchs E (1996) Keratinocyte growth factor is required for hair development but not for wound healing. Genes Dev 10:165–175

65. Sekine K, Ohuchi H, Fujiwara M et al (1999) Fgf10 is essential for limb and lung formation. Nat Genet 21:138–141. doi:10.1038/5096

66. Ngan ESW, Ma Z-Q, Chua SS et al (2002) Inducible expression of FGF-3 in mouse mammary gland. Proc Natl Acad Sci U S A 99:11187–11192. doi:10.1073/pnas.172366199

67. Lu P, Ewald AJ, Martin GR, Werb Z (2008) Genetic mosaic analysis reveals FGF receptor 2 function in terminal end buds during mammary gland branching morphogenesis. Dev Biol 321:77–87. doi:10.1016/j.ydbio.2008.06.005

68. Yant J, Buluwela L, Niranjan B et al (1998) In vivo effects of hepatocyte growth factor/scatter factor on mouse mammary gland development. Exp Cell Res 241:476–481. doi:10.1006/excr.1998.4028

69. Yang Y, Spitzer E, Meyer D et al (1995) Sequential requirement of hepatocyte growth factor and neuregulin in the morphogenesis and differentiation of the mammary gland. J Cell Biol 131:215–226

70. Garner OB, Bush KT, Nigam KB et al (2011) Stage-dependent regulation of mammary ductal branching by heparan sulfate and HGF-cMet signaling. Dev Biol 355:394–403. doi:10.1016/j.ydbio.2011.04.035

71. Niranjan B, Buluwela L, Yant J et al (1995) HGF/SF: a potent cytokine for mammary growth, morphogenesis and development. Development 121:2897–2908

72. Delehedde M, Lyon M, Sergeant N et al (2001) Proteoglycans: pericellular and cell surface multireceptors that integrate external stimuli in the mammary gland. J Mammary Gland Biol Neoplasia 6:253–273. doi:10.1023/A:1011367423085

73. Liu BY, Kim YC, Leatherberry V et al (2003) Mammary gland development requires syndecan-1 to create a β-catenin/TCF-responsive mammary epithelial subpopulation. Oncogene 22:9243–9253. doi:10.1038/sj.onc.1207217

74. Wu Z-ZZ, Sun N-KN, Chao CC-KC (2011) Knockdown of CITED2 using short-hairpin RNA sensitizes cancer cells to cisplatin through stabilization of p53 and enhancement of p53-dependent apoptosis. J Cell Physiol 226:2415–2428. doi:10.1002/jcp.22589

75. Zcharia E, Jia J, Zhang X et al (2009) Newly generated heparanase knock-out mice unravel co-regulation of heparanase and matrix metalloproteinases. PLoS One 4, e5181. doi:10.1371/journal.pone.0005181.t002

76. Huebner RJ, Ewald AJ (2014) Cellular foundations of mammary tubulogenesis. Semin Cell Dev Biol 31:124–131. doi:10.1016/j.semcdb.2014.04.019

77. Shackleton M, Vaillant F, Simpson KJ et al (2006) Generation of a functional mammary gland from a single stem cell. Nature 439:84–88. doi:10.1038/nature04372

78. Stingl J, Smalley M, Glukhova M, Bentires-Alj M (2010) Methods in mammary gland

development and cancer: the second ENDBC meeting – intravital imaging, genomics, modeling and metastasis. Breast Cancer Res 12:311. doi:10.1186/bcr2630

79. Phillips S, Prat A, Sedic M et al (2014) Cell-state transitions regulated by SLUG are critical for tissue regeneration and tumor initiation. Stem Cell Rep 2:633–647. doi:10.1016/j.stemcr.2014.03.008

80. Van Keymeulen A, Rocha AS, Ousset M et al (2011) Distinct stem cells contribute to mammary gland development and maintenance. Nature 479:189–193. doi:10.1038/nature10573

81. Kouros-Mehr H, Slorach EM, Sternlicht MD, Werb Z (2006) GATA-3 maintains the differentiation of the luminal cell fate in the mammary gland. Cell 127:1041–1055. doi:10.1016/j.cell.2006.09.048

82. Asselin-Labat M-L, Sutherland KD, Barker H et al (2007) Gata-3 is an essential regulator of mammary-gland morphogenesis and luminal-cell differentiation. Nat Cell Biol 9:201–209. doi:10.1038/ncb1530

83. Asselin-Labat ML, Sutherland KD, Vaillant F et al (2011) Gata-3 negatively regulates the tumor-initiating capacity of mammary luminal progenitor cells and targets the putative tumor suppressor caspase-14. Mol Cell Biol 31:4609–4622. doi:10.1128/MCB.05766-11

84. Pei X-H, Bai F, Smith MD et al (2009) CDK inhibitor p18INK4c is a downstream target of GATA3 and restrains mammary luminal progenitor cell proliferation and tumorigenesis. Cancer Cell 15:389–401. doi:10.1016/j.ccr.2009.03.004

85. Pietersen AM, Evers B, Prasad AA et al (2008) Bmi1 regulates stem cells and proliferation and differentiation of committed cells in mammary epithelium. Curr Biol 18:1094–1099. doi:10.1016/j.cub.2008.06.070

86. Kurpios NA, MacNeil L, Shepherd TG et al (2009) The Pea3 Ets transcription factor regulates differentiation of multipotent progenitor cells during mammary gland development. Dev Biol 325:106–121. doi:10.1016/j.ydbio.2008.09.033

87. Jia Y (2005) Peroxisome proliferator-activated receptor-binding protein null mutation results in defective mammary gland development. J Biol Chem 280:10766–10773. doi:10.1074/jbc.M413331200

88. Jiang P, Hu Q, Ito M et al (2010) Key roles for MED1 LxxLL motifs in pubertal mammary gland development and luminal-cell differentiation. Proc Natl Acad Sci U S A 107:6765–6770. doi:10.1073/pnas.1001814107

89. Hasegawa N, Sumitomo A, Fujita A et al (2012) Mediator subunits MED1 and MED24 cooperatively contribute to pubertal mammary gland development and growth of breast carcinoma cells. Mol Cell Biol 32:1483–1495. doi:10.1128/MCB.05245-11

90. Parashurama N, Lobo NA, Ito K et al (2012) Remodeling of endogenous mammary epithelium by breast cancer stem cells. Stem Cells 30:2114–2127. doi:10.1002/stem.1205

91. Booth BW, Boulanger CA, Anderson LH et al (2010) Amphiregulin mediates self-renewal in an immortal mammary epithelial cell line with stem cell characteristics. Exp Cell Res 316:422–432. doi:10.1016/j.yexcr.2009.11.006

92. Santos SJ, Haslam SZ, Conrad SE (2010) Signal transducer and activator of transcription 5a mediates mammary ductal branching and proliferation in the nulliparous mouse. Endocrinology 151:2876–2885. doi:10.1210/en.2009-1282

93. Yamaji D, Na R, Feuermann Y et al (2009) Development of mammary luminal progenitor cells is controlled by the transcription factor STAT5A. Genes Dev 23:2382–2387. doi:10.1101/gad.1840109

94. Vafaizadeh V, Klemmt PA, Groner B (2012) Stat5 assumes distinct functions in mammary gland development and mammary tumor formation. Front Biosci 17:1232–1250

95. Vafaizadeh V, Klemmt P, Brendel C et al (2010) Mammary epithelial reconstitution with gene-modified stem cells assigns roles to Stat5 in luminal alveolar cell fate decisions, differentiation, involution, and mammary tumor formation. Stem Cells 28:928–938. doi:10.1002/stem.407

96. Rios AC, Fu NY, Lindeman GJ, Visvader JE (2014) In situ identification of bipotent stem cells in the mammary gland. Nature 506:322–327. doi:10.1038/nature12948

97. Mani SA, Guo W, Liao M-J et al (2008) The epithelial-mesenchymal transition generates cells with properties of stem cells. Cell 133:704–715. doi:10.1016/j.cell.2008.03.027

98. Bai F, Chan HL, Scott A et al (2014) BRCA1 suppresses epithelial-to-mesenchymal transition and stem cell dedifferentiation during mammary and tumor development. Cancer Res 74(21):6161–6172

99. Deng C-X, Xu X, Wagner K-U et al (1999) Conditional mutation of Brca1 in mammary epithelial cells results in blunted ductal morphogenesis and tumour formation. Nat Genet 22:37–43. doi:10.1038/8743

100. Bai F, Smith MD, Chan HL, Pei X-H (2013) Germline mutation of Brca1 alters the fate of mammary luminal cells and causes luminal-to-basal mammary tumor transformation. Oncogene 32:2715–2725. doi:10.1038/onc.2012.293

101. Lamber EP, Horwitz AA, Parvin JD (2010) BRCA1 represses amphiregulin gene expression. Cancer Res 70:996–1005. doi:10.1158/0008-5472.CAN-09-2842

102. Lee K, Gjorevski N, Boghaert E et al (2011) Snail1, Snail2, and E47 promote mammary epithelial branching morphogenesis. EMBO J 30:2662–2674

103. Nelson CM, Vanduijn MM, Inman JL et al (2006) Tissue geometry determines sites of mammary branching morphogenesis in organotypic cultures. Science 314:298–300. doi:10.1126/science.1131000

104. Kouros-Mehr H, Werb Z (2006) Candidate regulators of mammary branching morphogenesis identified by genome-wide transcript analysis. Dev Dyn 235:3404–3412. doi:10.1002/dvdy.20978

105. Nelson CM, Inman JL, Bissell MJ (2008) Three-dimensional lithographically defined organotypic tissue arrays for quantitative analysis of morphogenesis and neoplastic progression. Nat Protoc 3:674–678. doi:10.1038/nprot.2008.35

106. Ewald AJ, Brenot A, Duong M et al (2008) Collective epithelial migration and cell rearrangements drive mammary branching morphogenesis. Dev Cell 14:570–581

107. Jennifer L, Andrews ACKJRH (2012) The role and function of cadherins in the mammary gland. Breast Cancer Res 14:203. doi:10.1186/bcr3065

108. Daniel CW, Strickland P, Friedmann Y (1995) Expression and functional role of E- and P-cadherins in mouse mammary ductal morphogenesis and growth. Dev Biol 169:511–519. doi:10.1006/dbio.1995.1165

109. Albergaria A, Ribeiro A-S, Vieira A-F et al (2011) P-cadherin role in normal breast development and cancer. Int J Dev Biol 55:811–822. doi:10.1387/ijdb.113382aa

110. Delmas V, Pla P, Feracci H et al (1999) Expression of the cytoplasmic domain of E-cadherin induces precocious mammary epithelial alveolar formation and affects cell polarity and cell–matrix integrity. Dev Biol 216:491–506. doi:10.1006/dbio.1999.9517

111. Boussadia O, Kutsch S, Hierholzer A et al (2002) E-cadherin is a survival factor for the lactating mouse mammary gland. Mech Dev 115:53–62. doi:10.1016/S0925-4773(02)00090-4

112. Gallagher RCJ, Hay T, Meniel V et al (2002) Inactivation of Apc perturbs mammary development, but only directly results in acanthoma in the context of Tcf-1 deficiency. Oncogene 21:6446–6457. doi:10.1038/sj.onc.1205892

113. Imbert A, Eelkema R, Jordan S et al (2001) Delta N89 beta-catenin induces precocious development, differentiation, and neoplasia in mammary gland. J Cell Biol 153:555–568

114. Incassati A, Chandramouli A, Eelkema R, Cowin P (2010) Key signaling nodes in mammary gland development and cancer: β-catenin. Breast Cancer Res 12:213. doi:10.1186/bcr2723

115. van Genderen C, Okamura R, Farinas I et al (1994) Development of several organs that require inductive epithelial-mesenchymal interactions is impaired in LEF-1-deficient mice. Genes Dev 8:2691–2703

116. Wheelock MJ, Shintani Y, Maeda M et al (2008) Cadherin switching. J Cell Sci 121:727–735. doi:10.1242/jcs.000455

117. Qian X, Karpova T, Sheppard AM et al (2004) E-cadherin-mediated adhesion inhibits ligand-dependent activation of diverse receptor tyrosine kinases. EMBO J 23:1739–1784. doi:10.1038/sj.emboj.7600136

118. Qian X, Anzovino A, Kim S et al (2013) N-cadherin/FGFR promotes metastasis through epithelial-to-mesenchymal transition and stem/progenitor cell-like properties. Oncogene 33:3411–3421. doi:10.1038/onc.2013.310

119. Baum B, Georgiou M (2011) Dynamics of adherens junctions in epithelial establishment, maintenance, and remodeling. J Cell Biol 192(6):907–917

120. Dow LE, Humbert PO (2007) Polarity Regulators and the Control of Epithelial Architecture, Cell Migration, and Tumorigenesis. International Review of Cytology. Elsevier, Berlin, pp 253–302

121. Bray K, Gillette M, Young J et al (2013) Cdc42 overexpression induces hyperbranching in the developing mammary gland by enhancing cell migration. Breast Cancer Res 15:91

122. Gillette M, Bray K, Blumenthaler A, Vargo-Gogola T (2013) P190B RhoGAP overexpression in the developing mammary epithelium induces TGFβ-dependent fibroblast activation. PLoS One 8, e65105. doi:10.1371/journal.pone.0065105

123. Nathan J et al (2014) Scribble modulates the MAPK/Fra1 pathway to disrupt luminal and ductal integrity and suppress tumour formation in the mammary gland. PLoS Genet. doi:10.1371/journal.pgen.1004323

124. Srinivasan K, Strickland P, Valdes A et al (2003) Netrin-1/neogenin interaction stabilizes multipotent progenitor cap cells during mammary gland morphogenesis. Dev Cell 4:371–382

125. Munarini N, Jäger R, Abderhalden S et al (2002) Altered mammary epithelial development, pattern formation and involution in transgenic mice expressing the EphB4 receptor tyrosine kinase. J Cell Sci 115:25–37

126. Silberstein G, Daniel C (1987) Reversible inhibition of mammary gland growth by transforming growth factor-beta. Science 237:291–293. doi:10.1126/science.3474783

127. Pierce DF, Johnson MD, Matsui Y et al (1999) Inhibition of mammary duct development but not alveolar outgrowth during pregnancy in transgenic mice expressing active TGF-beta 1. Genes Dev 7(12):2308–2317

128. Ewan KB, Shyamala G, Ravani SA et al (2002) Latent transforming growth factor-beta activation in mammary gland: regulation by ovarian hormones affects ductal and alveolar proliferation. Am J Pathol 160:2081–2093

129. Silberstein GB, Flanders KC, Roberts AB, Daniel CW (1992) Regulation of mammary morphogenesis: evidence for extracellular matrix-mediated inhibition of ductal budding by transforming growth factor-β1. Dev Biol 152:354–362. doi:10.1016/0012-1606(92)90142-4

130. Roarty K, Serra R (2007) Wnt5a is required for proper mammary gland development and TGF-mediated inhibition of ductal growth. Development 134:3929–3939. doi:10.1242/dev.008250

131. Vogel WF, Aszódi A, Alves F, Pawson T (2001) Discoidin domain receptor 1 tyrosine kinase has an essential role in mammary gland development. Mol Cell Biol 21:2906–2917. doi:10.1128/MCB.21.8.2906-2917.2001

132. Roarty K, Baxley S, Crowley M et al (2009) Loss of TGF-β or Wnt5a results in an increase in Wnt/β-catenin activity and redirects mammary tumour phenotype. Breast Cancer Res 11:R19. doi:10.1186/bcr2244

133. Silberstein GB (2001) Tumour-stromal interactions. Role of the stroma in mammary development. Breast Cancer Res 3:218–223

134. Paulsson M (1992) Basement membrane proteins: structure, assembly, and cellular interactions. Crit Rev Biochem Mol Biol 27:93–127. doi:10.3109/10409239209082560

135. Wiseman BS, Sternlicht MD, Lund LR et al (2003) Site-specific inductive and inhibitory activities of MMP-2 and MMP-3 orchestrate mammary gland branching morphogenesis. J Cell Biol 162:1123–1133. doi:10.1083/jcb.200302090

136. Sympson CJ, Talhouk RS, Alexander CM et al (1994) Targeted expression of stromelysin-1 in mammary gland provides evidence for a role of proteinases in branching morphogenesis and the requirement for an intact basement membrane for tissue-specific gene expression. J Cell Biol 125:681–693

137. Thomasset N, Lochter A, Sympson CJ et al (1998) Expression of autoactivated stromelysin-1 in mammary glands of transgenic mice leads to a reactive stroma during early development. Am J Pathol 153:457–467. doi:10.1016/S0002-9440(10)65589-7

138. Fata JE, Leco KJ, Moorehead RA et al (1999) Timp-1 is important for epithelial proliferation and branching morphogenesis during mouse mammary development. Dev Biol 211:238–254. doi:10.1006/dbio.1999.9313

139. Taddei I, Deugnier M-A, Faraldo MM et al (2008) Beta1 integrin deletion from the basal compartment of the mammary epithelium affects stem cells. Nat Cell Biol 10:716–722. doi:10.1038/ncb1734

140. Klinowska TC, Soriano JV, Edwards GM et al (1999) Laminin and beta1 integrins are crucial for normal mammary gland development in the mouse. Dev Biol 215:13–32. doi:10.1006/dbio.1999.9435

141. Nagy T, Wei H, Shen T-L et al (2007) Mammary epithelial-specific deletion of the focal adhesion kinase gene leads to severe lobulo-alveolar hypoplasia and secretory immaturity of the murine mammary gland. J Biol Chem 282:31766–31776. doi:10.1074/jbc.M705403200

142. van Miltenburg MHAM, Lalai R, de Bont H et al (2009) Complete focal adhesion kinase deficiency in the mammary gland causes ductal dilation and aberrant branching morphogenesis through defects in Rho kinase-dependent cell contractility. FASEB J 23:3482–3493. doi:10.1096/fj.08-123398

143. Gomez EW, Chen QK, Gjorevski N, Nelson CM (2010) Tissue geometry patterns epithelial-mesenchymal transition via intercellular mechanotransduction. J Cell Biochem 110:44–51. doi:10.1002/jcb.22545

144. Gjorevski N, Nelson CM (2010) Endogenous patterns of mechanical stress are required for branching morphogenesis. Integr Biol (Camb) 2:424–434. doi:10.1039/c0ib00040j

145. Gouon-Evans V, Rothenberg ME, Pollard JW (2000) Postnatal mammary gland development requires macrophages and eosinophils. Development 127:2269–2282

146. Sferruzzi-Perri AN, Robertson SA, Dent LA (2003) Interleukin-5 transgene expression and eosinophilia are associated with retarded mammary gland development in mice. Biol Reprod 69:224–233. doi:10.1095/biolreprod.102.010611

147. Lilla JN, Werb Z (2010) Mast cells contribute to the stromal microenvironment in mammary gland branching morphogenesis. Dev Biol 337:124–133. doi:10.1016/j.ydbio.2009.10.021

148. Van Nguyen A, Pollard JW (2002) Colony stimulating factor-1 is required to recruit macrophages into the mammary gland to facilitate mammary ductal outgrowth. Dev Biol 247:11–25. doi:10.1006/dbio.2002.0669

149. O'Brien J, Martinson H, Durand-Rougely C, Schedin P (2011) Macrophages are crucial for epithelial cell death and adipocyte repopulation during mammary gland involution. Development 139:269–275. doi:10.1242/dev.071696

150. Gyorki DE, Asselin-Labat M-L, van Rooijen N et al (2009) Resident macrophages influence stem cell activity in the mammary gland. Breast Cancer Res 11:R62. doi:10.1186/bcr2353

151. Doyle A, McGarry MP, Lee NA, Lee JJ (2011) The construction of transgenic and gene knockout/knockin mouse models of human disease. Transgenic Res 21:327–349. doi:10.1007/s11248-011-9537-3

152. Ristevski S (2005) Making better transgenic models: conditional, temporal, and spatial approaches. Mol Biotechnol 29:153–164. doi:10.1385/MB:29:2:153

153. Medina D (2010) Of mice and women: a short history of mouse mammary cancer research with an emphasis on the paradigms inspired by the transplantation method. Cold Spring Harb Perspect Biol 2(10):004523

154. SHILLINGFORD J, Hennighausen L (2001) Experimental mouse genetics – answering fundamental questions about mammary gland biology. Trends Endocrinol Metab 12:402–408. doi:10.1016/S1043-2760(01)00471-4

155. Fantl V, Stamp G, Andrews A et al (1995) Mice lacking cyclin D1 are small and show defects in eye and mammary gland development. Genes Dev 9(19):2364–2372

156. Sicinski P, Donaher JL, Parker SB et al (1995) Cyclin D1 provides a link between development and oncogenesis in the retina and breast. Cell 82:621–630. doi:10.1016/0092-8674(95)90034-9

157. Casimiro MC, Wang C, Li Z et al (2013) Cyclin D1 determines estrogen signaling in the mammary gland in vivo. Mol Endocrinol 27:1415–1428. doi:10.1210/me.2013-1065

158. Xu J, Liao L, Ning G et al (2000) The steroid receptor coactivator SRC-3 (p/CIP/RAC3/AIB1/ACTR/TRAM-1) is required for normal growth, puberty, female reproductive function, and mammary gland development. Proc Natl Acad Sci U S A 97:6379–6384. doi:10.1073/pnas.120166297

159. Yang Y-A, Tang B, Robinson G et al (2002) Smad3 in the mammary epithelium has a nonredundant role in the induction of apoptosis, but not in the regulation of proliferation or differentiation by transforming growth factor-beta. Cell Growth Differ 13:123–130

160. Ucar A, Vafaizadeh V, Jarry H et al (2010) miR-212 and miR-132 are required for epithelial stromal interactions necessary for mouse mammary gland development. Nat Genet 42:1101–1108. doi:10.1038/ng.709

161. Remenyi J, van den Bosch MWM, Palygin O et al (2013) miR-132/212 knockout mice reveal roles for these mirnas in regulating cortical synaptic transmission and plasticity. PLoS One 8:62509. doi:10.1371/journal.pone.0062509

162. Kayo H, Kiga K, Fukuda-Yuzawa Y et al (2014) miR-212 and miR-132 are dispensable for mouse mammary gland development. Nat Genet 46:802–804. doi:10.1038/ng.2990

163. Ucar A, Erikci E, Ucar O, Chowdhury K (2014) miR-212 and miR-132 are dispensable for mouse mammary gland development. Nat Genet 46:804–805. doi:10.1038/ng.3032

164. Torres-Ruiz R, Rodriguez-Perales S (2015) CRISPR-Cas9: a revolutionary tool for cancer modelling. Int J Mol Sci 16:22151–22168. doi:10.3390/ijms160922151

165. Blasco RB, Karaca E, Ambrogio C et al (2014) Simple and rapid in vivo generation of chromosomal rearrangements using CRISPR/Cas9 technology. Cell Rep 9:1219–1227. doi:10.1016/j.celrep.2014.10.051

166. Maddalo D, Manchado E, Concepcion CP et al (2014) In vivo engineering of oncogenic chromosomal rearrangements with the CRISPR/Cas9 system. Nature 516:423–427. doi:10.1038/nature13902

167. Sánchez-Rivera FJ, Papagiannakopoulos T, Romero R et al (2014) Rapid modeling of cooperating genetic events in cancer through somatic genome editing. Nature 516:428–431. doi:10.1038/nature13906

168. Mou H, Kennedy Z, Anderson DG et al (2015) Precision cancer mouse models through genome editing with CRISPR-Cas9. Genome Med 7:53. doi:10.1186/s13073-015-0178-7

169. Yin H, Xue W, Chen S et al (2014) Genome editing with Cas9 in adult mice corrects a disease mutation and phenotype. Nat Biotechnol 32:551–553. doi:10.1038/nbt.2884

170. Metser G, Shin HY, Wang C et al (2016) An autoregulatory enhancer controls mammary-specific STAT5 functions. Nucleic Acids Res 44(3):1052–1063. doi:10.1093/nar/gkv999

171. Zhao H, Pearson EK, Brooks DC et al (2012) A humanized pattern of aromatase expression is associated with mammary hyperplasia in mice. Endocrinology 153:2701–2713. doi:10.1210/en.2011-1761

172. Gallego MI, Binart N, Robinson GW et al (2001) Prolactin, growth hormone, and epidermal growth factor activate Stat5 in different compartments of mammary tissue and exert different and overlapping developmental effects. Dev Biol 229:163–175. doi:10.1006/dbio.2000.9961

173. Xu J, Qiu Y, DeMayo FJ et al (1998) Partial hormone resistance in mice with disruption of the steroid receptor coactivator-1 (SRC-1) gene. Science 279:1922–1925

174. Kouros-Mehr H, Kim J-W, Bechis SK, Werb Z (2008) GATA-3 and the regulation of the mammary luminal cell fate. Curr Opin Cell Biol 20:164–170. doi:10.1016/j.ceb.2008.02.003

175. Zcharia E, Metzger S, Chajek-Shaul T et al (2004) Transgenic expression of mammalian heparanase uncovers physiological functions of heparan sulfate in tissue morphogenesis, vascularization, and feeding behavior. FASEB J 18(2):252–263

176. Hathaway HJ, Shur BD (1996) Mammary gland morphogenesis is inhibited in transgenic mice that overexpress cell surface beta1,4-galactosyltransferase. Development 122(9):2859–2872

177. Steffgen K, Dufraux K, Hathaway H (2002) Enhanced branching morphogenesis in mammary glands of mice lacking cell surface beta1,4-galactosyltransferase. Dev Biol 244:114–133. doi:10.1006/dbio.2002.0599

178. Crowley MR, Head KL, Kwiatkowski DJ et al (2000) The mouse mammary gland requires the actin-binding protein gelsolin for proper ductal morphogenesis. Dev Biol 225:407–423. doi:10.1006/dbio.2000.9844

179. Shillingford JM, Miyoshi K, Flagella M (2002) Mouse mammary epithelial cells express the Na-K-Cl cotransporter, NKCC1: characterization, localization, and involvement in ductal development and morphogenesis. Mol Endocrinol 16(6):1309–1321

180. Okolowsky N, Furth PA, Hamel PA (2014) Oestrogen receptor-alpha regulates noncanonical Hedgehog-signalling in the mammary gland. Dev Biol 391:219–229. doi:10.1016/j.ydbio.2014.04.007

181. Vapola MH, Rokka A, Sormunen RT et al (2014) Peroxisomal membrane channel Pxmp2 in the mammary fat pad is essential for stromal lipid homeostasis and for development of mammary gland epithelium in mice. Dev Biol 391:66–80. doi:10.1016/j.ydbio.2014.03.022

182. Michalak EM, Nacerddine K, Pietersen A et al (2013) Polycomb group gene Ezh2 regulates mammary gland morphogenesis and maintains the luminal progenitor pool. Stem Cells 31:1910–1920

183. Stairs DB, Notarfrancesco KL, Chodosh LA (2005) The serine/threonine kinase, Krct, affects endbud morphogenesis during murine mammary gland development. Transgenic Res 14:919–940. doi:10.1007/s11248-005-1806-6

Chapter 4

Analysis of Mammary Gland Phenotypes by Transplantation of the Genetically Marked Mammary Epithelium

Duje Buric and Cathrin Brisken

Abstract

The mammary gland is the only organ to undergo most of its development after birth and therefore particularly attractive for studying developmental processes. In the mouse, powerful tissue recombination techniques are available that can be elegantly combined with the use of different genetically engineered mouse models to study development and differentiation in vivo.

In this chapter, we describe how epithelial intrinsic gene function can by discerned by grafting mammary epithelial cells of different genotypes to wild-type recipients. Either pieces of mammary epithelial tissue or dissociated mammary epithelial cells are isolated from donor mice and subsequently transplanted into recipients whose mammary fat pads were divested of their endogenous epithelium. This is followed by phenotypic characterization of the epithelial outgrowth either by fluorescence stereomicroscopy for the fluorescently marked grafts or carmine alum whole mount for the unmarked epithelia.

Key words Mammary epithelium, Single cells, Transplantation, Engraftment, Injection, Donor tissue, Recipient mice, Cleared fat pad, Epithelial outgrowth

1 Introduction

The mouse mammary gland is a very attractive experimental system. Most of its development occurs after birth making it easy to study. As mammary glands are skin appendages that can be found on the back of the skin, they are readily accessible by surgery. The glands are paired organs and contralateral glands can to the best of our current knowledge be directly compared.

The mammary gland's major components are a mammary fat pad and an epithelial structure that invades it. During embryonic development, a mammary bud forms from a placode in the ventral skin around E12.5. This develops into a small ductal system that grows into the underlying specialized fatty stroma by E18.5. During the first 3 weeks of life (prepubertal stage), the rudimentary ductal system grows isometrically with the rest of the body.

Finian Martin et al. (eds.), *Mammary Gland Development: Methods and Protocols*, Methods in Molecular Biology, vol. 1501, DOI 10.1007/978-1-4939-6475-8_4, © Springer Science+Business Media New York 2017

During puberty, between 3 and 8 weeks of age, the ductal system expands and invades the fat pad, driven by ovarian estrogens. With the onset of adulthood, around 8 weeks of age, the ovaries begin to secrete estrogens and progesterone regularly, and estrous cycles are established. Next, the milk duct system becomes more complex through a process called side branching driven by progesterone [1]. When pregnancy occurs, progesterone levels rise further, the estrous cycles are suppressed and side branching continues. Finally, during the last third of pregnancy little saccular outpouchings sprout from the ducts, which will produce milk. This process called alveologenesis is under the influence of prolactin.

DeOme was the first to show that the endogenous ductal tree can be surgically removed from prepubertal females leaving behind approximately half the mammary gland as "cleared" fat pad, in which an epithelium fragment from another mouse can be engrafted [2]. It will grow out to form a new ductal tree that behaves like the endogenous epithelium without establishing a link to the teat. Even dissociated mammary epithelial cells, injected into the "cleared" fat pad, were able to do so [3]. Initially, the approach was used to characterize the properties of hyperplastic and malignant lesions in different mouse strains [4–8]. Subsequent reports on the engraftment efficiency and the growth potential of normal mammary tissue established that mammary epithelium can serially engraft [9–13].

With the advent of targeted gene deletion in the mouse germ line. The transplantation of mammary epithelium was used to reveal mammary phenotypes secondary to systemic effects of the genetic change and to discern intrinsic epithelial phenotypes [14–16]. Additionally, the transplantation approach can be used to rescue epithelium from mouse mutants that are lethal by engrafting embryonic mammary buds into wild-type mice as early as E12.5 [17, 18]. More recently, engraftment of single cells obtained by limiting dilution [19] or of a specific single-cell population obtained by fluorescence-activated cell sorting [20, 21] became a standard tool in mammary stem cell research.

A potential problem of the fat pad clearing approach is that endogenous epithelium may not be completely removed and compete with the graft. When the engrafted gland is analyzed prior to pregnancy and/or up to mid-pregnancy the implanted graft can readily be distinguished from endogenous epithelium because of its radial versus the uni-directional growth pattern of the endogenous ductal system (Fig. 1). However, the ductal growth pattern can be impossible to see when the fat pad is filled with mammary epithelium during late pregnancy. To unequivocally distinguish engrafted from endogenous epithelium the use of a marker is advisable. Initially, we utilized donor mice systemically expressing LacZ [22]. This required 5-bromo-4-chloro-3-indolyl-β-D-galactopyranoside (X-gal) staining followed by carmine alum counterstaining and

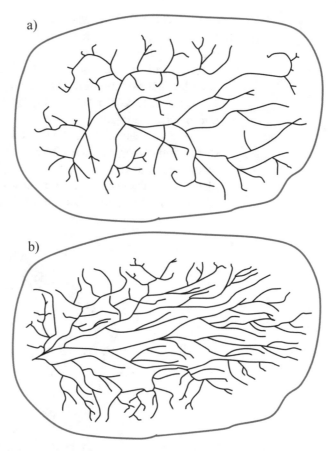

Fig. 1 Difference in growth pattern between engrafted and endogenous mammary epithelium. Scheme of radial epithelial growth pattern in transplanted mammary gland (**a**) and uni-directional growth pattern in the endogenous mammary gland (**b**)

whole mounting of the glands [14]. Nowadays, the availability of mice expressing different fluorescent proteins [23, 24] in the mammary gland has made the discrimination of epithelium from adipose tissue by fluorescent stereomicroscopy more convenient and applicable to live tissue [14, 15]. The genetic markers are also useful for normalizing gene expression to the transplant outgrowth when contralateral glands are processed for Western blotting or quantitative RT-PCR analysis.

In this chapter, we describe the preparation of the graft material, whether it is a piece of epithelium or a suspension of single cells, the preparation of the recipient animals, the engraftment into the cleared fat pad, and the analysis of the epithelial outgrowth.

As engrafting material we use either pieces of epithelium excised directly from the mammary gland of the donor mouse or suspensions of single cells. To obtain single cells from the mammary gland epithelium, we use a shortened and slightly modified version of the protocol from Matthew J. Smalley [25] where mammary glands are minced and treated with collagenase A and trypsin, washed with red blood cell lysis buffer, and at the end shortly digested with trypsin and DNase-1. For injection of single cells, we use Matrigel as a medium for the engraftment. This was shown to increase the success rate possibly by preventing cell dispersal from the injection site [26].

To obtain as much material as possible for the preparation of the suspension of single cells we isolate four out of the five pairs of mammary glands in the mouse (Fig. 2). The cervical pair of mammary glands is usually not collected because of their position: they are both difficult to access and difficult to distinguish from the salivary glands.

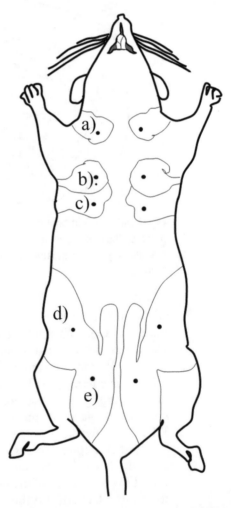

Fig. 2 Anatomic location of mouse mammary glands. Scheme shows a female mouse lying on its back. The position of the mammary glands is depicted with *thin lines*, and *teats* are represented by *black dots*. (**a**) Cervical, (**b**) first thoracic, (**c**) second thoracic, (**d**) abdominal, (**e**) inguinal

2 Materials

2.1 Media and Buffers

1. Phosphate-buffered saline (PBS): 1.0% w/v Sodium chloride, 0.025% w/v potassium chloride, 0.025% w/v disodium hydrogen orthophosphate, and 0.1437% w/v potassium dihydrogen orthophosphate (prepared in the laboratory, filtered, and autoclaved).

2. PBS/10% FCS medium: Phosphate-buffered saline plus 10% v/v heat-inactivated filtered fetal calf serum (FCS).

3. Leibowitz L15 medium with L-glutamine (L15) with no additives.

4. Trypsin solution 1: 15 mg/ml Trypsin from bovine pancreas is in serum-free L15. Stored at –20 °C in 1 ml aliquots (*see* **Note 1**).

5. Collagenase A solution: 100 mg/ml Collagenase A in PBS. Stored at –20 °C in 1.2 ml aliquots (*see* **Note 2**).

6. Digestion solution: 1.2 ml of Collagenase A solution, 1 ml of trypsin solution 1, 37.8 ml of L15 medium with L-glutamine.

7. Trypsin solution 2: 0.25% Trypsin, 0.02% EDTA in Hanks' balanced salt solution.

8. DNase-1 solution (5 μg/ml): 5 μg/ml Bovine pancreatic DNase-1 in serum-free L15. Store at –20 °C in 5 ml aliquots.

9. Matrigel solution: Falcon matrigel basement membrane matrix is mixed with sterile PBS in a 1:8 ratio.

2.1.1 Preparation of Donor Tissue

1. For tissue collection: Neoprene cork dissection board wrapped in aluminum foil and autoclaved, 70% ethanol in spray bottle to disinfect the animals, dissection tools (dissection needles for fixing the donor (dead) animal to the board, round-nosed scissors, and two pairs of forceps), 100 ml beaker containing 70% ethanol for instrument sterilization, and 15 ml Falcon tube containing sterile PBS for the tissue collection (kept on ice).

2. For preparation of tissue fragments: Two sterile 10 cm Petri dishes containing 10 ml sterile PBS (see Note 3), two pairs of forceps, round-nosed scissors, angled scissors (Vanna Scissors, angled-on-flat blades, 0.1 mm tip), 5/45 jewellers' forceps, clips for stitching, paper tissues, and fluorescent dissection stereo microscope.

3. For preparation of single cells: A small beaker containing 70% ethanol for sterilizing tools, a 50 ml Falcon tube containing sterile PBS for collection of mammary glands, scale to weigh the isolated tissue, 40 μl of digestion solution per batch of glands, red blood cell lysis buffer, 2 ml per batch of glands of trypsin solution 2 and 5 ml per batch of glands of DNase 1 solution, 40 μm cell strainers, Matrigel solution, and two #22 scalpels.

2.1.2 Surgery

Anesthesia is performed under the guidance of the Federal Veterinary Office of Switzerland. To minimize side effects, it is advised to use isoflurane anesthesia (5 % of isoflurane in the atmosphere for the induction period until the mouse becomes ataxic, and 2 % of isoflurane during the maintenance period with oxygen supply at the rate of 5 l/min during induction and 1 l/min during maintenance period) with addition of analgesics chosen in accord with local veterinary guidance (buprenorphine at the rate of 0.1 mg/kg of mouse) (*see* **Note 4**).

Required is a heating pad to warm the animals during the surgery, double-sided tape, Betadine standardized solution for the sterilization of the surgery spot (*see* **Note 5**), two pairs of forceps, round-nosed scissors, angled spring scissors (Vanna Scissors, angled-on-flat blades, 0.1 mm tip), 5/45 jewellers' forceps, battery-operated cauterizer, 100 μl Hamilton's syringe, clips for stitching, clip-removing forceps, cotton pads, sterile physiological solution, and analgesic (buprenorphine) for postoperative treatment (as advised by the Federal Veterinary Office of Switzerland) (*see* **Note 6**).

2.1.3 Preparation for Subsequent Analysis

Fluorescent stereomicroscope, a suitable camera, dissection tools, histological glass slides, plastic clips, glass beaker (size depending on the number of samples) with 4 % w/v paraformaldehyde in PBS, glass beaker with 70 % ethanol, and container with liquid nitrogen are required.

3   Methods

3.1 Dissection of the Donor Mice

1. Mice are sacrificed in the CO_2 chamber and fixed on their back with pins to the dissection board and the ventral side is thoroughly sprayed with 70 % ethanol to disinfect the skin. A ventral incision is carefully made with round-nosed scissors pulling up the skin with forceps to avoid puncturing the muscle wall and the incision is extended to the top of the rib cage. Two further incisions are made to generate a Y-shaped opening extending down the lower limbs and up the upper limbs (Fig. 3). The skin is carefully pulled back from the body wall with forceps to expose the abdominal and the thoracic mammary glands, which stay attached to the skin. Thoracic glands are carefully detached from the skin by scissors and forceps. The connection between the abdominal and the inguinal gland is carefully cut with scissors (Fig. 5b). A small incision is made above the subiliacal lymph node; the node is isolated and removed using forceps. The fourth mammary gland is carefully removed by forceps and scissors and placed into the Falcon tube with sterile PBS.

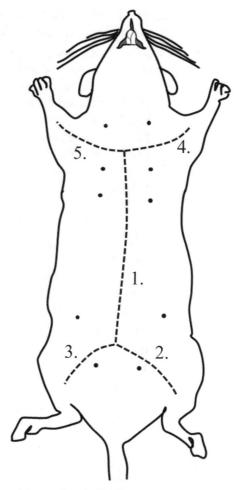

Fig. 3 Surgical field for dissection of the donor mouse. Scheme shows a female mouse lying on its back. The *incision lines* are shown by *dashed lines*, order of incisions follows the numbering

3.2 Preparation of the Engraftment Material

3.2.1 Pieces of the Tissues

1. Mammary gland from the donor mouse expressing fluorescent protein is placed into the 10 cm plastic dish filled with sterile PBS and observed under a dissecting fluorescence stereo-microscope. Pieces of approximately 1 mm³ in size are cut with jewellers' forceps and angled scissors, placed in another petri dish containing PBS, which is kept on ice until the surgery is performed.

3.2.2 Single Cells

1. A batch of collected mammary glands is transferred into the tissue culture hood and placed on the previously autoclaved dissection board using forceps. Excess PBS is aspirated by lifting the board. Mammary glands are finely chopped with #22 scalpels until they become a fine semi-liquid slurry (*see* **Note 7**).

2. The tissue slurry is weighed and transferred into the 50 ml Falcon tubes, 2 g per tube.

3. Forty milliliter of digestion mix is added per tube and tubes are placed on the rotator at 37 °C for 2 h (*see* **Notes 8** and **9**).

4. After incubation, check that solution is homogeneous and fragments are smaller than 1–2 mm in size. The Falcon tube is placed in the centrifuge and spun at $300 \times g$ for 5 min. The pellet will be enriched for epithelial fragments. Supernatant containing digestion medium and the layer of fat is transferred to a new Falcon tube and spun down again at same speed and time (*see* **Note 10**). Supernatant from the second Falcon tube is removed by aspiration. The pellet from the first Falcon tube is resuspended in 10 ml and from the second Falcon tube in 5 ml of PBS/10% FCS; they are pooled in a 15 ml Falcon tube and spun down again followed by aspiration of supernatant.

5. The pellet is resuspended in 5 ml of red blood cell lysis buffer and incubated for 5 min at room temperature. Suspension is again spun down at $300 \times g$ for 5 min and supernatant is aspirated.

6. The pellet is washed with 5 ml of PBS/10% FCS and spun down. Supernatant is aspirated.

7. The pellet is resuspended in 2 ml of trypsin solution 2 and incubated for 2 min at 37 °C. Following incubation 5 ml of DNase 1 solution is added and suspension is incubated for another 5 min at 37 °C.

8. To inactivate trypsin, 8 ml of PBS/10% FCS is added. Suspension is filtered through a 40 µM cell strainer, spun down, and resuspended into minimal amount of PBS/10% FCS. Cells are counted.

9. Just before the engraftment, 50,000 cells are transferred into an Eppendorf tube. They are spun down in the tabletop centrifuge at $10,000 \times g$ for 1 min at room temperature, resuspended in 10 µl Matrigel solution, and kept on ice.

3.3 Transplantation

Regarding the choice of host for the transplantation a few points need to be considered. First choice, whenever possible, is an isogenic recipient. When in doubt, isogenicity can be checked by skin grafts between randomly selected mice [27]. When a mutation of interest is in a mixed genetic background of two distinct strains, frequently 129SV/C57BL/6J F-1 hybrids generated by parents of either background are suitable hosts [28, 29]. Not only will the F1 daughter accept any mixture of alleles from the two strains but in addition the experiment benefits from the hybrid vigor that results from crossing two inbred strains and makes the F1 generation particularly healthy. However, not all the strains show histocompatibility with either the F-1 hybrids or the hosts from the same strains, requiring extensive backcrossing. In particular, with the advent of conditional deletions many mouse strains now contain elements of more than two genetic backgrounds. These complexities require the use of immunocompromised mice.

A widely used model for transplantation experiments were *nude mice* in which *foxn1* gene is disrupted. As a result the mice are athymic and lack thymus-derived T-cells; these are important in allograft rejection [30]. However, nude females have abnormally low levels of circulating estrogens which may influence the growth of transplanted mammary epithelium [31, 32].

Better recipients are mice lacking recombination-activating genes 1 or 2 (Rag1$^{-/-}$) [33] or (Rag2$^{-/-}$) [34]. The two genes are required for recognizing and cleaving signal-specific sequences for somatic rearrangement of B- and T-cell receptors. As a result Rag1$^{-/-}$ and Rag2$^{-/-}$ mice have neither B nor T cells. Transplantation to these mice gives very good and reproducible results. However, the recent discoveries of the important role of immune cells in development and carcinogenesis of the mammary gland [35] suggest that results from this system need to be carefully interpreted.

3.3.1 Transplantation Procedure

1. Mice are anesthetized in the incubation chamber with 5 % isoflurane in the atmosphere and 5 l/min of oxygen supply and then transferred to the heating pad (*see* **Note 11**) at 37 °C with mask on the nose supplying 2 % of isoflurane and 1 l/min of oxygen. Buprenorphine at the rate of 0.1 mg/kg bodyweight is administered subcutaneously for analgesia.

2. Mice are fixed on the heating pad with double-sided tape. The inguinal area is disinfected with Betadine standardized solution.

3. A ventral incision is made carefully with round-nosed scissors and two other incisions perpendicular to the ventral, one on each side of the mouse, finishing half way between teat #4 and teat #5 being careful not to puncture the peritoneum (*see* **Note 12**). Skin is carefully peeled off the peritoneum with forceps and the abdominal gland is exposed (Fig. 4).

4. To stop bleeding, the cauterizer is applied to the blood vessel near the junction by the lymph node (Fig. 5a) and the blood vessel on the fat pad connection between the fourth and fifth glands (Fig. 5b) (*see* **Note 13**).

5. Using angle spring scissors a cut is made at the peritoneal side of the subiliacal lymph node (Fig. 5c) and the teat side half of the gland containing the rudimental ductal tree is excised leaving behind the cleared fat pad. The same procedure is applied to the contralateral side.

6. (a) For the tissue fragment: Using only one side of the 5/45 jewellers' forceps a small pocket is made in the middle of the cleared fat pad (*see* **Note 14**). The tissue piece is placed on the top of the pocket and gently pushed inside using one side of the 5/45 jewellers' forceps. (b) For the single-cell injections: 50,000 cells in 10 μl of the Matrigel solution are taken up with

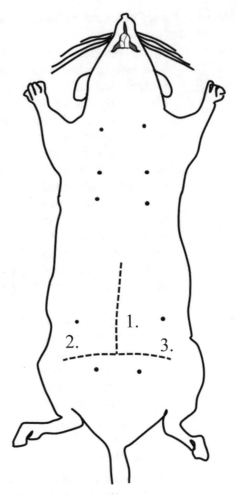

Fig. 4 Surgical field for skin incision in the recipient mouse. Scheme shows a female mouse lying on its back. The incisions positions are indicated with *dashed lines. Numbers* indicate order of incisions

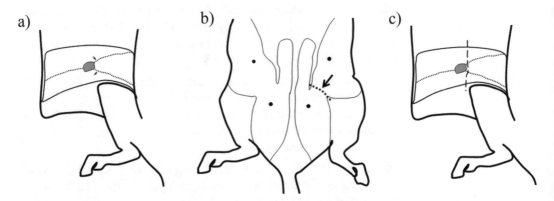

Fig. 5 Surgical field for fat pad clearance in the recipient mouse. Scheme of inguinal area with abdominal and inguinal gland (mammary gland is depicted with *thin line* and blood vessels with *dotted line*). (**a**) *Arrow* indicates where the blood vessels near the junction by the lymph node have to be cauterized. (**b**) *Dotted line* shows a connection between abdominal and inguinal gland that has to be cauterized. (**c**) *Dashed line* shows the position of the incision required to clear the fat pad

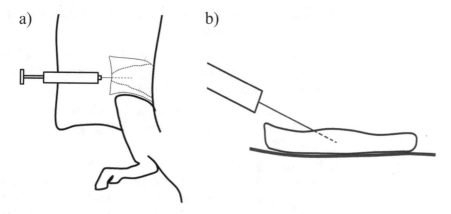

Fig. 6 Cell injection into cleared fat pad of the recipient mouse. (**a**) Scheme of inguinal area with fat pad after clearing being injected with cell suspension. (**b**) Schematic cross section of inguinal fat pad; *dashed line* represents part of needle inside the fat pad

100 µl Hamilton syringe and this suspension is carefully injected into the middle of the cleared fat pad (Fig. 6a, b).

7. Mice are left on the heating pad while incisions are closed with metal clips (approx. 5 mm between each clip) (*see* **Note 15**). Five-hundred microliter of physiological saline is injected into the mouse intraperitoneally (*see* **Note 16**) and tape, used for holding the mouse during surgery, is removed. Buprenorphine (0.1 mg/kg) is administered subcutaneously every 8–12 h for 3 subsequent days.

8. Metal clips are removed with clip-removing forceps 10 days after the surgery.

3.4 Phenotypic Characterization

The timing of analysis of the transplanted glands is a parameter that has to be determined in light of the working hypothesis. As early as 3 days after surgery, limited growth can be observed in transplanted glands. As a rule of thumb, up to 21 days after surgery the terminal end buds can be seen and after 28–35 days the pubertal development in the recipient is over so the mammary gland should have reached its adult stage [29]. There is some variation depending on donor and host genetic backgrounds. To observe side branching, mice have to be examined at least 10 weeks after surgery. A phenotype related to alveologenesis can be observed between 14.5 and 18.5 days of pregnancy; lactation is best examined immediately after birth. As the milk cannot leave the engrafted epithelium because there is no connection to the teat, involution will start within a day after giving birth.

Mice are sacrificed in the CO_2 chamber, fixed on their back to the dissection board with four pins through their paws. The ventral side is sprayed with 70 % ethanol to disinfect the skin. A Y-shaped incision is made and the mammary gland is dissected from the mouse, as described above.

1. For analysis of fluorescent epithelial grafts a mammary gland is placed between the two glass slides. The slides are held together with two plastic clips (*see* **Note 17**).

2. The mammary gland between the two slides is observed under the fluorescence microscope and photographed for analysis (*see* **Note 18**).

3. For subsequent histological analysis mammary glands are placed into the plastic cassettes in PFA solution and left o/n at 4 °C. Next day they are washed in PBS and transferred to a beaker containing 70% ethanol. After this, the histological analysis can be performed at any time.

4. For the whole-mount analysis the dissected mammary gland is spread on the glass slide and left to dry for several hours before fixation.

5. For any analysis requiring protein, RNA or DNA extraction mammary glands are placed in Eppendorf tubes, flash-frozen in liquid nitrogen, and stored at −80 °C (*see* **Note 19**).

4   Notes

1. To avoid weighing of the trypsin, which is a very light charged powder that easily sticks to metal surfaces, the best way to prepare the solution is to add 16.7 ml of L15 medium directly into the original packaging and to vortex immediately. The trypsin dissolves easily and is instantly ready to use or to aliquot.

2. Collagenase A does not readily dissolve in PBS. Hence, it is recommended to prepare aliquots in advance. Briefly, add 25 ml of PBS directly to original packaging and mix it on a horizontal orbital shaker for several hours at 4 °C until dissolved.

3. Usually, in the transplantation experiment mammary epithelium from a genetically modified mouse and a control wild-type mice are engrafted. The experimental, genetically modified tissue is engrafted on one side, and the wild-type is engrafted contralaterally. When this experimental design is followed, it is recommended to use 2 cm tissue culture dishes or 6-well plates to prepare the epithelial fragments for transplantation.

4. We use isoflurane gas anesthesia because it is well tolerated; we observe few side effects and recovery time is short facilitating postoperative care. Alternatively, injected anesthetics (xylazine 10–15 mg/kg bodyweight+ketamine 80–100 mg/kg bodyweight) can be used; however, recovery time is much longer. We have noticed that *RAG 1−/−* mice are particularly sensitive to the injected anesthetics. Therefore, for any strain

the dose may have to be adjusted. When using injected anesthetics additional analgesia during surgery is not required.

5. In accord with the advice from the Federal Veterinary Office of Switzerland we use Betadine (standardized solution) to disinfect the mouse skin. Seventy percent ethanol should be avoided because evaporation of ethanol may lead to cooling down of the mouse and may increase mortality. Cotton is soaked with Betadine solution and rubbed into the mouse skin against the direction of the hair growth to optimize skin sterilization.

6. Buprenorphine at a dose of 0.1 mg/kg body weight is used during surgery and for postoperative analgesia and the same dose is given daily for 3 days after surgery. Additionally, paracetamol can be provided in the drinking water at 1 mg/ml

7. To mince the mammary gland tissue, two #22 scalpels are taped together side by side. If different experimental groups of mice are used for the same experiment it is advisable to mince the glands in as similar a way as possible, meaning for the same amount of time. Usually around 3 min is adequate for up to two mice (four glands per mouse); count an extra 30 s for each additional mouse (maximum five mice per batch). While cutting/mincing the tissue bring it together with scalpels to the center of the board from time to time.

8. Twenty milliliter of the digestion mix is optimal for 1 g of tissue.

9. It is advisable to gently shake the Falcon tubes with the digestion mix every 15 min during collagenase A/trypsin digestion to ensure homogeneous digestion.

10. Shake the transferred supernatant vigorously to release the leftover organoids from the fat on the top.

11. If the heating pad does not have a precise electronic regulator, place a thermometer on it in order to follow the temperature. Slightest increases can cause dehydration and be fatal.

12. If the peritoneum is punctured, it is possible to stitch it with absorbable suture, placing a stitch every 0.5 mm.

13. Cauterization has to be very gentle and careful; otherwise fat pad can get damaged and necrosis can prevent the graft taking.

14. It is extremely important that the pocket is well centered in the middle of the fat pad and that the fragment is not placed underneath it. Outside the fat pad the graft will not grow.

15. When stitching the two perpendicular incisions, it is important to remove the tape fixing the lower limbs of the mouse to the board so that the stitches are made in a physiological position to avoid interfering with the mobility of the animal after surgery.

16. Sterile physiological solution is injected to accelerate recovery because the surgery can cause dehydration.

17. The mammary glands should be as dry as possible to avoid it slipping between the slides. Extra liquid can give rise to a false border that may cause problems with subsequent analyses, such as determining the extent of fat pad filling, as the size of fat pad may be overestimated.

18. EGFP transgenic mice [23] whose use is proposed in this chapter have a high-intensity fluorescence signal in the epithelium and a very low fluorescence signal in the stroma. This makes it difficult to distinguish the fat pad borders from the epithelium in the GFP channel. In the DsRed channel, stroma has higher auto-fluorescence making it convenient to take the picture of the fat pad in red and epithelium in green.

19. To preserve RNA quality, mammary glands have to be isolated from live anesthetized mice.

Acknowledgements

The authors thank all current and former members of the Brisken laboratory, who contributed to the development of these techniques, and Gisèle Ferrand for advice on anesthesia procedures.

References

1. Knight CH, Peaker M (1982) Development of the mammary gland. J Reprod Fertil 65:521–536

2. Faulkin LJ, DeOme KB (1958) The Effect of estradiol and cortisol on the transplantability and subsequent fate of normal, hyperplastic, and tumorous mammary tissue of C3H Mice. Cancer Res 18:51–56

3. Daniel CW, Deome KB (1965) Growth of mouse mammary glands in vivo after monolayer culture. Science 149:634–636

4. Deome KB, Faulkin LJ Jr, Bern HA, Blair PB (1959) Development of mammary tumors from hyperplastic alveolar nodules transplanted into gland-free mammary fat pads of female C3H mice. Cancer Res 19:515–520

5. Hoshino K, Gardner WU, Pawlikowski RA (1965) The incidence of cancer in quantitatively transplanted mammary glands and its relation to age and milk agent of the donor and host mice. Cancer Res 25:1792–1803

6. Muhlbock O (1956) The hormonal genesis of mammary cancer. Adv Cancer Res 4:371–391

7. Prehn RT (1953) Tumors and hyperplastic nodules in transplanted mammary glands. J Natl Cancer Inst 13:859–871

8. Shimkin MB, Wyman RS, Andervont HB (1946) Mammary tumors in mice following transplantation of mammary tissue. J Natl Cancer Inst 7:77

9. Hoshino K (1962) Morphogenesis and growth potentiality of mammary glands in mice. I. Transplantability and growth potentiality of mammary tissue of virgin mice. J Natl Cancer Inst 29:835–851

10. Hoshino K (1963) Morphogenesis and growth potentiality of mammary glands in mice. II. Quantitative transplantation of mammary glands of normal male mice. J Natl Cancer Inst 30:585–591

11. Hoshino K (1964) Regeneration and growth of quantitatively transplanted mammary glands of normal female mice. Anat Rec 150:221–235

12. Hoshino K (1967) Transplantability of mammary gland in brown fat pads of mice. Nature 213:194–195

13. Hoshino K, Gardner WU (1967) Transplantability and life span of mammary gland during serial transplantation in mice. Nature 213:193–194

14. Brisken C, Park S, Vass T, Lydon JP, O'Malley BW, Weinberg RA (1998) A paracrine role for

the epithelial progesterone receptor in mammary gland development. Proc Natl Acad Sci U S A 95:5076–5081

15. Mallepell S, Krust A, Chambon P, Brisken C (2006) Paracrine signaling through the epithelial estrogen receptor alpha is required for proliferation and morphogenesis in the mammary gland. Proc Natl Acad Sci U S A 103:2196–2201

16. Ciarloni L, Mallepell S, Brisken C (2007) Amphiregulin is an essential mediator of estrogen receptor alpha function in mammary gland development. Proc Natl Acad Sci U S A 104:5455–5460

17. Brisken C, Heineman A, Chavarria T, Elenbaas B, Tan J, Dey SK et al (2000) Essential function of Wnt-4 in mammary gland development downstream of progesterone signaling. Genes Dev 14:650–654

18. Heckman-Stoddard BM, Vargo-Gogola T, Herrick MP, Visbal AP, Lewis MT, Settleman J et al (2011) P190A RhoGAP is required for mammary gland development. Dev Biol 360:1–10

19. Pond AC, Bin X, Batts T, Roarty K, Hilsenbeck S, Rosen JM (2013) Fibroblast growth factor receptor signaling is essential for normal mammary gland development and stem cell function. Stem Cells 31:178–189

20. Shackleton M, Vaillant F, Simpson KJ, Stingl J, Smyth GK, Asselin-Labat ML et al (2006) Generation of a functional mammary gland from a single stem cell. Nature 439:84–88

21. Sleeman KE, Kendrick H, Robertson D, Isacke CM, Ashworth A, Smalley MJ (2007) Dissociation of estrogen receptor expression and in vivo stem cell activity in the mammary gland. J Cell Biol 176:19–26

22. Friedrich G, Soriano P (1991) Promoter traps in embryonic stem cells: a genetic screen to identify and mutate developmental genes in mice. Genes Dev 5:1513–1523

23. Okabe M, Ikawa M, Kominami K, Nakanishi T, Nishimune Y (1997) 'Green mice' as a source of ubiquitous green cells. FEBS Lett 407:313–319

24. Vintersten K, Monetti C, Gertsenstein M, Zhang P, Laszlo L, Biechele S et al (2004) Mouse in red: red fluorescent protein expression in mouse ES cells, embryos, and adult animals. Genesis 40:241–246

25. Smalley MJ (2010) Isolation, culture and analysis of mouse mammary epithelial cells. Methods Mol Biol 633:139–170

26. LaMarca HL, Visbal AP, Creighton CJ, Liu H, Zhang Y, Behbod F et al (2010) CCAAT/enhancer binding protein beta regulates stem cell activity and specifies luminal cell fate in the mammary gland. Stem Cells 28:535–544

27. Daniel CW, Deome KB, Young JT, Blair PB, Faulkin LJ Jr (2009) The in vivo life span of normal and preneoplastic mouse mammary glands: a serial transplantation study. J Mammary Gland Biol Neoplasia 14:355–362

28. Faulkin LJ Jr, Deome KB (1960) Regulation of growth and spacing of gland elements in the mammary fat pad of the C3H mouse. J Natl Cancer Inst 24:953–969

29. Williams MF, Hoshino K (1970) Early histogenesis of transplanted mouse mammary glands. I. Within 21 days following isografting. Z Anat Entwicklungsgesch 132:305–317

30. Flanagan SP (1966) 'Nude', a new hairless gene with pleiotropic effects in the mouse. Genet Res 8:295–309

31. Seibert K, Shafie SM, Triche TJ, Whang-Peng JJ, O'Brien SJ, Toney JH et al (1983) Clonal variation of MCF-7 breast cancer cells in vitro and in athymic nude mice. Cancer Res 43:2223–2239

32. Soule HD, McGrath CM (1980) Estrogen responsive proliferation of clonal human breast carcinoma cells in athymic mice. Cancer Lett 10:177–189

33. Mombaerts P, Iacomini J, Johnson RS, Herrup K, Tonegawa S, Papaioannou VE (1992) RAG-1-deficient mice have no mature B and T lymphocytes. Cell 68:869–877

34. Shinkai Y, Rathbun G, Lam KP, Oltz EM, Stewart V, Mendelsohn M et al (1992) RAG-2-deficient mice lack mature lymphocytes owing to inability to initiate V(D)J rearrangement. Cell 68:855–867

35. Reed JR, Schwertfeger KL (2010) Immune cell location and function during post-natal mammary gland development. J Mammary Gland Biol Neoplasia 15:329–339

Pubertal Ductal Morphogenesis: Isolation and Transcriptome Analysis of the Terminal End Bud

Joanna S. Morris and Torsten Stein

Abstract

The terminal end bud (TEB) is the growing part of the ductal mammary epithelium during puberty, enabling the formation of a primary epithelial network. These highly proliferative bulbous end structures that drive the ductal expansion into the mammary fat pad comprise an outer cap cell layer, containing the progenitor cells of the ductal myoepithelium, and the body cells, which form the luminal epithelium. As TEB make up only a very small part of the whole mammary tissue, TEB-associated factors can be easily missed when whole-tissue sections are being analyzed. Here we describe a method to enzymatically separate TEB and ducts, respectively, from the surrounding stroma of pubertal mice in order to perform transcriptomic or proteomic analysis on the isolated structures and identify potential novel regulators of epithelial outgrowth, or to allow further cell culturing. This approach has previously allowed us to identify novel TEB-associated proteins, including several axonal guidance proteins. We further include protocols for the culturing of isolated TEB, processing of mammary tissue into paraffin and immunohistochemical/fluorescent staining for verification, and localization of protein expression in the mammary tissue at different developmental time points.

Key words Puberty, Mammary gland, TEB, Ducts, Stroma

1 Introduction

The mammary gland is a unique organ in that it mainly develops postnatally [1]. At birth only a rudimentary epithelial structure can be found at the nipple within a specialized stromal environment, the mammary fat pad [2]. In response to hormonal changes at ~3 weeks of age the mammary epithelium grows out and branches within the surrounding fat pad to form the primary ductal network. This dramatic epithelial expansion and branching is driven by highly specialized proliferative structures, the terminal end buds (TEB) [1, 3]. The epithelium expands and branches to form the primary and secondary ducts until it reaches the outer borders of the fat pad (Fig. 1a), when the TEB regress by an as yet unknown mechanism(s) to form terminal end ducts. During each pregnancy

Finian Martin et al. (eds.), *Mammary Gland Development: Methods and Protocols*, Methods in Molecular Biology, vol. 1501, DOI 10.1007/978-1-4939-6475-8_5, © Springer Science+Business Media New York 2017

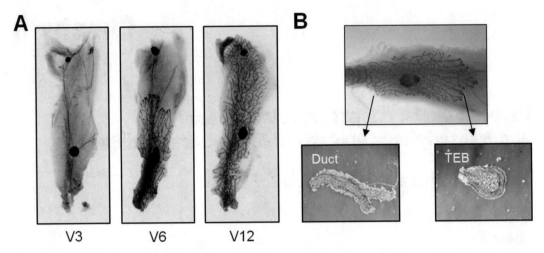

Fig. 1 Pubertal mammary epithelial development. (a) Carmine-alum-stained whole-mounts showing the pubertal mammary epithelial development at (V3) 3 weeks, (V6) 6 weeks, (V12) and 12 weeks of age in C75/Bl6 mice. The branching epithelial tree grows out from the nipple end past the central lymph node to fill the mammary fat pad. (b) Mammary gland whole-mount at 6 weeks of age with examples of isolated TEB and duct

cycle these ducts bud and form tertiary ducts and secretory alveoli, which regress after lactation to form a mammary epithelium that resembles the pre-pregnancy state.

The exact mechanism and physiological control of TEB outgrowth are still only partly understood, though some important regulators have been identified using several mouse models. These have been extensively described by McBryan and Howlin within this book (chapter 3). Proteomic and transcriptomic analyses can be very efficient tools to identify factors involved in such processes, but to date there have been relatively few studies focusing on genome-wide RNA expression in the pubertal mammary gland. The power of this approach to study mammary gland RNA expression changes was first shown by Master et al. [4] using the first generation of Affymetrix mouse transcriptome arrays, covering just over 4000 transcripts. This study identified brown fat degradation in the stroma of 4-week-old mouse mammary glands, not previously seen. McBryan et al. [5] also studied the early changes during pubertal growth induction using whole glands and the next generation of Affymetrix RNA arrays, which covered ~22,000 transcripts. By identifying RNAs that were abundant when TEB were detectable in the mammary gland by whole-mount analysis they defined a "TEB-profile." However, since the TEB only constitute a small proportion of the total number of cells in the mammary gland it is challenging to study specific TEB gene expression in whole mammary glands, as these RNAs are strongly diluted by the rest of the gland. In 2006, Kouros-Mehr and Werb therefore compared RNA expression profiles of laser micro-dissected TEB and ductal epithelium with that obtained from epithelium-free mammary stroma. Using this approach they identified potential regulators of mammary branching, including several members of the Wnt pathway [6].

We have taken a different approach and further developed and optimized a method based on the one described by Richard et al. [7] to mechanically and enzymatically isolate TEB and ducts from freshly dissected pubertal mouse mammary glands to enable more precise analysis. The advantage of this method is that TEB and ducts can be freed gently and relatively quickly from their stromal environment and can be used for either proteome or transcriptome analysis. By comparing the transcriptomes and proteomes of TEB and of ducts from the pubertal gland we established gene signatures associated with either outgrowing or resting ductal epithelium, and thereby identified novel factors associated with the controlled invasive growth of this multicellular epithelial structure into its stromal environment, including the expression of axonal guidance protein-encoding genes [8, 9]. Comparison of our transcriptome data from isolated TEB and ducts with those obtained from mammary gland tissue strips that were enriched for TEB or ducts showed that many identified TEB-associated RNAs would have been missed due to either a very low abundance in TEB or dilution with RNAs from other epithelial and stromal cells if the whole tissues had only been used.

Here we describe this easy method for the isolation of TEB and ducts (Fig. 1b), which produces high-quality RNA or protein for transcriptome and qRT-PCR or proteomic analysis, and facilitates the search for potential new regulators of mammary epithelial outgrowth [8, 9]. We had previously used the MG-U74Av2 microarray from Affymetrix and have since then followed this up with mouse whole-genome exon arrays (GeneChip-Mouse-Exon-1.0-ST-Array, manuscript in preparation). 5 µg of total RNA was necessary in each case, requiring the pooling of over 1000 TEB and ducts in order to obtain enough RNA for replicate experiments. The necessity of pooling the tissue samples meant that biological variation was reduced and replicates reflected technical variability of the experiment. Improved reliability of RNA amplification kits allows the reduction of the starting material to a few nanograms or even less (see chapter 6), dramatically reducing the number of mice required.

Since microarray technology has developed so considerably in recent years and is in fact increasingly replaced by RNA sequencing technology or outsourced to companies or genomics facilities, we will not cover this technology itself, but focus on the TEB isolation. The described approach has the additional advantage over methods like micro-dissection in that the isolated TEB and ducts can also be cultured and the primary cells analyzed further in vitro. We include in this chapter a brief description of a method for culturing these organoids on fibronectin-coated plastic for further immunofluorescent staining. We also include protocols for the processing of tissues into paraffin (FFPE) blocks and immunohistochemical or immunofluorescent analysis of FFPE-mammary gland sections for verification purposes and cellular localization.

2 Materials

2.1 Mice

The protocols below have been used successfully with Balb/C and C57BL/6 mice in our laboratories, but other strains can be used as well. For optimal isolation, mice of ~6–7 weeks of age and 16–18 g weight are advisable, though this may need adjustment when using other mouse strains to allow for differences in weights and the speed of ductal elongation.

2.2 Media and Reagents

1. L15 serum-free medium: Leibovitz L15 medium.

2. L15/0.1% FCS medium: Leibovitz L15 medium supplemented with 2 mM L-glutamine supplemented with 0.1% v/v heat-inactivated fetal calf serum (FCS), 100 IU/ml penicillin, 100 μg/ml streptomycin.

3. DMEM/F12 10% FCS medium: Dulbecco's modified Eagle's medium (DMEM) supplemented with 2 mM L-glutamine and 10% v/v heat-inactivated FCS, 100 IU/ml penicillin, 100 μg/ml streptomycin.

4. D-PBS: Dulbecco's phosphate-buffered saline; 1.0% w/v sodium chloride, 0.025% w/v potassium chloride, 0.025% w/v disodium hydrogen orthophosphate, and 0.1437% w/v potassium dihydrogen orthophosphate.

5. Collagenase: *Clostridium histolyticum* Collagenase Type II solution: stock 10 mg/ml in L15 medium.

6. TRI-solution or equivalent.

7. RNase decontamination solution.

8. 10% Neutral buffered formalin.

9. Autoclaved DEPC-treated water.

10. TBS/Tween buffer: 50 mM Tris–HCl, 150 mM NaCl, pH7.4 with HCl, 0.1% Tween-20 (v/v).

11. 2.5% Normal horse serum (in TBS/Tween) blocking solution.

12. Primary antibodies.

13. 3% Hydrogen peroxide solution.

14. 1 mM EDTA buffer pH 8: 1 mM EDTA disodium salt, 4.5 mM Tris–HCl, pH 8 in dH_2O.

15. 10 μg/ml Fibronectin or fibronectin-like protein in D-PBS solution.

16. 4 % Paraformaldehyde solution (PFA) in PBS for TEB fixation on glass slides: Dissolve 4 % PFA (w/v) in an appropriate amount of PBS in a glass beaker, heating it up and stirring on a hot plate (to ~60 °C) in a fume hood. Add a few drops of 1 N NaOH if necessary (if solution remains cloudy) to help to dissolve the paraformaldehyde. Once the solution is clear, let it cool down to room temperature and filter. Test and adjust the pH 7.4 with HCl if required and adjust the final volume. Freeze at –20 °C or store at 4 °C for short-term storage (up to 1 month).

17. ImmPRESS HRP link reagent.

18. DAB reagent.

19. Hematoxylin.

20. Fluorescent-labeled secondary antibodies for IF.

21. 1 % Acid alcohol: 99.5 ml 75 % aqueous ethanol, 0.5 g sodium chloride, 0.5 g HCl.

22. Scott's tap water substitute: 81.14 mM Anhydrous $MgSO_4$, 41.66 mM $NaHCO_3$ in dH_2O.

23. Mounting medium for IHC.

24. Image-iT FX Signal Enhancer Solution.

25. Mounting medium with DAPI for IF.

2.3 Tools and Instruments

2.3.1 Mammary Gland Dissection

1. One or more cork dissection boards.

2. 70 % Ethanol solution for sterilization of the animal.

3. Dissection needles or 23 G injection needles for securing the animal and the skin flap onto the cork board.

4. Dissection scissors (preferably with rounded end) for opening the skin.

5. Microdissecting spring scissors to remove the mammary gland.

6. Two pairs of fine blunt-ended forceps (all autoclaved or cleaned thoroughly with ethanol).

7. One universal tube for tissue collection, containing ice-cold L15 medium.

2.3.2 TEB and Duct Preparation

1. One tissue culture hood for handling mammary glands and isolating TEB and ducts.

2. One 90 mm sterile petri dish for cutting the mammary glands.

3. Two scalpel blades for cross-cutting of mammary glands.

4. One universal tube with 9 ml pre-warmed (37 °C) L15 medium without serum and 1 ml of 10 mg/ml stock collagenase type II in serum-free L15 medium (final concentration 1 mg/ml).

5. A 37 °C shaking incubator set at 100 rpm (or a roller in a hot room (37 °C)) for incubation with collagenase.

6. One refrigerated centrifuge.

7. One gridded 6 cm contact dish within a 9 cm petri dish.

8. A P2 or P10 pipette with sterile filter tips for collection of the TEB and ducts.

9. Two Eppendorf tubes filled with 0.5–1 ml of TRI solution in which to collect the TEB and ducts.

10. One dissection microscope with 4× magnification for identification of TEB and ducts.

2.3.3 TEB and Duct In Vitro Culture

1. 8-Well chamber slides or 12-well plate with sterile glass cover slip for further IF studies.

2.3.4 Immunohistochemistry

1. Universal tubes.

2. Plastic cassette for wax embedding.

3. Paraffin wax.

4. One heated paraffin dispensing module.

5. One rotary microtome.

6. One tissue flotation water bath at 40 °C.

7. Electrostatically charged glass microscope slides.

8. Glass staining dishes with lids and slide rack for ethanol series (70, 95, 99, 100 %) and xylene.

9. One micro wave-proof pressure cooker.

10. One microwave with at least 600 W power setting.

11. One ImmunoEdge pen.

12. One humidification incubation box for antibody incubation.

13. Cover glass, 22 × 22 mm.

3   Methods

The following protocols describe how to dissect the mouse mammary glands from pubertal mice followed by TEB/duct isolation and RNA or protein extraction. It will yield TEB that still retain the cap cell layer (Fig. 1b), which itself can be isolated further by hyaluronidase treatment. We further describe the preparation of the mammary glands for wax embedding and immunohistochemistry for protein localization within the pubertal mammary gland tissue for verification of any identified gene/protein. In order to do transcriptome analysis, proteomic analysis, or high-throughput sequencing one needs a large number of TEB and ducts to gain enough material for RNA or protein. The protocol described here yields on average of ~4–6 TEB per mouse gland. The use of new technologies like linear RNA amplification will reduce the requirement to only nanograms of RNA, thereby reducing the number of mice needed.

3.1 Extraction of Fourth Inguinal Glands

1. Sacrifice the mouse by cervical dislocation or increasing concentrations of CO_2 according to schedule one killing and place it on its back on the cork dissection board. Stretch the legs out laterally and push the needles through each of the 4 feet into the cork board to hold the mouse in place. Squirt the fur with 70% alcohol to dampen it down and sterilize it.

2. Grasp the skin of the midline using blunt forceps approximately half way between front and back legs and using sterile/alcohol-cleaned scissors make an incision through the skin alone, taking care not to penetrate through into the abdominal cavity. Extend the incision cranially to the neck and caudo-laterally along each hind leg to create an inverted Y-shape incision.

3. Push the skin away from the underlying muscle with either blunt-ended scissors or sterile cotton swabs, and stretch the skin flaps out on each side of the mouse, pinning into place with 1–2 needles. The inguinal mammary gland should be obvious on top of each skin flap. Push the body wall away from the skin flap medially to expose the entire gland.

4. Using sharp dissection scissors, held flat against the skin flap, separate the mammary gland from the skin, moving laterally to medially, and once the whole gland is dissected free, place it in medium for TEB isolation, store it in formalin, or snap freeze it in liquid nitrogen, as required.

5. Repeat the dissection for the inguinal mammary gland on the other side.

3.2 Isolation of TEB and Ducts from Pubertal Mouse Mammary Glands

1. Harvest the glands into a universal tube containing approximately 10 ml of chilled serum-free L15 medium and kept on ice (see **Note 1**). Take the chilled glands into a tissue culture facility for further dissection and processing as quickly as possible to reduce the time for RNA degradation and cell death to occur. Pour the dissected mammary glands with the medium from the universal tube into a sterile 90 mm petri dish, and carefully aspirate the medium, leaving the glands in the dish.

2. Using two new sharp scalpel blades, chop the glands coarsely by the crossed scalpels technique into smaller pieces (Fig. 2a; see **Note 2**). Using the same scalpels, scrape the minced tissue into a new sterile universal tube and 9 ml of pre-warmed (37 °C) serum-free L15 medium, and add 1 ml of 10 mg/ml stock collagenase type II in serum-free L15 medium.

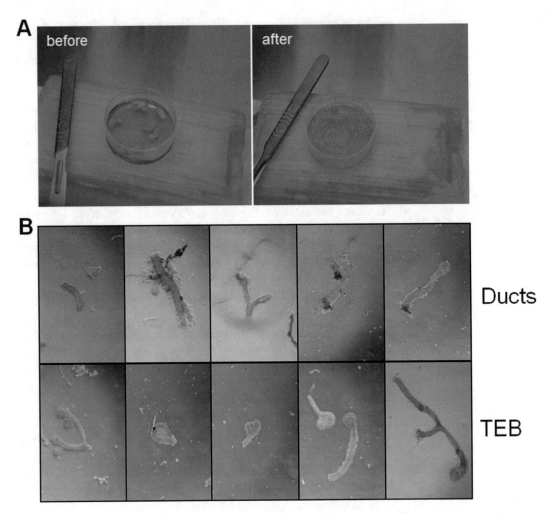

Fig. 2 Mammary gland preparation with examples of isolated TEB and ducts. (a) Mammary glands before and after cutting with scalpels. (b) Examples of isolated TEB and ducts as found after the isolation procedure. If TEB are still attached to long ducts these can be carefully broken off with a pipette tip

3. Place the tube into a shaking incubator at 37 °C for 20–30 min (TEB) or 30–45 min (ducts) with mild agitation (100 rpm) to digest the mammary tissue and free the epithelium of the surrounding collagenous stroma. Alternatively, leave the tube on a roller in a hot room set at 37 °C for the indicated times (*see* **Note 3**).

4. After incubation, shake the universal tubes vigorously by hand to encourage release of the epithelial structures from the stroma. Evenly split the contents of the universal between two 50 ml Falcon tubes and dilute the tissue soup tenfold to 50 ml with fresh ice-cold L15 medium to suppress the collagenase activity (*see* **Note 4**). Centrifuge the Falcon tubes at $300 \times g$ for 5 min at 4 °C and carefully remove the fatty scum on the top

and the supernatant without disturbing the cell pellet, which contains the TEB and ducts.

5. Resuspend each pellet in 0.5–1 ml ice-cold L15 medium and combine them before transferring them to a gridded 60 mm contact dish inside a 90 mm petri dish in case of medium spillage (*see* **Note 5**).

6. Examine the dish under a dissecting microscope (4× magnification) to identify individual TEB or duct segments (Fig. 2b, *see* **Note 6**). Using a 2 or 10 μl micropipette set at 1 μl, transfer a single TEB or duct segment at a time to separate microcentrifuge tubes, containing 50–100 μl TRI solution (*see* **Notes 7** and **8**) and record the number transferred to each tube (one tube for TEB, one for ducts). The samples can be further homogenized using a plastic 1.5 ml Eppendorf pellet pestle by carefully plunging the pestle up and down with a rotating motion 15–20 times or until no more fragments are visible. Snap freeze these samples and keep them frozen at –80 °C until sufficient numbers have been collected to batch for RNA extraction (*see* **Note 9**).

7. For protein extraction, collect the TEB into 100–150 μl of protein lysis buffer and homogenize them with an Eppendorf pellet pestle. Incubate the tube with the lysis buffer (*see* **Note 10**) at 10 °C for 10 min, centrifuge the tube at 20,000 × g for 5 min, and store the supernatant frozen at –80 °C until a sufficient number of TEB or ducts have been collected (450–500 TEB in 500 μl to a protein concentration of ~0.7 mg/ml).

3.3 TEB/Duct In Vitro Cultures

1. The day before the TEB isolation, coat sterile glass cover slips (in 12-well plates) or chamber slides with fibronectin or fibronectin-like polymer by covering the area with a 10 μg/ml fibronectin in D-PBS solution overnight at 4 °C. The next day, aspirate any excess solution and wash the cover slip/well with sterile D-PBS. The coated cover slips/slides can be used immediately or dried and stored under sterile conditions at 4 °C for several weeks.

2. Pipette the TEB or ducts into the sterile fibronectin-coated well of a chamber slide or onto the cover slip in the 12-well dish and carefully add just enough DMEM/F12/10%FCS medium with added antibiotics and anti-mycotics (*see* **Note 11**) to the side of the well, so that the medium just touches the organoid, leaving the TEB or ductal fragment in the liquid-air interphase to allow for efficient initial adhesion. Transfer the dish or chamber slide carefully to a 37 °C/5% CO$_2$ incubator and *do not move it for 24 h* (*see* **Note 12**).

3. After 24 h, carefully remove the dish from the incubator and very carefully add just enough medium to the side of the well, without disturbing the TEB so that it is just covered (~2 mm)

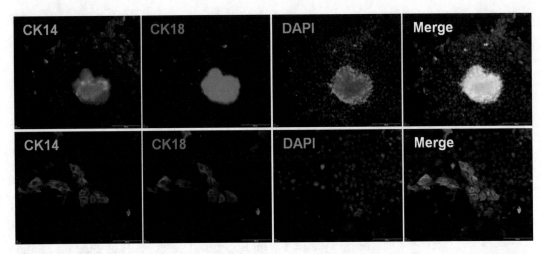

Fig. 3 Immuno-fluorescent staining of a TEB culture. Low- (200×; *top row*) and high-power magnifications (400×; *bottom row*) of a TEB grown in a fibronectin-like protein coated well. The TEB was fixed and stained for cytokeratin (CK) 14 (*green*) and CK18 (*red*) as described. The seeded TEB can be seen as the large bright structure from which the cells have grown out

and the TEB does not float off. It is best to check under a microscope that the TEB/duct is still in the desired place. Return the dish to the incubator and do not move it for four days to allow for cell attachment and outgrowth (*see* **Note 12**). The TEB/ducts will attach to the fibronectin-coated dish and the epithelial cells will grow out to form a distinct colony with cytokeratin-18 and cytokeratin-14 double-positive primary epithelial cells (Fig. 3) surrounded by smooth-muscle actin-positive cells.

For further immunohistochemical or fluorescent staining fix the cells in 4 % PFA for 15 min at room temperature and/or ice-cold absolute ethanol and keep them at 4 °C for up to a week under ethanol, making sure that the cells never dry out.

3.4 RNA Isolation from TEB and Ducts

1. Thaw the frozen samples fully at room temperature and then vortex them thoroughly (10–15 s on highest setting) before combining the samples to get adequate total numbers for one RNA sample. For efficient RNA isolation use a phenol-based guanidine thiocyanate system (TRI solution) and follow the recommended manufacturer's procedures (*see* **Note 13**).

2. Add 0.2 ml of chloroform per 1 ml TRI solution. Shake the tube vigorously by hand for 15 s, and incubate it at room temperature for 2–3 min. For separation of the aqueous and organic phases, centrifuge the samples for 15 min at 4 °C at 12,000×*g*.

3. Transfer the upper RNA containing aqueous phase carefully to a fresh tube without disturbing the protein-containing interphase or lower organic phase. Precipitate the RNA by adding 0.5 ml of isopropanol. Shake the samples briefly by hand and incubate them for 10 min at room temperature before centrifugation at $12,000 \times g$ for 10 min at 4 °C.

4. Carefully remove the supernatant without disturbing the translucent pellet and wash it once by adding ~1 ml of 75 % ethanol (in nuclease-free water) to remove salts, mix the sample briefly by vortexing, and centrifuge them at $7500 \times g$ for 5 min at 4 °C to pellet the RNA.

5. Carefully remove the ethanol by pipetting, again trying not to disturb the pellet (which can be very loose by now and easily aspirated), and allow the RNA pellet to air-dry (*see* **Note 14**). Dissolve the RNA in approx. 15–40 µl of nuclease-free water depending on the size of the pellet.

6. Place the RNA samples directly on ice and quantify it by spectrophotometer. For long-term storage snap freeze the RNA in dry ice or liquid nitrogen and store it at –80 °C. If necessary, pool the RNA from several isolates for further downstream applications such as microarray analysis (Fig. 4).

7. Small RNAs and DNA may interfere with some downstream applications. To remove these, purify the RNA extracts further using a sepharose bead-based RNA extraction system with on-column RNase-free DNase I digestion (*see* **Note 15**). The RNA pellet from the RNA extraction can either be directly resuspended in the lysis buffer or the resuspended RNA can be diluted in the appropriate amount of lysis buffer solution according to the manufacturer's protocol (*see* **Note 16**).

8. Elute the RNA in nuclease-free water, keep on ice, and immediately quantify it by spectrophotometer or snap freeze the RNA for long-term storage at -80 °C.

3.5 Immuno-histochemistry on Formalin-Fixed Paraffin-Embedded (FFPE) Tissue Sections

3.5.1 Fixation of Mammary Gland Tissue

1. Dissect the fourth inguinal mammary gland from the mouse as described under Subheading 3.1 without removal of the central lymph node, and transfer it into a sterile universal tube containing enough 10 % neutral buffered formalin. A 10:1 ratio v/v of formalin to tissue is required to enable efficient fixation of the whole gland.

2. Fix the gland in the formalin for at least 4 h at room temperature (or overnight), remove it with forceps from the formalin solution, and place it into a plastic cassette for further processing (*see* **Note 17**).

3. Dehydrate the tissue slowly through a graded concentration of ethanol (50 min in 50 % (v/v), 50 min in 80 % (v/v) ethanol,

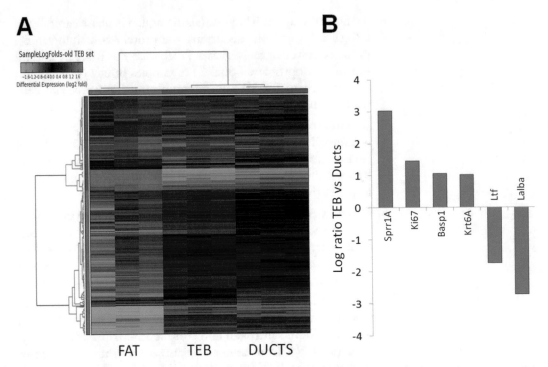

Fig. 4 Heat-map from TEB and duct microarray data. (**a**) The heat-map shows the differences between the TEB and duct transcriptome profiles in comparison to empty fat pat after hierarchical clustering using Altanalyze software [10]. (**b**) Log ratio (TEB vs. ducts) of normalized signal intensities of axonal guidance proteins small-proline rich protein 1a (Sprr1a) and brain acid-soluble protein 1 (Basp1) as well as the proliferation marker Ki67, TEB-associated protein cytokeratin 6a (Krt6a), and ductal-associated proteins lactotransferrin (Ltf) and lactalbumin α (Lalba)

and 1 h in 100% ethanol (v/v)) and incubate in xylene for 2 h at 37 °C.

4. Transfer the fixed and dehydrated gland into the paraffin wax for 4 h at 60 °C to allow for even penetration and coating.

5. Place the paraffin wax-embedded tissue at the bottom of metal pans and add additional wax to embed the tissue into paraffin wax blocks using a heated paraffin-dispensing module. Leave the paraffin tissue blocks to solidify for 30 min on a cold plate, chilled to –5 °C, and store at room temperature (RT) until required.

3.5.2 Immuno-Staining of FFPE-Mammary Gland Sections

1. Cut paraffin sections with a microtome and place these on the surface of a 40 °C water bath (*see* **Note 18**). A scalpel blade can be used to separate the wax sections, which will form a ribbon when cut. Pick the sections up from the water bath using electrostatically charged slides and leave them to dry in a slide rack for 1 h at 62 °C. If consecutive sections are being used, it is important to pick the sections up in the order that they have been cut and to label them accordingly.

2. De-wax the sections by moving the slides in a slide rack through (a) xylene 3×5 min, followed by (b) immersion in 99% denatured alcohol (IMS) twice for 2 min each, and finally (c) rinse them in running tap water for at least 2 min to remove all traces of ethanol.

3. Quench any endogenous peroxidase activity by incubating the sections in 3% hydrogen peroxide solution for 10 min and rinse them in running tap water (see **Note 19**).

4. For antigen retrieval with EDTA buffer bring 1 l of 1 mM EDTA buffer to the boil in a microwave pressure cooker (see **Note 20**). Add the slides to the boiling EDTA buffer and heat them in the microwave at full pressure for a further 5 min, of which 1–2 min has to be precisely timed for the slides to be held at the retrieval pressure (see **Note 21**). Remove the pressure cooker from the microwave, release the pressure quickly, and place the slides in running tap water for 10 min to cool.

5. Place the slides in a humid staining chamber. Using an Immunoedge pen circle the tissue section so that the reagents only cover the area of the tissue and do not run off. To enable the antibodies to penetrate the cells cover the sections with TBS/Tween buffer and leave them for 1 min (see **Note 22**). Drain off the TBS/Tween buffer and incubate the section with 2.5% normal horse serum for 10 min at room temperature to block any nonspecific antibody-binding sites.

6. Afterwards, rinse the slides with TBS/Tween and flick any excess off the slide. Dilute the primary antibody in antibody diluent solution (or TBS) to the required concentration (see **Note 23**). Cover the section with as much primary antibody as needed to cover the tissue and incubate it for at least 30 min at room temperature (see **Note 24**).

7. Wash the slides three times for 10 min with TBS/Tween buffer to remove any excess antibody and incubate each section with 2–4 drops of ready-to-use ImmPRESS HRP link reagent per slide for 30 min at room temperature (see **Note 25**). Afterwards, wash the sections thoroughly three times for 5 min each in TBS/Tween buffer.

8. For a brown chromagen reaction, incubate the section with 3,3-diaminobenzidine tetrahydrochloride (DAB—see **Note 26**) for 5 min (one drop of DAB in 1 ml of antibody diluent—see **Note 27**).

9. Finally, wash the stained sections under running tap water and counterstain the tissue section with hematoxylin solution (stains the nuclei blue) and wash them for 2 min under running tap water. Dip the slides in 1% acid alcohol (see **Note 28**)

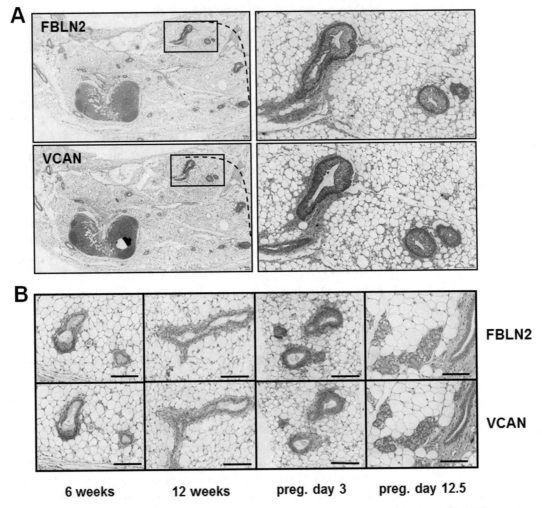

Fig. 5 Versican and fibulin 2 detection in mammary gland sections. (a) Immunohistochemical staining for two TEB-associated proteins, fibulin 2 (FBLN2) and versican (VCAN), in consecutive sections of a pubertal mammary gland with close-up (*box*). The *dashed line* shows the growing front of the epithelium. (b) FBLN2 and VCAN staining in consecutive sections from a mouse mammary gland during puberty (V6), adulthood (V12), as well as early- (P3) and mid-pregnancy (P12.5) (adapted from [11])

once or twice, rinse them again in running tap water, wash in Scotts Tap Water substitute to increase the pH for 1 min, and rinse them again. The intensity of blue hematoxylin staining can be examined under a light microscope. Dehydrate the slides through increasing concentrations of ethanol (1 min in 30% (v/v), 50% (v/v), 70% (v/v) ethanol each followed by xylene for 40 s) and mount them using a suitable histology mounting medium and cover slips (Fig. 5).

4   Notes

1. For speed of collection and to keep tissues as fresh as possible and minimize RNA degradation, use up to six or seven mice per digest (12–14 inguinal glands). Balb/C mice or C57BL/6 mice should be 16–18 g (approx. 6–7 weeks) to ensure that the TEB have grown beyond the central lymph node, and mice should be well settled in the Biological Services unit if bought in to ensure normal estrus cycles and mammary development.

2. It is very important that the glands are not cut too thoroughly at this stage (less than 30 s), as otherwise, the TEB and ducts themselves will be chopped up and only cell debris will be left after collagenase treatment. It is important to remember that everybody has a slightly different cutting technique and will therefore cut the mammary glands with different efficiencies, so some optimization of the technique may be required. A tissue chopper should not be used as it will mince the tissue too efficiently.

3. It is advisable to optimize the collagenase treatment time necessary for the release of the TEB and ducts from the stromal compartment, according to strain of mouse and experimental requirements, since extended digestion will destroy TEB and ducts. In general, TEB are released twice as quickly as ducts, since they are not as tightly embedded in the surrounding stroma. Although TEBs can be extracted with 15-min incubation, a greater yield is obtained after 20–30 min. For ducts, 30–45 min is necessary. It can therefore be advisable to cut the pubertal mammary gland in half at the lymph node to obtain TEB-enriched and ductal-enriched tissues before further treatment. This allows one to treat the TEB and ductal preparations separately.

4. If the TEB and ducts are to be used for proteomic analysis then no serum must be added to the L15 medium for gland removal and incubation, as serum proteins will interfere with the analysis. For RNA analysis, serum could be added to the medium at this point to assist inhibition of the collagenase.

5. If the TEB are used for RNA or protein isolation, it will not be necessary to work under sterile conditions, though the usual requirements for clean and nuclease-free working apply.

6. It is sometimes difficult to decide whether a segment should be counted as a TEB or a duct, for example if the TEB is attached to a fragment of duct (Fig. 2, image 9). In such cases, with careful manipulation of the micropipette, the fragment can sometimes be broken in two parts to separate the TEB from

the duct. In other cases a large branching duct may have small end buds at the end (Fig. 2, image 10). Such fragments would just be counted as ducts. In general, a small fragment of duct within a large pool of TEB or vice versa will not affect the end analysis.

7. Alternative RNA lysis buffer may be used if different RNA extraction methods are preferred; similarly protein lysis buffer is substituted for protein analysis.

8. If the tip touches the TRI solution it needs to be changed since it becomes sticky and difficult to pick up further structures.

9. The precise number of structures required will depend on their end use. For microarray studies, we collected ~1100–1200 TEB or 700–1000 duct fragments for three replicates; however with amplification kits, fewer numbers could be collected.

10. The lysis buffer may vary according to the use of the protein after extraction, e.g., 1D or 2D electrophoresis.

11. To culture the TEB after isolation it is important that sterile conditions are maintained throughout, although the recommended addition of antibiotics and anti-mycotics in the medium will help to reduce any contamination that may occur during the isolation process. DMEM/F12 is preferential to L15 here because of its enhanced buffering capacity.

12. The organoids float off very easily, so avoid looking at them during this time.

13. Ensure that all work surfaces and equipment, e.g., Gilson's and forceps, are treated with RNase-Zap before use to remove RNases.

14. It is important here to remove as much of the ethanol as possible, but not to over-dry the RNA as it will otherwise be difficult to resuspend the RNA again.

15. Small RNAs (<200–300 nucleotides, including 5S rRNA and tRNAs), which can make up ~20% of the total RNA, can have a negative effect on cDNA synthesis efficiency, where they can act as primers and lead to a production of shortened cDNA fragments. It is therefore strongly advisable to remove these together with any potential genomic DNA contamination.

16. The key for the binding of RNAs to the sepharose is the ethanol concentration in the first step, which should be reduced to ~30%, as small RNAs do not bind efficiently under these conditions.

17. It is advisable to orientate the gland in such a way that the whole length of the gland will be cut.

18. The thickness of the sections may need to be adjusted for the protein of interest. Three to four μm thick sections are adequate for good localization of most proteins without having

several cell layers on top of each other. Ten micrometer sections can be used for 3D reconstruction and confocal microscopy.

19. This step is not necessary when using immunofluorescence, though it should be included if a signal amplification step is necessary, which includes a peroxidase step.

20. The fixation and embedding process can block antibody access for epitopes and antigen retrieval is frequently required. The method of choice is dependent on the antibody and antigen and needs to be optimized, but usually boiling of the samples in citrate buffer pH 6.0 or in EDTA buffer at pH 8.0 or 9.0 is used.

21. The exact timing of the pressure needs to be optimized for each antibody as the accessibility of the epitopes will vary. However, 1–2 min at the highest pressure, which can be heard by a hissing sound, is a good starting point.

22. For IF, tissue sections are rinsed in TBS-Tween 20 pH 7.4 after antigen retrieval and incubated with Image-iT FX Signal Enhancer for 30 min to block the nonspecific background staining that can result from tissue auto-fluorescence.

23. It is always advisable to perform a dilution series to determine the optimal concentration of the primary antibody. It is also important to include a non-primary antibody control and, if possible, a positive control tissue with known protein localization.

24. Sometimes it is necessary to incubate longer or overnight, in which case the antibody needs to be incubated at 4 °C in a humidified atmosphere to avoid evaporation.

25. For IF, incubation with a 1:500–1:1000 dilution of a fluorescent dye-conjugated secondary antibody is performed instead and in the dark for 30 min. When using several antibodies, the secondary antibodies can be incubated at the same time, as long as they do not cross-react with one another. Otherwise, the cross-reacting antibody needs to be incubated first, and washed thoroughly after incubation, and any free binding sites are then blocked by incubation with unlabeled target IgG for 30 min before the next secondary antibody is used. It is important to include single-antibody incubations as control.

26. As DAB solution cannot be washed down the sink it must be collected into a tube after use and disposed of separately.

27. For IF, tissue sections are mounted directly using mounting medium that includes anti-fade reagents with DAPI to counterstain the nuclei. If a mounting medium is used which hardens, then cover slips do not need to be sealed. Otherwise, the cover slips need to be sealed with nail varnish to stop them from drying out.

28. The sections are initially over-stained and the staining procedure stopped with acid alcohol. If the hematoxylin staining is still too strong after incubation in Scott's Tap Water, the staining can be reduced by briefly dipping the slides again into acid alcohol.

Acknowledgements

We would like to thank Dr. Daria Olijnyk for the images of the whole mounts, Rod Ferrier for critical reading of the IHC protocol, Stephanie Stengel for culturing and IF staining of TEBs, and Ayman Ibrahim for photographs of the isolation procedure.

References

1. Silberstein GB (2001) Postnatal mammary gland morphogenesis. Microsc Res Tech 52(2):155–162. doi:10.1002/1097-0029(20010115)52:2<155::AID-JEMT1001>3.0.CO;2-P

2. Propper AY, Howard BA, Veltmaat JM (2013) Prenatal morphogenesis of mammary glands in mouse and rabbit. J Mammary Gland Biol Neoplasia 18(2):93–104. doi:10.1007/s10911-013-9298-0

3. Silberstein GB, Daniel CW (1982) Glycosaminoglycans in the basal lamina and extracellular matrix of the developing mouse mammary duct. Dev Biol 90(1):215–222, doi:0012-1606(82)90228-7 [pii]

4. Master SR, Hartman JL, D'Cruz CM, Moody SE, Keiper EA, Ha SI, Cox JD, Belka GK, Chodosh LA (2002) Functional microarray analysis of mammary organogenesis reveals a developmental role in adaptive thermogenesis. Mol Endocrinol 16(6):1185–1203. doi:10.1210/mend.16.6.0865

5. McBryan J, Howlin J, Kenny PA, Shioda T, Martin F (2007) ERalpha-CITED1 co-regulated genes expressed during pubertal mammary gland development: implications for breast cancer prognosis. Oncogene 26(44):6406–6419. doi:10.1038/sj.onc.1210468

6. Kouros-Mehr H, Werb Z (2006) Candidate regulators of mammary branching morphogenesis identified by genome-wide transcript analysis. Dev Dyn 235(12):3404–3412. doi:10.1002/dvdy.20978

7. Richards J, Guzman R, Konrad M, Yang J, Nandi S (1982) Growth of mouse mammary gland end buds cultured in a collagen gel matrix. Exp Cell Res 141(2):433–443

8. Davies CR, Morris JS, Griffiths MR, Page MJ, Pitt A, Stein T, Gusterson BA (2006) Proteomic analysis of the mouse mammary gland is a powerful tool to identify novel proteins that are differentially expressed during mammary development. Proteomics 6(21):5694–5704. doi:10.1002/pmic.200600202

9. Morris JS, Stein T, Pringle MA, Davies CR, Weber-Hall S, Ferrier RK, Bell AK, Heath VJ, Gusterson BA (2006) Involvement of axonal guidance proteins and their signaling partners in the developing mouse mammary gland. J Cell Physiol 206(1):16–24

10. Emig D, Salomonis N, Baumbach J, Lengauer T, Conklin BR, Albrecht M (2010) AltAnalyze and DomainGraph: analyzing and visualizing exon expression data. Nucleic acids research 38 (Web Server issue):W755-762. doi:10.1093/nar/gkq405

11. Olijnyk D, Ibrahim AM, Ferrier RK, Tsuda T, Chu ML, Gusterson BA, Stein T, Morris JS (2014) Fibulin-2 is involved in early extracellular matrix development of the outgrowing mouse mammary epithelium. Cell Mol Life Sci 71(19):3811–3828. doi:10.1007/s00018-014-1577-4

Chapter 6

RNA Profiling of Non-cultured Fibroblasts Isolated from Pubertal Mouse Mammary Gland Sections

Ayman M. Ibrahim, Claire Cairney, Joanna S. Morris, and Torsten Stein

Abstract

The epithelium of the pubertal mouse mammary gland grows and invades the mammary fat pad to form a primary ductal network. This outgrowth is tightly controlled by epithelial and stromal factors that are present in the environment around the terminal end buds (TEB) at the growth front and the newly formed ducts. Identifying the contribution that each cell type makes to this regulation is a major challenge. To identify the role that fibroblasts play during this process we have optimised a fibroblast isolation procedure, followed by cell cleanup, RNA extraction, and amplification from non-cultured, freshly isolated fibroblasts from around the TEB as well as the subtending ducts. This was facilitated by the use of mice that constitutively expressed EGFP, which allowed the visualization of the growth front of the pubertal mammary tree under UV light. The isolated RNA is of sufficiently high quality, giving reproducible qRT-PCR results, for transcriptome analysis after RNA amplification.

Key words Mammary gland, Puberty, Fibroblasts, RNA amplification, Epithelial-stromal interaction

1 Introduction

Ductal elongation during pubertal mouse mammary gland development occurs when oestrogen stimulation leads to the formation of highly proliferative terminal ends buds (TEB) (for review *see* [1]). These TEB lead the growth from the rudimentary ductal network close to the nipple into the surrounding mammary fat pad, being responsible for laying down a primary network of epithelial ducts through dichotomous branching and lateral side branching. The TEB "migrate"/grow very quickly into the fat pad, elongating the ducts by ~0.5 mm/day [2]. Both epithelial and mesenchymal factors influence this controlled invasive growth [3], and this has been reviewed elsewhere in this book (introductory chapter and the Chapter 3 by McBryan & Howlin). The important role, which macrophages and eosinophils play during this process was identified using knockout mouse models that were deficient in the cytokine CSF1 [4, 5] or the

Finian Martin et al. (eds.), *Mammary Gland Development: Methods and Protocols*, Methods in Molecular Biology, vol. 1501, DOI 10.1007/978-1-4939-6475-8_6, © Springer Science+Business Media New York 2017

chemokine eotaxin [6]. Fibroblasts also play a major role in this epithelial growth control [7]. However, our knowledge about the direct contribution they make during pubertal ductal outgrowth is limited. Fibroblasts can be detected within the neck region of the TEB, interspersed between adipocytes, and along the subtending ducts [8]. Koledova and Lu (Chapter 10 this book) describe a co-culture model in which cultured isolated fibroblasts can be used to study their effect on ductal branching. However, the process of culturing isolated fibroblasts itself has the potential to greatly alter their gene expression profiles. Thus, these may not completely capture the transcriptome of the fibroblasts when associated with the growing and subtending ducts in vivo. We have therefore optimised a fibroblast isolation and cleanup procedure which allows the recovery of RNA from very small numbers of freshly isolated fibroblasts without the need for culturing. Global EGFP-expressing mice (under actin-promoter control) were used to allow identification of the epithelial growth front in the dissected mammary glands. The RNA isolation procedure gives highly reproducible results in qRT-PCR analysis for housekeeping RNAs, and following a linear RNA amplification step, provides enough high-quality RNA for transcriptome analysis.

2   Materials

2.1   Mice

The protocols described below have been used successfully with C57BL/6 mice expressing EGFP under the control of the β-actin promoter/CMV enhancer [9]. For optimal RNA isolation from pubertal ducts at the peak of TEB growth, mice of ~5–6 weeks of age and 16–18 g weight should be used (*see* Chapter 5 by Morris and Stein on TEB isolation).

2.2   Media and Reagents

2.2.1   For Fibroblast Isolation

1. DMEM/F12 serum-free medium: Dulbecco's modified Eagle's medium and Ham's F-12 nutrient mixture medium (1:1).

2. DMEM/F12/5% FCS medium: Dulbecco's modified Eagle's medium and Ham's F-12 nutrient mixture medium supplemented with 2 mM L-glutamine supplemented with 5% v/v heat-inactivated fetal calf serum (FCS), 100 IU/ml penicillin, 100 μg/ml streptomycin.

3. DMEM/F12 10% FCS medium: Dulbecco's modified Eagle's medium and Ham's F-12 nutrient mixture supplemented with 2 mM L-glutamine and 10% v/v heat-inactivated FCS, 100 IU/ml penicillin, 100 μg/ml streptomycin.

4. D-PBS: Dulbecco's phosphate-buffered saline; 1.0% w/v sodium chloride, 0.025% w/v potassium chloride, 0.025% w/v disodium hydrogen orthophosphate, and 0.1437% w/v potassium dihydrogen orthophosphate.

5. Collagenase: *Clostridium histolyticum* collagenase type II solution: stock 10 mg/ml in DMEM/F12 medium.

6. 10× Trypsin stock (2% w/v in DMEM/F12 with 5% FCS.)

7. Trypsin 0.05%/0.5 mM EDTA solution.

8. DNase I.

2.2.2 For Leukocyte Removal

1. Anti-mouse CD45-biotin antibody (rat IgG2b, 0.5 mg/ml).

2. Fc blocking antibody.

3. Biotin selection cocktail.

4. Magnetic nanoparticles.

2.2.3 For RNA Isolation

1. RNA isolation micro-kit (kits from several manufacturers are available).

2.2.4 For RNA Amplification

1. First-strand synthesis kit.

2. Second-strand synthesis kit.

3. SPIA reaction kit.

4. Paramagnetic beads for cDNA purification.

5. PCR purification kit.

6. RNase decontamination solution.

7. Autoclaved DEPC-treated water.

2.3 Tools and Instruments

2.3.1 Mammary Gland Dissection

1. One or more cork dissection boards.

2. 70% Ethanol solution for sterilization of the animal's skin.

3. Dissection needles or 23 G injection needles for securing the animal and the skin flap onto the cork board.

4. Dissection scissors (preferably with rounded end) for opening the skin.

5. Micro dissecting spring scissors to remove the mammary gland.

6. Two pairs of fine blunt-ended forceps (all instruments should be autoclaved or cleaned thoroughly with ethanol).

7. Clean glass slides.

8. Dissecting microscope with UV light source or UV-light goggles.

9. Universal tubes for tissue collection, containing ice-cold DMEM/F12 with 5% serum medium.

2.3.2 For Fibroblast Preparation

1. One tissue culture hood for handling mammary glands and isolating fibroblasts.

2. One 90 mm sterile petri dish for cross-cutting/mincing mammary gland fragments.

3. Two scalpel blades for cross-cutting/mincing mammary gland fragments.

4. One tube with 4-5 ml pre-warmed (37 °C) DMEM/F12 medium with 5 % serum with collagenase type II (final concentration 2.5 mg/ml).

5. A 37 °C shaking incubator set at 130 rpm (or a roller in a hot room (37 °C)) for incubation with collagenase.

6. One refrigerated centrifuge.

2.3.3 For Leukocyte Removal

1. A tissue culture hood for handling cells under aseptic conditions.

2. A suitable magnet.

2.3.4 For RNA Isolation

1. A refrigerated centrifuge.

2. A microfluidic analysis system.

2.3.5 For RNA Amplification

1. A thermocycler.

2. A 96-well magnet.

3. A microcentrifuge.

4. A microfluidic analysis system.

3   Methods

The following protocols describe how to harvest tissue fragments from dissected pubertal mouse mammary glands, how to isolate fibroblasts from the mammary gland fragments and how to remove contaminating leukocytes from the fibroblast isolates. The fibroblast isolation procedure is a modification of the protocol for primary mammary epithelial cell (MEC) isolation developed in Mina Bissell's laboratory. The isolated fibroblasts can either be cultured (as described by Koledova and Lu, Chapter 10) or used directly for micro-level RNA extraction, RNA amplification, cDNA synthesis and subsequent transcriptome analysis. Because of the small number of cells harvested per animal, tissue fragments need to be pooled from several mice.

3.1 Dissection of TEB-Enriched and Ductal-Enriched Pubertal Mammary Tissue Strips from the Fourth Inguinal Mammary Glands

1. Dissect out the mammary glands, following the procedure described by Morris and Stein (Chapter 5), using a mouse strain which allows the in situ identification of the mammary epithelium within the fat pad, e.g., *B6 ACTb-EGFP* mice (*see* **Note 1**).

2. To visualize the TEB growth front, spread the dissected glands out on a clean glass slide and observe under UV light (*see* **Note 2**).

3. Using clean, sterile scalpel blades, cut the regions close to the nipple and before the inguinal lymph node (pre-LN) and the region after the LN extending to just beyond the growth front of the TEB (post-LN) avoiding inclusion of the lymph nodes (*see* **Note 3**).

4. Collect the tissue strips directly into 10 ml of ice-cold DMEM/F12 medium. Separately pool the post-LN and pre-LN strips from the inguinal mammary glands of 5–7 mice (*see* **Note 4**) and keep on ice until processing can be carried out.

3.2 Isolation of Fibroblasts from Mouse Mammary Gland Tissue Strips

1. In a tissue culture cabinet, pour the pooled dissected post-LN or the pre-LN strips into a sterile 90 mm petri dish.

2. Using two new sharp scalpel blades and on ice, chop the sections finely by the crossed scalpels technique (*see* **Note 5**). Using the same scalpels, scrape the minced tissue into a new sterile universal tube and add 9 ml of pre-warmed (37 °C) DMEM/F12 with 5% serum medium containing collagenase type II at 2.5 mg/ml and 0.2% trypsin.

3. Place the tube into a shaking incubator at 37 °C for 30 min with mild agitation (130 rpm) to digest the mammary tissue and free the cells from the surrounding collagenous stroma (*see* **Note 6**).

4. Spin down the homogenate in a tissue culture centrifuge at $250 \times g$ for 10 min.

5. Transfer the supernatant with the fatty layer to a fresh tube and spin down again (*see* **Note 7**).

6. Discard supernatant.

7. Suspend pellets from **steps 4** and **6** in serum-free DMEM/F12 media and combine them into one tube.

8. Add DNase to the cell suspension at 2 units/ml and incubate at room temperature with hand shaking for 5 min (*see* **Note 8**).

9. Spin the suspension down at $250 \times g$ for 10 min.

10. Discard supernatant and thoroughly suspend cells in fresh serum-free DMEM/F12 (4–5 ml or according to pellet size).

11. Perform differential centrifugation by spinning cell suspension in a tissue culture centrifuge and stop the centrifuge immediately when it reaches $250 \times g$ (*see* **Note 9**).

12. Collect the supernatant into a fresh tube and resuspend pellet in 2–3 ml of serum-free DMEM/F12.

13. Repeat this washing step at least five times, and collect supernatant from each wash in the same tube (*see* **Note 10**).

14. Spin the collected supernatants at $250 \times g$ for 10 min.

15. Discard supernatant and suspend cells in DMEM/F12 media with 10% serum.

16. Plate cell suspension in 6-well plates and incubate at 37 °C and 5% CO_2 for 1 h (*see* **Note 11**).

17. After 1 h incubation of stromal cells, aspirate off media with suspended cells.

18. Wash adherent cells using PBS at least three times (*see* **Note 12**).

19. Treat cells with 0.05% trypsin/EDTA for 7 min at 37 °C and 5% CO_2.

20. Collect the released fibroblast/leukocyte cell mixture in DMEM/F12 with 10% serum, and spin down at $250 \times g$ for 10 min.

21. Suspend pellet thoroughly in 200 µl DMEM/F12 with 10% serum. The cells should be used immediately for leukocyte subtraction.

3.3 Removal of CD45-Positive Cells (Leukocytes)

As some leukocytes will also stick to the plastic it is important to remove these prior to analysis of the mammary stromal fibroblasts. This can be achieved using a negative selection with biotinylated-CD45 antibodies and magnetic streptavidin beads as described below. The successful removal of CD45-positive cells should be optimized beforehand and can be followed by FACS analysis (Fig. 1):

1. Add 2 µl (10 µl/ml of cells) Fc-receptor blocking antibody to the fibroblasts, mix well, and incubate for 1 min at RT.

2. Add a CD45-biotin antibody at 2 µg/ml and incubate for 15 min at RT.

3. Add 20 µl of a biotin selection cocktail to the cell/antibody mix and incubate for 15 min at RT (*see* **Note 13**).

4. Add 10 µl dextran-coated magnetic nanoparticles and incubate these for 10 min at RT (*see* **Note 14**).

5. Subsequently, subject the labeled cells to negative magnetic selection for 5 min within a magnetic separation rack, and collect the fibroblast-containing supernatant.

6. Spin the fibroblasts for 10 min at $250 \times g$ and use them directly for RNA extraction.

The procedure can be optimized using isolated leukocytes from mouse blood as positive controls (Fig. 2b), which can be obtained following the protocol below. Use cultured fibroblasts as negative control (Fig. 2a).

1. Collect blood in a tube with EDTA (1.5 mg/ml) to prevent cell clotting.

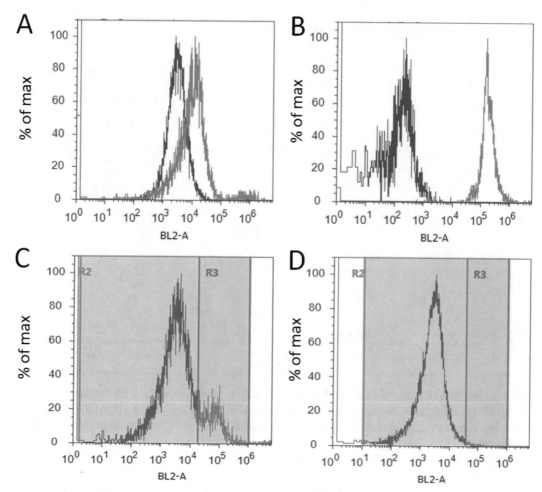

Fig. 1 The percentage of CD45pos cells was measured using a CD45-PE antibody. Cultured fibroblasts (**a**) and CD45-pos white blood cells (**b**) were used as negative and positive controls, respectively. For analysis of the freshly isolated fibroblasts, cells were suspended thoroughly in 500 µl 10 % DMEM/F12 and incubated for 15 min at RT with anti CD45-PE antibody at 2 µg/ml before analysis (**c**); (**d**) shows the FACS analysis after bead treatment

2. For each 1 ml of blood use 14 ml 1× RBCs lysis buffer (0.826 g NH_4Cl, 0.119 g $NaHCO_3$, 20 µl EDTA in 100 ml H_2O (pH 7.3)).

3. Incubate with rocking for 10 min at RT. The mixture will turn clear red after incubation.

4. Spin down at 4 °C for 10 min at 250×g.

5. Discard supernatant and wash cell pellet with ice cold PBS with 2 % fetal calf serum.

6. Spin down at 4 °C for 10 min at 250×g and repeat washing two more times.

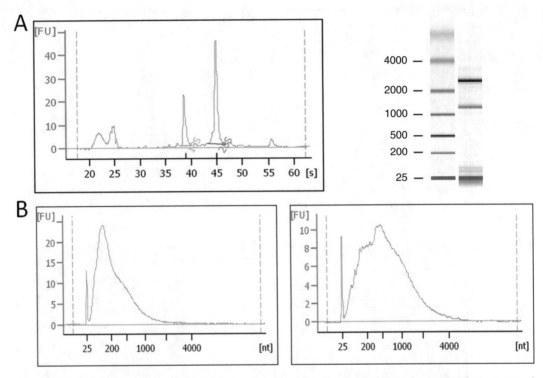

Fig. 2 (**a**) Typical RNA profile of isolated RNA measured with a pico-green kit on a microfluidic system and virtual RNA gel of the same samples; (**b**) cDNA profiles after RNA amplification using a high RNA volume (4 μl; *left*) or low volume (1 μl; *right*) show that a higher volume leads to an over-representation of very small amplification products

7. Put cells on ice and use them directly for anti-CD45 treatment and further FACS analysis.

3.4 RNA Isolation from Post-LN and Pre-LN Fibroblasts

Because of the extremely small sample size (*see* **Note 4**) the use of an optimized commercial RNA isolation micro-kit is strongly advisable. These should be used when working in a range of 10^1–5×10^5 cells or less than 5 mg of tissue.

1. Mix the cells in a suitable cell lysis buffer (*see* **Note 15**).

2. Mix these well and transfer them to a spin column within a collection tube.

3. Spin these for 30 s in a microfuge at top speed and discard the flow-through.

4. Wash the column with 400 μl wash buffer, spin again for 30 s, and discard the flow-through.

5. Perform an on-column DNase treatment at this step. Add 40 μl DNase I reaction mix directly to the column, and incu-

bate at room temperature (RT) for 15 min. Centrifuge for 30 s and discard the flow-through.

6. Wash the column once with 400 μl wash buffer followed by a 700 μl wash as above.

7. Wash with another 400 μl of wash buffer and centrifuge for 2 min at RT.

8. Transfer the dry column to a fresh nuclease-free reaction tube.

9. Add 10–15 μl of nuclease-free water directly onto the column matrix and spin for 30 s at RT to recover the RNA (*see* **Note 16**).

10. Use the RNA immediately or store at –80 °C until further use.

11. Analyse the RNA on a microfluidic system, using a Pico-kit, to measure RNA integrity and concentration (Fig. 2a) (*see* **Note 17**).

3.5 RNA Amplification Using Single-Primer Isothermal Amplification (SPIA) Method

With an average total yield of up to 15 ng of RNA from our pooled cell samples, the resulting RNA yield is not high enough to carry out a microarray hybridization experiment, which requires a minimum of 1.5 μg of cDNA (i.e., ~1.5 μg RNA if the RT reaction was working at 100% efficiency). Therefore, a linear amplification step is required. Several kits are commercially available. Here we describe the single-primer isothermal amplification (SPIA) method [10], which results in excess of 1000-fold amplification, producing single-stranded cDNA suitable for microarray analysis.

This method uses chimeric DNA/RNA primers to produce multiple copies of cDNA in three steps: (1) cDNA synthesis, (2) double-strand synthesis, and (3) cDNA amplification. In the first step, cDNA is produced by a typical reverse-transcriptase reaction, using a mixture of chimeric DNA/RNA random and oligo-dT primers, which produce a cDNA with a unique 5' RNA tag (SPIA tag). In the second step, the RNA of the newly formed cDNA/RNA complex is fragmented, creating a priming site for double-stranded cDNA synthesis, followed by a DNA polymerase reaction. This DNA polymerase also has reverse transcriptase activity so that the RNA section of the DNA/RNA primer becomes reverse-transcribed. The resulting double-stranded cDNA includes DNA which is complementary to the 5' SPIA tag of the chimeric primers, thereby creating a DNA/RNA hetero-duplex at one end. In the amplifying third step, RNase H digests the RNA part of the tag, thereby unmasking the binding site for the SPIA primer. A DNA polymerase with strand displacement activity then synthesizes cDNA from the 3' end of the primer, during which the existing forward strand becomes displaced. New SPIA primer can then bind to the tag region, creating a new substrate for RNase H digestion and cDNA synthesis initiation. This sequence of primer binding, replication, forward strand displacement, and RNA cleavage is repeated again and again, leading to a strong amplification of the original cDNA by more than 1000-fold (Fig. 3). This amplified cDNA can be used for qPCR or array-based gene expression analysis, using a range of microarray platforms:

1. Dilute equal volumes of RNA samples (minimum concentration 500 pg in 1 μl of maximum volume) (*see* **Note 18**) to 5 μl using nuclease-free water in nuclease-free tubes (0.2 ml PCR tubes are advisable).

2. Add 2 μl of first-strand synthesis primer mix (*see* **Note 19**) to diluted RNA samples, mix thoroughly, and spin down briefly in a microfuge.

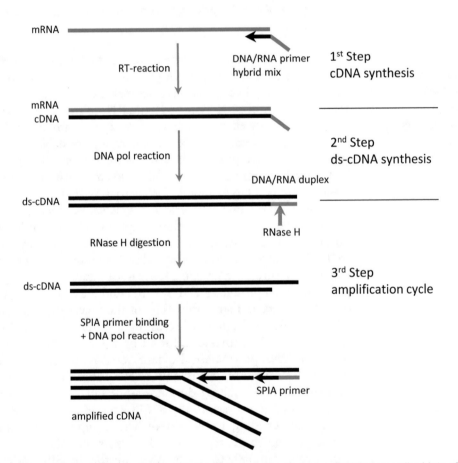

Fig. 3 Diagram of the amplification method in three steps. In the first step, RNA is transcribed into cDNA by reverse transcriptase, using a mix of random and oligo-dT primers, which contain a 5′ SPIA DNA/RNA tag. In the second step, the RNA is fragmented and double-stranded cDNA produced using a DNA polymerase with reverse transcriptase activity, thereby producing a DNA/RNA complex at one end. In step 3, RNase H recognizes this complex, digesting the RNA and unmasking a SPIA primer-binding site. A DNA polymerase displaces the cDNA strand and produces new cDNA. This final step is repeated again and again to amplify the cDNA in a linear fashion

3. Place the tubes in a thermocycler at 65 °C for 5 min.

4. Then, immediately place tubes on ice and add 2.5 μl of first-strand buffer and 0.5 μl of reverse transcriptase to each tube.

5. Place in a precooled thermocycler and run the following program:

   4 °C—2 min.

   25 °C—30 min.

   42 °C—15 min.

   70 °C—15 min.

   Hold at 4 °C.

6. Place tubes on ice and 10 μl of a master mix containing 9.7 μl of second-strand buffer and 0.3 μl of DNA polymerase per reaction to each tube to synthesize the second cDNA strand.

7. Place in a precooled thermocycler and run the following program:

   4 °C—1 min.

   25 °C—10 min.

   50 °C—30 min.

   80 °C—20 min.

   Hold at 4 °C.

8. In a microfuge, briefly spin down any condensate that may have formed in the lid.

9. Prior to the amplification step, purify the double-stranded product from excess primers, by using, e.g., a magnetic bead system as described below (*see* **Note 20**).

   (a) Thoroughly mix 32 μl of well-suspended magnetic beads in binding buffer with each reaction tube by pipetting at least ten times up and down.

   (b) Incubate these at RT for 10 min.

   (c) Place the tubes in a 96-well magnet and let it stand for 5 min at RT to completely clear the solution.

   (d) With the tubes in the magnet, carefully remove only 45 μl of binding buffer from each tube, leaving some buffer behind (*see* **Note 21**).

   (e) For washing, add 200 μl of 70% ethanol in each tube and incubate for 1 min at RT (*see* **Note 22**).

   (f) Repeat washing twice (*see* **Note 23**).

   (g) Finally, remove tubes from magnet and let them dry for 15 min at RT.

10. To amplify the double-stranded cDNA, add 40 μl of master mix containing 20 μl of SPIA buffer, 10 μl SPIA primer, and 10 μl SPIA enzyme mix (RNase H and DNA polymerase), to each tube (*see* **Note 24**).

11. Place tubes in thermocycler and run the following program:

    4 °C—1 min.

    47 °C—75 min.

    95 °C—5 min.

    Hold at 4 °C.

12. Transfer the tubes to the 96-well magnet and let them stand for 5 min to remove the beads from the product.

13. With the tubes in the magnet, remove all of the cleared supernatant that contains the amplified product to fresh nuclease-free tubes (*see* **Note 25**).

14. Purify the amplified product with a column-based PCR product cleaning kit as follows:

    (a) Add five times of nucleic acid binding buffer to each tube, mix thoroughly, and incubate for 1 min at RT.

    (b) Transfer the mixture to a spin column with a collection tube and spin down in a microfuge at top speed for 1 min.

    (c) Add 700 μl of 70% ethanol to each column to wash and spin down at top speed for 1 min.

    (d) Add 15–20 μl of nuclease-free water to each column placed in a 1.5 ml fresh nuclease-free tube incubate for 2 min at RT, and spin down at top speed for 1 min.

15. Test size distribution of cDNA on a microfluidic system (Fig. 4) (*see* **Note 26**).

16. Store the products at –20 °C until further use.

17. Use 1.5–2 μg cDNA for further microarray hybridization experiments (Fig. 5) (*see* **Note 27**).

4   Notes

1. The B6 ACTb-EGFP mice will express EGFP in all cells and therefore it can be difficult to see the epithelium clearly within the bright stromal background. It is therefore important that the TEB have grown past the very bright lymph node. It may be easier to visualize the epithelium using mice which express EGFP or other fluorescent proteins under the control of epithelial markers including CK14; however, the authors have no experience with these.

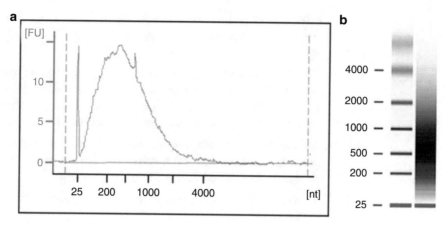

Fig. 4 cDNA profile and virtual RNA gel of amplified RNA before microarray analysis

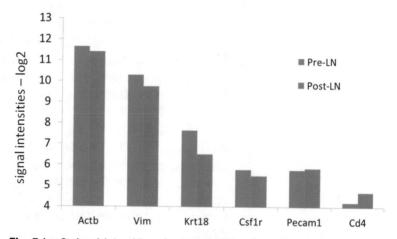

Fig. 5 Log2-signal intensities of selected RNAs after microarray hybridization show the high enrichment of fibroblast RNAs (vimentin) compared to RNAs associated with epithelial cells (*Krt18*), macrophages (*Csf1r*, *Emr1*), or endothelial cells (*Pecam1*). B-Actin (*Actb*) expression levels show the similarity in overall RNA levels

2. Good visualisation is achieved with UV goggles without the use of a dissecting microscope; however, the use of such a microscope is still advised if available.

3. As the fibroblasts are not cultured it is crucial at this point not to include any lymph node tissue, as this could easily contaminate the fibroblast isolate. When cutting the post-LN strip care should be taken that the lymph node at the end of gland, close to the thorax is not included.

4. Five to six pairs of mammary gland tissue strips will provide ~500–600 fibroblasts per isolation procedure. As mammalian

cells contain on average 10–30 pg of RNA per cell one can only expect up to 15–18 ng recoverable RNA per pooled sample.

5. The chilled tissue sections should be processed as quickly as possible in order to reduce the time for possible RNA degradation or cell death. A tissue chopper could be useful though we have used this protocol successfully without it.

6. Although other collagenases can be used, collagenase II has given us the best survival results. The length of the treatment should be optimised and is for guidance only.

7. This step is for making sure there are no cells lost when trapped in the fatty layer.

8. The DNase treatment is necessary to break up any potential cell clumps as the genomic DNA from damaged cells will stick the cells together.

9. This step separates the relatively dense epithelial cells and organoids from the stromal cells (containing fibroblasts) and this process can be assigned as a washing step.

10. For collecting epithelial cells as controls, the pellet can be plated in medium favouring epithelial growth (5 μg/ml EGF, 50 μg/ml gentamycin, 5 ml of FBS, Pen/Strep (50 U/ml penicillin; 50 μg/ml streptomycin) in DMEM/F12 media with FUNGIZONE at 1 μg/ml).

11. During this time fibroblasts will bind stronger to the plastic compared to epithelial cells or adipocytes. Some leukocytes may also bind, but the number of these can be reduced by properly cleaning the glands in medium at the time of collection, reducing the amount of blood, and by avoiding lymph tissue.

12. Adherent cells (fibroblasts) will still be rounded and not stretching out at this stage.

13. This cocktail includes bispecific tetrameric antibody complexes that recognize the dextran and the biotin of the biotinylated antibody.

14. These beads need to be well suspended before use.

15. Some RNA isolation kits contain ~50% ethanol in the cell lysis buffer. We strongly advise that this is reduced to 30% to prevent small RNAs, including miRNAs and tRNAs, from binding to the column as these are present in abundance and are likely to interfere with the RNA amplification and cDNA synthesis.

16. It is advisable to use as small an elution volume as possible and to apply this twice to the column for maximum yield and the highest RNA concentration.

17. As the concentration of the RNA will be very low, a spectrophotometer cannot be used for measuring the concentration

accurately. A microfluidic system allows the concentration measurement, using a pico-kit, while at the same time testing the RNA quality. The amount of small RNAs, including tRNAs and miRNAs, should be kept to a minimum, while the 26S/16SrRNA ration should be ~2.

18. Low volumes (1 µl) of RNA solution are advised as our experience has shown that higher volumes can strongly affect the size distribution of the amplified products, possibly due to their salt content. Equal volumes among samples will ensure the consistency of their product size distribution (Fig. 4b).

19. The first strand primer mix is a DNA/RNA (chimeric) primer mix that hybridizes uniformly across the input RNA (Fig. 3).

20. These are paramagnetic beads in an optimized binding buffer, which selectively bind to cDNA molecules in the sample mixture leaving oligonucleotides, nucleotides, salts, and enzymes to be washed off and discarded when placed in the suitable magnet. However, other PCR product clean-up kits are also suitable.

21. Leaving some buffer behind will reduce the danger of losing beads during this procedure. Removal of beads during the binding procedure or the subsequent washes can greatly affect cDNA yields and must therefore be avoided.

22. The beads should remain on the sides of the tube and should not be dispersed, as this could lead to significant bead loss.

23. Remove as much ethanol as possible, using two or more pipetting steps. Allow any excess ethanol to first collect at the bottom of the tubes before removing it by subsequent pipetting.

24. The SPIA reaction mixture contains RNase which will degrade the original RNA added in the first-strand synthesis step, DNA polymerase to initiate replication at the 3′ end of the primer and to displace the existing forward strand. The RNA sequence at the 5′ end of the newly synthesized strand is again removed by RNase H to expose the priming sequence for the next round of SPIA primer attachment.

25. As even small amounts of bead carryover could interfere with the quantification of the sample, it must be removed carefully without disturbing the beads.

26. The distribution pattern of the cDNA should show a round bell-shaped curve with a shallow peak of at least ~500 nt as shown in Fig. 2b (right panel). From our experience, distribution patterns as those shown in Fig. 2b (left panel) will give poor reproducibility in qPCR experiments and very high Ct values and high standard variation.

27. When using Illumina bead gene expression arrays the hybridization temperature needs to be reduced from 58 to 48 °C.

Acknowledgements

AI was funded by a grant from the Egyptian Ministry for Higher Education. CC and TS were funded by a project grant from Breast Cancer Now.

References

1. Howlin J, McBryan J, Martin F (2006) Pubertal mammary gland development: insights from mouse models. J Mammary Gland Biol Neoplasia 11(3-4):283–297. doi:10.1007/s10911-006-9024-2

2. Hinck L, Silberstein GB (2005) Key stages in mammary gland development: the mammary end bud as a motile organ. Breast Cancer Res 7(6):245–251. doi:10.1186/bcr1331

3. Parmar H, Cunha GR (2004) Epithelial-stromal interactions in the mouse and human mammary gland in vivo. Endocr Relat cancer 11(3):437–458

4. Van Nguyen A, Pollard JW (2002) Colony stimulating factor-1 is required to recruit macrophages into the mammary gland to facilitate mammary ductal outgrowth. Dev Biol 247(1):11–25. doi:10.1006/dbio.2002.0669

5. Pollard JW, Hennighausen L (1994) Colony stimulating factor 1 is required for mammary gland development during pregnancy. Proc Natl Acad Sci U S A 91(20):9312–9316

6. Gouon-Evans V, Pollard JW (2001) Eotaxin is required for eosinophil homing into the stroma of the pubertal and cycling uterus. Endocrinology 142(10):4515–4521. doi:10.1210/endo.142.10.8459

7. Haslam SZ (1986) Mammary fibroblast influence on normal mouse mammary epithelial cell responses to estrogen in vitro. Cancer Res 46(1):310–316

8. Richert MM, Schwertfeger KL, Ryder JW, Anderson SM (2000) An atlas of mouse mammary gland development. J Mammary Gland Biol Neoplasia 5(2):227–241

9. Okabe M, Ikawa M, Kominami K, Nakanishi T, Nishimune Y (1997) 'Green mice' as a source of ubiquitous green cells. FEBS Lett 407(3):313–319, doi:S0014-5793(97)00313-X [pii]

10. Kurn N, Chen P, Heath JD, Kopf-Sill A, Stephens KM, Wang S (2005) Novel isothermal, linear nucleic acid amplification systems for highly multiplexed applications. Clin Chem 51(10): 1973–1981.doi:10.1373/clinchem.2005.053694

Chapter 7

Analysis of the Involuting Mouse Mammary Gland: An In Vivo Model for Cell Death

Bethan Lloyd-Lewis, Timothy J. Sargeant, Peter A. Kreuzaler, Henrike K. Resemann, Sara Pensa, and Christine J. Watson

Abstract

Involution of the mammary gland occurs at the end of every period of lactation and is an essential process to return the gland to a pre-pregnant state in readiness for the next pregnancy. Involution is a complex process of regulated alveolar cell death coupled with tissue remodeling and requires exquisite control of transcription and signaling. These processes can be investigated using a variety of molecular and morphological approaches.

In this chapter we describe how to initiate involution and collect mammary glands, measure involution morphologically, and quantify lysosomal leakiness in mammary tissue and in cultured mammary epithelial cells. These procedures encompass a range of microscopy and molecular biology techniques.

Key words Involution, Cell death, Electron microscopy, Lysosome, Caspase 3, Immunofluorescence, Mammary gland

1 Introduction

The primary function of the mammary gland is to produce milk. This life-giving substance is produced by alveolar epithelial cells, a specific lineage which arises during pregnancy from stem cells/progenitors in the ducts. Although the hierarchy of alveolar lineage commitment has not been clearly defined, it has been shown that there are at least two different types of luminal alveolar cells: those that are ERα/PR/Gata3 expressing and those that are Stat5/Elf5 expressing [1–4]. At the cessation of lactation, alveolar cells become redundant and most undergo cell death concomitant with tissue remodeling and re-appearance of adipocytes [5]. The involuting mammary gland is one of the most spectacular examples of cell death in a physiological context and involution has been used to investigate the mechanisms of programmed cell death. Recent work from our laboratory has shown that the first wave of cell death, which is characterised by

Finian Martin et al. (eds.), *Mammary Gland Development: Methods and Protocols*, Methods in Molecular Biology, vol. 1501, DOI 10.1007/978-1-4939-6475-8_7, © Springer Science+Business Media New York 2017

detachment and shedding of dying cells into the alveolar lumen and occurs within 48 h of synchronous (forced) weaning, is not apoptosis but is mediated by leakage of cathepsins from lysosomes [6, 7] a process named lysosomal-mediated programmed cell death (LM-PCD). These cysteine proteases cleave cellular components and bring about the demise of the cell. A second wave of cell death occurs after 48 h when involution becomes irreversible. The mechanism(s) of cell death in this phase has not been defined but probably involves classical apoptosis and is associated with cell death in situ, i.e., cells die without detaching from the alveolar wall and are TUNEL positive. In this chapter, we will describe the various methods that can be used to measure the progress of involution including activation of the LM-PCD pathway both in mammary gland tissues and in the mammary epithelial cell line EpH4 which we have shown to be a good model for Stat3-induced LM-PCD [6, 7]. These protocols include: (1) measuring the number of dead/shed cells; (2) evaluating the extent of remodeling; and (3) quantifying lysosomal leakiness.

Many factors affect the initiation and kinetics of involution and the remodeling process and we suggest that several methods are utilised in order to rigorously determine if a particular genetic modification or treatment affects the involution process.

2   Materials

2.1   Forced Involution and Mammary Gland Harvest

1. 4% Paraformaldehyde.
2. Modified RIPA buffer (50 mM Tris, 1% NP40, 0.25% sodium deoxycholate, 150 mM NaCl, 1 mM EGTA, at pH = 7.4).
3. Complete protease inhibitor.
4. $NaVO_3$.
5. NaF.
6. Pefabloc.
7. TRIZOL.
8. Liquid N_2 in a thermos flask.

2.2   Measuring Involution Morphologically

2.2.1   Visualizing and Quantifying Dead Cells via Cleaved Caspase 3/E-Cadherin Staining

1. Xylene.
2. Ethanol series (100, 90, 70, 50, 30%).
3. Phosphate-buffered saline (PBS).
4. Sodium citrate.
5. PAP pen or Vaseline.
6. Goat serum.
7. Fluorescent secondary antibodies.

8. Cleaved caspase 3 antibody.

9. E-cadherin antibody.

10. Hoechst dye.

11. Glycerol.

2.2.2 Measurement of the Area Occupied by Adipocytes to Quantify the Extent of Remodeling

1. ImageJ software (imagej.nih.gov).

2. Photoshop.

3. Perilipin antibody.

2.2.3 Perfusion Fixation for Cryosectioning, Paraffin Embedding, or Transmission Electron Microscopy

1. Fume hood.

2. Eye protection, gloves, labcoat.

3. 50 ml Syringes.

4. 1 ml Syringe.

5. Low-gauge needles (approx 18 G).

6. High-gauge needles (27 G).

7. Butterfly needles (25G).

8. Sodium pentobarbital (Euthatal).

9. Wash solution.

10. Fixative:

   4-6% Formaldehyde in PBS (used for standard immunohisto-chemical staining).

   3% Glutaraldehyde/1% formaldehyde in 0.1 M PIPES pH 7.4 (used for TEM).

11. Cork board with pins.

12. Large collection tray that fits the cork board.

13. Scissors.

14. Forceps.

15. 70% Ethanol in spray bottle.

16. Bijou tubes (or similar sample containers).

2.3 Measuring Lysosomal Leakiness in Mammary Tissue

2.3.1 Cytosolic Cathepsin Activity Assays

1. Fractionation buffer: Hepes-KOH pH 7.5 20 mM, sucrose 250 mM, KCl 10 mM, $MgCl_2 \times 6H_2O$ 1.5 mM, sodium EDTA 1 mM, sodium EGTA 1 mM, dithiothreitol 8 mM (DTT, add fresh), Pefabloc 1 mM (add fresh) (*see* **Note 1**).

2. Reaction buffer (sodium acetate 50 mM, EDTA 8 mM, DTT 8 mM, Pefabloc 1 mM).

3. Cathepsin substrate: zFR-AMC (final concentration 50 μM, stock in DMSO 5 mM).

4. Cathepsin B inhibitor: Ca-074 (final concentration 5 μM, stock in DMSO 0.5 mM).

5. 5 ml Tight-fitting handheld tissue homogenizer.

6. Scalpel blade.

7. Petri dishes.

8. 96-Well plate.

9. Ultracentrifuge.

10. Pre-chilled fixed-angle rotor (e.g., Type 50 Ti).

11. Ultracentrifugation tubes.

12. Fluorophotometer (able to read at excitation 380 nM, emission 442 nM).

13. BCA assay for protein concentration.

14. Liquid N_2 in a thermos flask.

2.3.2 Lysosome Leakiness Assays

1. Fractionation buffer (Hepes-KOH pH 7.5 20 mM, sucrose 250 mM, KCl 10 mM, $MgCl_2 \times 6H_2O$ 1.5 mM, sodium EDTA 1 mM, sodium EGTA 1 mM, dithiothreitol 8 mM (DTT, add fresh), complete protease inhibitor (Roche, add fresh according to the manufacturer's instructions)) (*see* **Note 1**).

2. 2. RIPA buffer (Tris–HCl 50 mM pH 7.4, NP40 1%, sodium deoxycholate 0.25%, NaCl 150 mM, EGTA 1 mM, glycerol 1%, complete protease inhibitor).

3. Centrifuge.

4. Cathepsin B antibody.

5. Cathepsin L antibody.

6. Lamp2 antibody.

7. Thermomixer.

2.3.3 Immuno-fluorescence for LAMP1/2 and Cathepsin B/D: Mander's Coefficient

1. Cathepsin B antibody.

2. Cathepsin L antibody.

3. Lamp2 antibody.

4. All standard materials for immunofluorescence (see above).

5. ImageJ with Mander's coefficient: plug-in.

2.4 Measuring Lysosomal Leakiness in an In Vitro Model: EpH4 Cell Line

1. EpH4 cells.

2. Dulbecco's modified Eagle medium (DMEM).

3. Fetal calf serum (FCS).

4. Trypsin-EDTA.

5. PBS.

6. Oncostatin M (OSM): Made up in PBS/0.1% BSA.

7. 6-Well tissue culture plates.

8. 15 ml Centrifuge tubes.

9. Hemocytometer counting chamber.

10. 1.5 ml Centrifuge tubes.

11. Digitonin.

12. 0.1 % Triton X-100 in PBS.

13. Fractionation buffer (HEPES-KOH 20 mM, sucrose 250 mM, KCl 10 mM, MgCl2 1.5 mM, EDTA 1 mM, EGTA 1 mM, dithiothreitol (DTT) 8 mM, Pefabloc 1 mM, at pH 7.5).

14. Cathepsin reaction buffer (sodium acetate 50 mM, EDTA 8 mM, dithiothreitol 8 mM, and Pefabloc 1 mM, at pH 6).

15. Synthetic cathepsin substrate Z-Phe-Arg-AMC.

16. Clear 96-well plates.

17. Fluorescent plate reader—read at excitation wavelength 380 nm, emission wavelength 442 nm.

18. Lysotracker Red DND-99.

3   Methods

3.1   Forced Involution and Mammary Gland Harvest

Females to be used for mammary gland involution studies should be virgins and ideally at least 8 weeks of age at the time of mating. Males should be removed before the birth of the pups to avoid a second pregnancy, which could affect the involution process. In order to maintain consistency we routinely normalise the number of pups per dam to 8 wherever possible by cross-fostering offspring if necessary, ideally around 3–4 days after birth (*see* **Note 2**). This is to avoid imbalanced suckling and variation between the different glands. Lactation is allowed to continue for 10 days to reach the peak of lactation and a synchronous involution induced by removal of all the pups from the dam. This should take place at the same time of day for all animals in an experimental cohort to avoid confounding effects of circadian rhythms. Involution can also be induced by sealing of the teats with veterinary glue. This latter procedure demonstrated that involution is initiated by local factors [8] and not by a reduction in circulating prolactin.

For a full characterization of the involution phenotype, glands should be harvested at 10 days of lactation, as well as 12, 24, 48, 72, 96, and 144 h of involution. If a delay in involution is observed and is particularly extensive, later time points such as 10 and 15 days may be informative in addition to a full wean at 21 days.

Due to potential right-left differences in upper/thoracic mammary glands, it is recommended to use numbers 4 and 5 (abdominal)

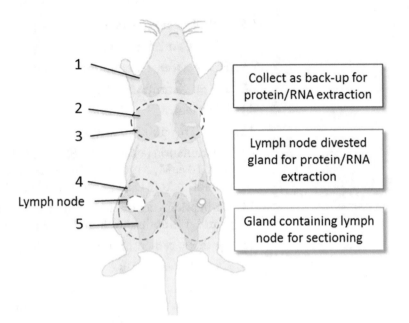

Fig. 1 Schematic of mouse mammary gland localization and tissue collection strategy

glands for all studies (*see* Fig. 1). For protein and RNA extraction, the lymph nodes should be removed from the glands prior to snap-freezing in liquid N_2 in a thermos flask. If the lymph node is not clearly visible, a larger area around the site of localisation of the lymph node should be removed. For sectioning, whole number 4 glands should be fixed in 4% paraformaldehyde or formalin overnight prior to paraffin embedding. Tissue may also be frozen, or fixed by perfusion under terminal anesthesia. For analysis of protein and RNA, numbers 2 and 3 glands are removed and snap frozen in liquid N_2.

3.2 Measuring Involution Morphologically

Involution can be assessed at a gross morphological level by examination of H&E sections of mammary tissue and immunofluorescence studies. The numbers of cells detached and shed into the alveolar lumen indicates the extent of cell death. It should be noted, however, that any changes in the efficiency or extent of phagocytosis of dead cells by the viable epithelium will affect the number of cells in the lumen [9] and could be misleading. After 48 h, when involution switches to an irreversible process, the gland begins to undergo remodeling which can be observed by the reappearance of adipocytes. This is a useful surrogate measure for the extent of cell death in the first phase of involution as abrogation of cell death results in retention of expanded alveoli and delayed reappearance of adipocytes.

We routinely carry out the following three procedures when assessing involution (*see* **Note 3**):

(a) Counting number of shed cells in the lumen

(b) Visualizing dead cells via cleaved caspase 3/E-cadherin staining

(c) Measuring the area occupied by adipocytes, and perilipin immunofluorescence microscopy.

A more detailed ultrastructural analysis can be carried out by transmission electron microscopy (TEM) and this also allows scoring of autophagy, uptake of milk fat globules (MFG), and other features such as mitochondrial structure. Cells from lactating glands are dramatically different to involuting glands as seen by TEM [7].

3.2.1 Counting Number of Shed Cells in the Lumen

Paraffin-embedded mammary tissue, collected as in protocol 1, is cut into 5–10 μm sections on glass slides and stained with hematoxylin and eosin (H&E).

H&E-stained slides are viewed with a light microscope and ten fields selected at random and photographed digitally. Fields are scored by counting nuclei of the alveolar cells and counting shed cells (which can be marked electronically for ease of counting) and dead/shed cells expressed as a percentage of total nuclei counted.

3.2.2 Visualizing and Quantifying Dead Cells via Cleaved Caspase 3/E-Cadherin Staining

Although executioner caspases are dispensable for cell death in the first 48 h of involution, once cells have been detached and shed into the alveolar lumen, they undergo a process (probably anoikis) that results in cleavage of caspase 3. Since this does not occur until cells are shed, immunofluorescence analysis for cleaved caspase 3 is a very useful marker for the number of shed cells. When coupled with staining for E-cadherin, to outline the viable luminal alveolar cells, this approach provides a relatively easy and accurate method to assess cell death (*see* Fig. 2).

1. Deparaffinize slides in xylene and rehydrate in ethanol series:

   (a) 3 × 5 min Xylene washes.

   (b) 1 × 5 min Washes in:
   - 100% Ethanol.
   - 90% Ethanol.
   - 70% Ethanol.
   - 50% Ethanol.
   - 30% Ethanol.
   - dH$_2$O.
   - PBS.

2. Antigen retrieval for optimal antibody staining:

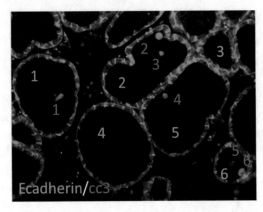

Number of cc3 +ve cells	6
Number of alveoli	6
Dead cells/alveolus	1

Fig. 2 Cleaved caspase 3/E-cadherin staining of involuting mammary gland to quantify cleaved caspase 3-positive cells/alveolus

(a) Heat in a pressure cooker under pressure in 1 l 0.01 M sodium citrate (pH = 6.0) for 11 min.

(b) Release pressure and allow slides to cool down slowly for at least 15 min (standing in cold water in the sink).

(c) Wash 1 × 5 min in ddH$_2$O.

(d) Wash 1 × 5 min in PBS.

3. Blocking to avoid unspecific antibody binding:

(a) Dry slides and circle samples with PAP pen or Vaseline.

(b) Add 10 % goat serum in PBS to each sample and leave to block in humid chamber (e.g., old tip box filled with wet tissues) for 90 min at room temperature.

4. Primary antibody incubation:

(a) Tip off blocking buffer.

(b) Replace with primary antibody diluted in 5 % goat serum in PBS:
    Rabbit-anti-cleaved caspase 3 (to stain dead cells).
    Mouse-anti-E-cadherin (to visualize alveoli).

(c) Leave at 4 °C overnight in a humid chamber.

5. Secondary antibody incubation:

(a) Tip off primary antibody.

(b) Wash 3 × 5 min in PBS.

(c) Add secondary antibody diluted 1:500 in 5 % goat serum in PBS, e.g.,
    Goat-anti-rabbit-594.
    Goat-anti-mouse-488.

(d) Incubate for 1 h in humid chamber at room temperature in the dark.

(e) Wash 2×5 min in PBS.

(f) Apply Hoechst dye (diluted in PBS) to each sample for nuclear staining.

(g) Incubate for 5 min at room temperature in the dark.

(h) Wash for 5 min in PBS.

(i) Apply 50:50 PBS:glycerol to each section and mount with a cover slip.

(j) Nail varnish around cover slip to avoid the sample drying out.

(k) Store dried slides at 4 °C away from light until viewing with a fluorescence microscope.

19. Quantification of cleaved caspase 3-positive cells per alveolus:

(a) Acquire several random images (at least ten) of multiple alveoli for each sample to use for quantification of cleaved caspase 3 positive cells per alveolus.

(b) In each picture, count the number of alveoli visible in their entirety and the number of cleaved caspase 3 cells shed in each alveolus.

(c) Calculate the average number of cleaved caspase 3-positive cells per alveolus for each sample (for example *see* Fig. 2).

3.2.3 Measurement of the Area Occupied by Adipocytes to Quantify the Extent of Remodeling

As alveolar cells die and alveoli collapse, the adipocytes in the gland re-differentiate and fill with fat. This can be measured morphologically and immunofluorescently, using an antibody for perilipin (also known as lipid droplet-associated protein) which is a marker for adipocytes.

1. Merging images for an entire mammary gland H&E sections.

(a) Multiple images covering the entirety of an H&E section should be taken with a light microscope at low magnification.

(b) To merge the images for a reconstruction of the section, open the images in Photoshop™ utilizing "File" > "Automate" > "Photomerge."

2. Quantification of the area occupied by adipocytes on H&E sections:

(a) In order to quantify adipocyte area, the NIH ImageJ software is used on the image obtained by merging H&E sections to reconstruct a full gland.

(b) Areas of the gland populated by adipocytes are outlined using the "polygon" tool and quantified using the "measure" function.

(c) The lymph node area is excluded.

(d) The total area occupied by adipocytes is expressed as a percentage of the total area of the reconstructed gland.

3. Perilipin staining to measure area occupied by adipocytes:

(a) Immunofluorescence staining of sections from paraffin-embedded tissues is performed as described above in Subheading 3.2.2.

(b) For the perilipin staining, a permeabilization step needs to be performed before antigen retrieval (between **step 1** and **2**) by incubating slides in PBS-Triton X-100 0.5 % for 5 min, followed by a wash with PBS for 5 min and a wash with H_2O for 5 min.

(c) Blocking is performed as described above. Perilipin antibody is diluted 1:50 in PBS 2 % NGS 0.3 % Triton X-100.

(d) Measurement of the area occupied by adipocytes can be performed on perilipin-stained sections with the method described above for H&E sections. Additional staining with an E-cadherin antibody facilitates this analysis. An example is shown in Fig. 3.

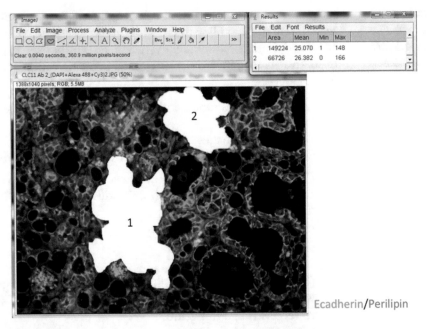

Fig. 3 Quantification of adipocyte area in ImageJ™ using E-cadherin/perilipin-stained involuting mammary gland sections

If a very high standard of fixation is required for immunohistological techniques or for transmission electron microscopy (TEM) of mammary gland tissue, perfusion-fixation may be used. During perfusion-fixation, aldehyde fixatives are delivered to tissues through the circulatory system via the heart. This ensures even and rapid fixation of the whole mammary gland when compared to immersion-fixation, which relies on diffusion for delivery of fixative and inherently fixes the tissue unevenly (*see* **Note 4**).

This method is routinely used on animals at the peak of lactation (10 days) or during mammary gland involution. Steps that involve the use of fixative or materials soaked in fixative should be performed in a fume hood as aldehydes are toxic. Eye protection is especially important (*see* **Note 5**).

1. Prepare by filling one syringe with 50 ml wash solution and one with 50 ml of your chosen fixative. Attach a butterfly needle to the 50 ml syringe containing wash solution and flush the line taking care to remove bubbles. Bubbles will block capillary beds and will impede fixation.

2. Fill the 1 ml syringe with 0.2 ml of sodium pentobarbital using the low-gauge needle. Swap the low-gauge needle for a new 27 G needle that will be used to pierce the abdomen of the mouse.

3. Restrain the mouse and administer sodium pentobarbital (0.2 ml Euthatal) by intraperitoneal injection. This is sufficient for terminal anaesthesia.

4. Wait for the mouse to lose consciousness. This typically takes 3–5 min and can be tested by pinching one toe of the lower paws. Do not proceed if the mouse still reacts to pinching by retracting its leg.

5. Pin the mouse to a cork board by its limbs and facing up. Make sure that the cork board has been placed in a tray to collect fluid and fixative.

6. Spray the mouse with ethanol to wet the fur. Cut along the midline to open the abdominal and thoracic cavities, taking particular care not to damage any of the organs, especially the heart and major blood vessels when cutting the rib cage. Make sure that you have cut away enough of the ribcage and sternum to allow easy manipulation of the heart.

7. Make one cut in the right atrium with scissors. This allows blood and fixative to exit the circulation as it returns to the heart via the vena cava.

8. Pierce the left ventricle with the butterfly needle that is connected to the syringe containing wash solution. Push the needle about 5 mm into the left ventricle—it is important that the needle remains in the left ventricle and does not damage the septum or break through to the left atrium.

9. Apply pressure to the syringe to wash the blood out of the mouse's circulatory system. The liver should have now turned from a deep red to a light brown colour. Fluid coming out of the nose indicates incorrect placement of the needle or too much pressure.

10. Taking care not to introduce bubbles into the butterfly syringe, swap the wash solution containing syringe for the fixative containing one and continue perfusing the mouse with fixative. During the initial phases of fixation, you will see the mouse twitch and contract. Perfuse with approximately 100 ml of fixative.

11. If the perfuse-fixation has been successful, the mouse's limbs, tail, and head should be stiff.

12. Separate the abdominal skin from the abdominal muscle and pin the skin back to the cork board to reveal the perfuse-fixed mammary glands. Remove the mammary glands and place this tissue into fresh fixative at 4 °C to allow post-fixation for 4 h.

13. After post-fixation, wash the mammary gland three times in PBS for immunohistochemistry or 0.1 M PIPES for TEM. Tissues can be stored in buffer at 4 °C until processing for cryosectioning, paraffin embedding, or processing for TEM.

Using this fixation technique we have achieved good preservation of immunohistochemical markers in both cryosectioned and paraffin-embedded tissue. We have also obtained good ultrastructural detail in TEM analysis such as clear observation of mammary epithelial apical cilia and mitochondrial matrix [7].

3.3 Measuring Lysosomal Leakiness in Mammary Tissue

Mammary gland involution is marked by a widespread loss of integrity of the lysosomal compartment, leading to a release of lysosomal proteases, and possibly other lysosomal components, into the cytosol [6]. Lysosomal membrane permeabilization (LMP) can be detected within hours of pup removal from the lactating dam and is the main driver leading to cellular demise, as inhibition of lysosomal proteases substantially delays mammary gland involution [6]. An interesting observation, which is crucial for the methods to follow, is that lysosomal leakage is widespread throughout the mammary gland, while cell death remains stochastic. Furthermore, LMP appears to be reversible, or at least it can be contained, as it already happens during the first, reversible phase of mammary involution during which the gland can reinitiate lactation if pups are returned.

We will discuss a number of methods to assess LMP during involution. Individually these methods have strong points and weak points, but when combined will give a clear picture about the extent and localization of LMP (*see* **Note 6**).

All of the methods are carried out with freshly dissected and lymph node-divested number 4 mammary glands.

3.3.1 Cytosolic Cathepsin Activity Assays

In this approach the mammary gland is dissected, homogenized, and fractionated by differential centrifugation. The fractions that are retained for further analysis are a crude organelle fraction containing mitochondria and lysosomes, a cytosolic fraction, and the input as a positive control. Other fractions can in theory be added, but will not be discussed in this protocol. The cathepsin B and L activity will subsequently be measured in a fluorimetric kinetic assay based on the fluorescent molecule AMC, which is released from the synthetic cathepsin substrate Z-Phe-Arg-AMC by lysosomal cysteine cathepsins to generate fluorescence [10]. As there are no cathepsin-specific individual substrates the individual activities are dissected by adding a specific cathepsin B inhibitor. The caveat for this assay is that cytosolic cathepsin inhibitors can potentially interfere with the activity measurements.

1. Collect lymph node-divested number 4 mammary gland in ice-cold fractionation buffer.

2. Cut it into very small pieces using the scalpel blade in a Petri dish.

3. Place into homogenizer with 1 ml of fractionation buffer.

4. Homogenize on ice in the hand held homogenizer (if a powered homogenizer is available, it can be used instead, but the conditions will need to be adjusted). Move the pestle up and down while twisting until the whole tissue is homogenized. After that, continue for another 10–15 strokes (*see* **Note 7**).

5. Transfer the homogenized mammary gland to a centrifugation tube.

6. Centrifuge for 10 min at $750 \times g$ in a fixed angle rotor (e.g., Type 50 Ti) at 4 °C to pellet the unlysed cells, nuclei, and debris.

7. Collect 150 μl of total lysate and snap freeze. This is the control.

8. Collect the remaining supernatant, transfer to a second tube, and spin at $10,000 \times g$ at 4 °C in the same rotor as before. This pellet contains the crude organelle fraction (mitochondria/lysosomes).

9. Resuspend the crude organelle fraction in 300 μl ice-cold fractionation buffer and snap freeze in liquid N_2.

10. The supernatant is then spun for 1 h at $100,000 \times g$ at 4 °C in the same rotor as before, to pellet microns. The supernatant from this fraction is the cytosolic fraction. Snap freeze it in liquid N_2. Take care to avoid the fat floating on top of the solution; do not collect it.

The protocol can be stopped here for an indefinite amount of time.

11. Freeze and thaw all fractions 2–3 times prior to BCA and activity assays.

12. Perform a BCA test on all fractions to measure protein content.

13. Prepare the samples in fractionation buffer at a concentration of 4 μg/10 μl.

14. Dispense 10 μl into the wells of a 96-well plate. Prepare two triplicates for each fraction.

15. Add 130 μl of reaction buffer to the extracts.

16. To one set of triplicates, add 2 μl of DMSO in 28 μl of buffer.

17. To the other set of triplicates add 2 μl of Ca-074 (cathepsin B inhibitor, final concentration 5 μM) in 28 μl of buffer.

18. Wait 5–10 min for the inhibitor to bind to its target, then add 2 μl of zFR-AMC (substrate) in 28 μl of buffer (total volume is now 200 μl/well).

19. Incubate for 15 min at 37 °C in the dark.

20. Measure at excitation 380 nm, and emission 442 nm. If the Fluorophotometer allows it, kinetics can be recorded.

21. The wells with Ca-074 represent the cathepsin L activity. The wells without inhibitor represent the combined B and L activity. Subtracting the cathepsin L activity from the total activity yields the cathepsin B activity.

3.3.2 Lysosome Leakiness Assays

As mentioned above, lysosomes fundamentally change their properties during mammary gland involution, becoming increasingly leaky. This leakiness can be assayed for in vitro. This is a very powerful method to assess qualitative differences between the lysosomes. The caveat here is that while the difference between a population of non-leaky lysosomes (e.g. during lactation) and one of leaky lysosomes (e.g. involution) becomes very clear, a quantitative assessment of leakiness is difficult. This is mainly due to the lengthy tissue preparation and heterogeneity of primary material. An overview of the procedure is shown in Fig. 4.

1. Homogenize the gland as previously described.

2. Spin at 3500 rpm ($=750 \times g$) for 10 min to pellet nuclei and unlysed cells.

3. Collect the supernatant. Ensure that the pellet is not disturbed as it is better to lose a bit of sample rather than contaminating the supernatant.

4. Spin at $10,000 \times g$ at 4 °C for 35 min to pellet the organelles.

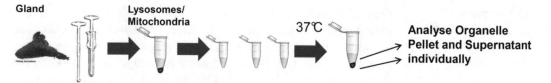

Fig. 4 Overview of the leakiness assay

    5. Resuspend in 400 μl fractionation buffer.

    6. Dispense into 6–8 tubes, 50 μl each.

> (a) Take one immediately without spinning down (Total 0). This fraction is also a control fraction for troubleshooting. It should contain all cathepsins, leaked or not, at the beginning of the procedure (*see* **Note 8**).

> (b) Take one tube and spin down immediately (Table top centrifuge, 4 °C, 12,000 rpm, 15 min). Then remove the supernatant very carefully without pellet contamination and snap freeze both the collected supernatant (*SN 0*) and Pellet (*P 0*). These fractions give an idea of the background leakage that occurs while processing the samples.

    7. The other tubes are incubated at 37 °C with agitation for 30, 60, and 90 min respectively.

    8. Spin down the fractions as above.

> (a) Remove and collect the supernatant very carefully without pellet contamination.

> (b) Snap freeze both the collected supernatants (*SN 30/60/90*) and pellets (*P 30/60/90*).

    9. Snap freeze one sample without spinning down (*Total 90*). This is another control fraction to assess the total remaining cathepsins after incubation. These should not differ much from *Total 0*, but any protein degradation should become apparent here.

  10. Resuspend the pellets in 150 μl of RIPA buffer. The supernatant fractions do not need further lysis.

  11. Lyse 35 min on ice with three pulses of vortexing (10 s).

  12. Spin down at 13,000 rpm at 4 °C, for 15 min in a tabletop centrifuge. The pellet is not solid but just a highly viscous mass. Take the supernatant without touching the soft pellet.

  13. For Western blotting load 4 μl of the supernatant/pellet fractions per well.

  14. Probing with cathepsin B and L antibodies will visualize the leakiness. LAMP2 can be used to confirm that the lysosomes did not disintegrate, as it is still recovered in the organelle fraction.

3.3.3 Immuno-
fluorescence for LAMP1/2
and Cathepsin B/D:
Mander's Coefficient

While the methods described above look at whole tissue extracts, immunofluorescence will give a better idea about the localisation and extent of lysosomal leakage. The very small size of lysosomes means that in standard fluorescence microscopes no sub-lysosomal distribution can be discerned. This fact is exploited, when assessing the amount of cathepsins within lysosomes by measuring the co-localization of a cathepsin stain with that of the lysosomal membrane marker LAMP1/2.

A good measure of co-localization is the so-called Mander's coefficient. In essence, this coefficient measures how may pixels in one channel overlap with pixels in another channel. A value of 1 means complete overlap, while a value of 0 means mutual exclusivity. Importantly, the values can be substantially different for the two channels. There can be a scenario, in which there is very little red stain, and a lot of green stain. If every red pixel overlaps with a green pixel, the value for the red channel equals 1. However, in this scenario there will be a substantial amount of green pixels that do not overlap with red pixels; the Mander's coefficient for the green channel will thus be <1 (*see* Fig. 5).

The advantage of this method is that clear numerical values can be generated, leading to a much more objective and unbiased way of displaying experimental data compared to an assessment of co-localization by eye. The only problem is the heavy image processing (background removal, choice of regions of interest etc.) needed to generate the coefficient, which can reintroduce an element of subjectivity. It is thus important to keep these processes as standardized as possible.

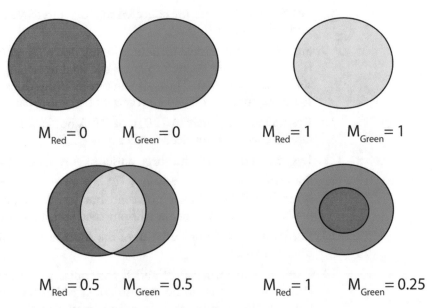

$M_{Red} = 0$ $M_{Green} = 0$ $M_{Red} = 1$ $M_{Green} = 1$

$M_{Red} = 0.5$ $M_{Green} = 0.5$ $M_{Red} = 1$ $M_{Green} = 0.25$

Fig. 5 Schematic representation of some extreme scenarios of pixel distribution and respective Mander's coefficients

This method consists of a standard immunofluorescence staining followed by image acquisition on a confocal or deconvolution microscope. It is important to use z-stacks, as a regular immunofluorescence with a deep focal plane will lead to false-positive readings. The coefficient can be generated using the Mander's coefficient plug-in for ImageJ (*see* **Note 9**). The Wright imaging facility offers this for free at: http://www.uhnresearch.ca/facilities/wcif/software/Plugins/Manders_Coefficients.html.

Thus in the following, the focus will not be on the image processing itself, but rather some points to keep in mind when working with mammary tissue.

1. Perform an immunofluorescence staining as described above (*see* **Note 10**).

2. Acquire z-stacks of the stained tissue sections. For every time point at least three z-stacks per mammary gland should be acquired and at least three individual mice should be analysed.

3. Remove background fluorescence using the manufacturer's software or other image processing programs (*see* **Note 11**).

4. Load z-stacks into ImageJ individually.

5. Choose regions of interest for every stack. This is important, since some structures within the mammary gland, such as red blood cells, show strong autofluorescence in both the red and the green channels. These need to be excluded, as they would give a false positive signal.

6. Run the Mander's coefficient plug-in (refer to website for a step-by-step guide).

3.4 Measuring Lysosomal Leakiness in an In Vitro Model: EpH4 Cell Line

EpH4 cells are a normal mammary epithelial cell line derived from a mid-pregnant BALB/c mouse [11] that has been previously shown to mimic Stat3-induced lysosomal mediated programmed cell death (LM-PCD) when stimulated with Oncostatin M (OSM) [6]. OSM treatment of EpH4 cells results in lysosomal membrane permeabilization and leakage of hydrolases into the cytosol, which can be measured quantitatively as described in Subheading 3.3.1 above by assessing cleavage of the fluorescent molecule AMC from the synthetic substrate Z-Phe-Arg-AMC by cathepsins. Briefly, the activity of lysosomal cysteine cathepsins is assessed in cytosolic preparations that are extracted from cells using a buffer containing 25 μg/ml digitonin, a glycosidic detergent. This concentration of digitonin was determined as optimal for the extraction of cytosolic proteins from EpH4 cells without damaging lysosomal membranes. The AMC fluorescence can be measured over time at 37 °C and the initial rate of fluorescence (corresponding to the initial rate of cathepsin activity—fluorescence/min) subsequently determined from the linear part of the resulting curve [10]. Lysotracker Red is an alternative approach to determine lysosomal integrity.

3.4.1 EpH4 Cell Culture

Cells are maintained in DMEM media supplemented with 10% FCS at 37 °C, 5% CO_2. Cells are grown to confluency and passaged by washing in PBS, prior to incubating with trypsin-EDTA at 37 °C until cells have detached. Trypsin is then inactivated by adding 4 volumes of medium to cells, which are then distributed into new tissue culture flasks as necessary.

3.4.2 Cell Seeding, Stimulation, and Extraction

1. Seed EpH4 cells at a density of a 100,000 cells/well in a plastic 6-well tissue culture plate in DMEM containing 10% FCS.

2. Next day, stimulate Stat3 signaling by treating cells with OSM (final concentration 25 ng/ml) or vehicle control (PBS/0.1%BSA) in DMEM containing 1% FCS.

3. 48 h later change medium with fresh OSM in 1% FCS/DMEM.

4. After 72 h of OSM treatment, remove media, wash cells in PBS, and detach using trypsin-EDTA. Inactivate trypsin in DMEM containing 10% FCS and harvest samples into 15 ml Falcon tubes.

5. Pellet cells by centrifuging at 1000 rpm for 5 min and resuspend in 1–2 ml serum-free DMEM.

6. Count cells using a hemocytometer.

7. Aliquot 175,000 cells into 1.5 ml centrifuge tubes—two aliquots per condition. One is for cytosolic extraction using digitonin, while the other is for total cell extraction using 0.1% Triton X-100 in PBS.

8. Pellet cells by centrifuging at $1000 \times g$ for 3 min at 4 °C.

9. Remove media and resuspend cell pellets in 300 μl fractionation buffer containing 25 μg/ml digitonin (for cytosolic extraction) or 0.1% Triton X-100/PBS (for total cell extraction).

10. Incubate cells on ice for 10 min, vortexing intermittently (5-s pulses every 2.5 min).

11. Spin samples at 13,000 rpm for 2 min at 4 °C. Remove supernatant and transfer into clean 1.5 ml tubes. For each condition there should be a digitonin and Triton X-100-extracted sample.

3.4.3 Cathepsin Activity Assay

1. Per well of a 96-well plate (samples run in triplicate) add:

   (a) 10 μl Sample.

   (b) 160 μl Cathepsin reaction buffer.

   (c) 30 μl Substrate (final concentration 50 μM) in reaction buffer (2 μl of 5 mM Z-Phe-Arg-AMC in 28 μl cathepsin reaction buffer).

Controls:

(d) Assay control—170 µl cathepsin reaction buffer + 30 µl substrate/reaction buffer.

(e) Blank—200 µl cathepsin reaction buffer.

2. Measure fluorescence at 1-min intervals for 1 h at 37 °C in a fluorescent plate reader (excitation: 380 nm, emission: 442 nm).

3. Subtract background (obtained from assay control) from all measurements and plot fluorescence over time for each sample.

4. From the linear part of the resulting curve determine the initial rate of fluorescence for each sample corresponding to initial rate of cathepsin activity (fluorescence/min).

5. Normalize cytosolic activity to total activity (measured in samples extracted with 0.1% TritonX-100) and display OSM treated values as fold over vehicle control. Expect a result of a 1.5-fold increase in cytosolic cathepsin activity in cells treated with OSM.

3.4.4 Lysosomal Leakiness Assay with Lysotracker Red

Carry out all procedures at room temperature unless otherwise specified.

1. EpH4 cells are cultured and stimulated with OSM as described above.

2. After 3 days of OSM treatment, wash cells twice with PBS and detach cells using trypsin-EDTA.

3. Spin down cells at 1500 rpm for 5 min.

4. Resuspend cells in 1 ml of culture medium and incubate them at 37 °C for 5 min.

5. Add Lysotracker Red (100 nM) to the suspension, vortex briefly at maximum speed, and incubate in the dark at 37 °C for 30 min.

6. Analyze samples by flow cytometry. Excitation/emission for Lysotracker Red: 577/590 nm.

Lysotracker Red is a weak base that can freely permeate membranes and concentrates in acidic compartments due to its protonation. LMP leads to leakage of this lysosomotropic dye from lysosomes to the cytosol. Cells with undamaged lysosomes are expected to absorb the dye equally and appear on a histogram as one peak of high fluorescence signal. LMP and leakage of lysosomal contents will result in the appearance of an additional population of low staining intensity. OSM treatment of EpH4 cells should result in a 30–50% of leakage [6].

4   Notes

1. If no cathepsin activity is to be measured, complete protease inhibitor can be used.

2. Care needs to be taken when cross-fostering pups in order not to disturb the dam such that she rejects the pups. Ideally, the dam should be allowed to suckle for 3 days before attempting to add new pups to the litter and no more than two or three should be added. The dam should be observed quietly to ensure that the fostered pups are accepted. This often happens immediately.

3. Involution can also be measured by analyzing expression of specific genes by immunoblotting or RT-PCR. Microarray analyses revealed that a huge number of genes are transcriptionally regulated at the onset, and during the first 4 days, of involution [12, 13]. A subset of these genes are also regulated by Stat3 which is a critical inducer of involution [6, 7, 14]. Thus, measuring the expression of a subset of these genes, at either the mRNA or protein level, is a useful indicator of the progress of involution [15]. For extraction of RNA, snap-frozen mammary glands should be pulverized to a fine powder in a freezing-cold mortar, e.g., standing in dry ice or cooled down with liquid nitrogen, to avoid thawing of the sample and RNA extracted using TRIZOL.

4. For perfusion fixation, although peristaltic pumps can be used, we routinely use large syringes with butterfly needles to obtain reliable results. With some practice, this is a cheap and accessible technique.

5. Full personal protective clothing should be worn when carrying out perfusion fixation.

6. Not all methods need to be combined at all times and certain aspects can be addressed satisfactorily with just one or two of these.

7. When preparing homogenates for cathepsin activity assays, the pestle might get stuck due to large pieces of connective tissue that can be removed manually.

8. For the lysosome leakiness assays, take a total lysate sample (*TL*). This fraction is useful for troubleshooting, but not necessary for the final assay.

9. Other repositories for Mander's coefficient can be found. Furthermore the same facility offers an excellent tutorial at http://www.uhnresearch.ca/facilities/wcif/imagej/colour_analysis.htm#colocalization.

10. Note that the LAMP2 antibody does not work following TritonX permeabilization, use saponin instead for immunofluorescence.

11. Note that milk is highly autofluorescent, particularly in the green channel. This needs to be taken into account when removing the background. Milk-filled areas in the alveolar lumen should appear black after background removal.

Acknowledgements

The Watson laboratory is supported by grants from the Medical Research Council and the NC3Rs. H.K.R. is supported by a Cambridge Cancer Centre CRUK Ph.D. studentship.

References

1. Oliver CH, Khaled WT, Frend H, Nichols J, Watson CJ (2012) The Stat6-regulated KRAB domain zinc finger protein Zfp157 regulates the balance of lineages in mammary glands and compensates for loss of Gata-3. Genes Dev 26(10):1086–1097

2. Oakes SR, Naylor MJ, Asselin-Labat ML, Blazek KD, Gardiner-Garden M, Hilton HN, Kazlauskas M, Pritchard MA, Chodosh LA, Pfeffer PL, Lindeman GJ, Visvader JE, Ormandy CJ (2008) The Ets transcription factor Elf5 specifies mammary alveolar cell fate. Genes Dev 22(5):581–586

3. Kouros-Mehr H, Slorach EM, Sternlicht MD, Werb Z (2006) GATA-3 maintains the differentiation of the luminal cell fate in the mammary gland. Cell 127(5):1041–1055

4. Asselin-Labat ML, Sutherland KD, Barker H, Thomas R, Shackleton M, Forrest NC, Hartley L, Robb L, Grosveld FG, van der Wees J, Lindeman GJ, Visvader JE (2007) Gata-3 is an essential regulator of mammary-gland morphogenesis and luminal-cell differentiation. Nat Cell Biol 9(2):201–209

5. Watson CJ, Kreuzaler PA (2011) Remodeling mechanisms of the mammary gland during involution. Int J Dev Biol 55(7-9):757–762

6. Kreuzaler PA, Staniszewska AD, Li W, Omidvar N, Kedjouar B, Turkson J, Poli V, Flavell RA, Clarkson RW, Watson CJ (2011) Stat3 controls lysosomal-mediated cell death in vivo. Nat Cell Biol 13(3):303–309

7. Sargeant TJ, Lloyd-Lewis B, Resemann HK, Ramos-Montoya A, Skepper J, Watson CJ (2014) Stat3 controls cell death during mammary gland involution by regulating uptake of milk fat globules and lysosomal membrane permeabilization. Nat Cell Biol 16(11):1057–1068

8. Li M, Liu X, Robinson G, Bar-Peled U, Wagner KU, Young WS, Hennighausen L, Furth PA (1997) Mammary-derived signals activate programmed cell death during the first stage of mammary gland involution. Proc Natl Acad Sci U S A 94(7):3425–3430

9. Sandahl M, Hunter DM, Strunk KE, Earp HS, Cook RS (2010) Epithelial cell-directed efferocytosis in the post-partum mammary gland is necessary for tissue homeostasis and future lactation. BMC Dev Biol 10:122–136

10. Jahreiss L, Renna M, Bittman R, Arthur G, Rubinsztein DC (2009) 1-O-Hexadecyl-2-O--methyl-3-O-(2'-acetamido-2'-deoxy-beta-D-glucopyranosyl)-sn-glycerol (Gln) induces cell death with more autophagosomes which is autophagy-independent. Autophagy 5(6):835–846

11. Reichmann E, Ball R, Groner B, Friis RR (1989) New mammary epithelial and fibroblastic cell clones in coculture form structures competent to differentiate functionally. J Cell Biol 108:1127–1138

12. Clarkson RW, Wayland MT, Lee J, Freeman T, Watson CJ (2004) Gene expression profiling of mammary gland development reveals putative roles for death receptors and immune mediators in post-lactational regression. Breast Cancer Res 6(2):R92–R109

13. Stein T, Morris JS, Davies CR, Weber-Hall SJ, Duffy MA, Heath VJ, Bell AK, Ferrier RK, Sandilands GP, Gusterson BA (2004) Involution

of the mouse mammary gland is associated with an immune cascade and an acute-phase response, involving LBP, CD14 and STAT3. Breast Cancer Res 6(2):R75–R91

14. Chapman RS, Lourenco PC, Tonner E, Flint DJ, Selbert S, Takeda K, Akira S, Clarke AR, Watson CJ (1999) Suppression of epithelial apoptosis and delayed mammary gland involu-tion in mice with a conditional knockout of Stat3. Genes Dev 13(19):2604–2616

15. Thangaraju M, Rudelius M, Bierie B, Raffeld M, Sharan S, Hennighausen L, Huang AM, Sterneck E (2005) C/EBPdelta is a crucial regulator of gene expression during mammary gland involution. Development 132(21): 4675–4685

Part II

In Vitro 2D & 3D-Models

Chapter 8

Contractility Assay for Established Myoepithelial Cell Lines

Stéphanie Cagnet, Marina A. Glukhova, and Karine Raymond

Abstract

The capacity of mammary myoepithelial cells to contract in response to suckling stimuli is essential for lactation. We describe here a protocol for studying the contractile activity of myoepithelial cells in vitro. This protocol includes the establishment of stable myoepithelial cell lines from mouse mammary glands and quantitative evaluation of the contraction and subsequent relaxation of cultured myoepithelial cells in response to oxytocin. It can be used for analyses of mouse mutants with gene deletions or overexpression altering myoepithelial cell function.

Key words Mammary myoepithelial cell, Contraction, Relaxation, Time-lapse video microscopy, Rho/ROCK pathway

1 Introduction

The mammary epithelium comprises two cell layers: the luminal and basal myoepithelial cell layers. In the functionally differentiated mammary gland, during lactation, luminal cells produce and secrete milk, whereas basal myoepithelial cells contract to eject the milk from the secretory alveoli into the ducts, towards the nipple and out of the body.

Mammary myoepithelial cells display the phenotypic characteristics of basal cells from stratified epithelia: they express basal-type cytokeratins 5, 14 (K5 and K14, respectively) and 17, P-cadherin and high levels of ΔNp63. However, differentiated myoepithelial cells are also contractile cells, and their ultrastructure is reminiscent of that of smooth muscle cells, as they contain large numbers of microfilaments and express smooth muscle-specific cytoskeletal and contractile proteins.

The mammary myoepithelium is organized differently in the ducts and alveoli. Elongated ductal myoepithelial cells are arranged in a more or less continuous monolayer, whereas alveolar myoepithelial cells are stellate and do not form a continuous layer between the secretory epithelium and the surrounding basement membrane (Fig. 1).

Finian Martin et al. (eds.), *Mammary Gland Development: Methods and Protocols*, Methods in Molecular Biology, vol. 1501, DOI 10.1007/978-1-4939-6475-8_8, © Springer Science+Business Media New York 2017

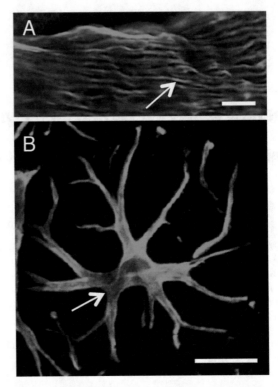

Fig. 1 Morphology of myoepithelial cells within ducts and alveoli. Immunofluorescence labeling of a 1-day lactating mouse mammary gland fragment (**A**) and section (**B**) with anti-αSMA antibody. *Arrows* indicate the longitudinally arranged ductal (**A**) and stellate alveolar (**B**) myoepithelial cells. Bars: 10 μm

The contractile activity of mammary myoepithelial cells is essential for lactation, and oxytocin (OT), a neuropeptide produced by the pituitary gland, is a major physiological regulator of myoepithelial cell contraction [1]. The contraction of myoepithelial cells, like that of smooth muscle, is induced by the phosphorylation of myosin light chains (MLC) by MLC kinase. The subsequent dephosphorylation of MLC by a specific phosphatase leads to relaxation [2]. Several other kinases, including RhoA-dependent kinase (ROCK), can also phosphorylate MLC. Moreover, ROCK is known to regulate smooth muscle contraction by inhibiting the MLC phosphatase [3]. We have recently shown that the RhoA/ROCK signaling cascade is essential for the OT-induced contraction of myoepithelial cells, and that ROCK inhibition completely prevents the contractile response of myoepithelial cells to OT [4]. As might be expected, the smooth muscle-like differentiation of mammary myoepithelial cells (i.e., the expression of smooth muscle-specific proteins) is required for their contractile activity, and deletion of the genes encoding α-smooth muscle actin (α-SMA) or the transcription factor MKL1, which has been implicated in the control of smooth muscle-specific protein expression, results in lactation failure [5–7].

Myoepithelial cells interact directly with the basement membrane, a special type of extracellular matrix (ECM) underlying the mammary epithelium. The integrins of the β1 family and α6β4 integrin are major ECM receptors expressed by mammary myoepithelial cells [8]. Mouse mutants lacking the laminin receptor, α3β1 integrin, in the myoepithelium, do not lactate normally due to defective myoepithelial cell contractile activity, resulting in the retention of milk within the gland [4]. We observed sustained MLC phosphorylation in the mammary epithelium of these mice, suggesting an impairment of postcontraction relaxation. Cultured mammary myoepithelial cells lacking α3β1 contract in response to oxytocin, but cannot sustain the post-contractile state of relaxation [4].

The contraction of myoepithelial cells in response to OT treatment is accompanied by major changes in cell shape that can be observed and recorded with a confocal microscope (Fig. 2). Myoepithelial cells from lactating α3β1-mutant mice have significantly shorter and thicker extensions than normal myoepithelial cells [4].

We describe here a protocol developed for studying the contractile activity of myoepithelial cells lacking α3β1 integrin in vitro. We obtained a stable mammary myoepithelial cell line from mice carrying conditional alleles of the gene encoding the α3 integrin chain (*itga3*), by modifying the protocol previously described by Dan Medina [9]. Further, the *itga3* gene was deleted in vitro, by treatment of the cells with Cre-recombinase-carrying adenoviruses, as described elsewhere [10]. Using this approach, we were able to study the signaling events underlying the impaired contractility of α3β1-deficient mammary myoepithelial cells. This protocol could also be used for analyses of other mouse mutants with gene deletions or overexpression leading to altered myoepithelial cell function.

2 Materials

2.1 For the Preparation and Culture of Mouse Mammary Epithelial Cells

1. Sterile scissors, forceps, and scalpels for tissue dissection.
2. Dulbecco's modified Eagle's medium/Ham's nutrient F-12 medium (DMEM/F12).
3. Collagenase A, hyaluronidase type IV-S.
4. Growth-factor reduced Matrigel, rat tail collagen (*see* **Note 1**).

2.2 For the Characterization of the Established Mouse Mammary Myoepithelial Cell Lines

1. Anti-β4 integrin antibody coupled to phycoerythrin.
2. Anti-keratin 5 antibody.
3. Anti-keratin 14 antibody.
4. Anti-α-SMA antibody.

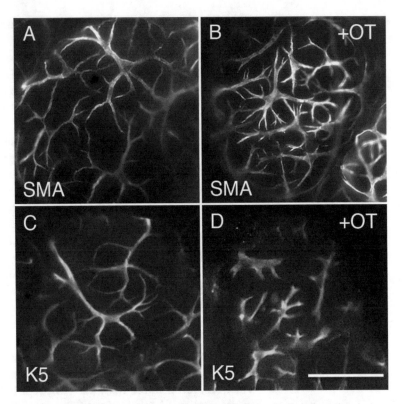

Fig. 2 Ex vivo contraction of mammary myoepithelial cells. To induce the contraction of myoepithelial cells, mammary gland fragments dissected from lactating mice were treated with OT, fixed and labeled with anti-αSMA (SMA) or anti-K5 antibodies, to visualize the changes in myoepithelial cell shape induced by contraction. (**A**) and (**C**), no treatment; (**B**) and (**D**), 50 nM OT, for 2 min. Alveolar myoepithelial cells have thin, elongated extensions, which become thicker and shorter after OT treatment. For more details, *see* [4]. Bar: 25 μm

2.3 In Vitro Cell Contractility Assay

1. Glass-bottomed dishes, uncoated.

2. Biostation IM time-lapse imaging system with an integrated incubation system (37 °C, 5 % CO_2).

3. Oxytocin.

4. Myosin light-chain kinase inhibitor ML7.

5. ROCK inhibitor Y27632.

3   Methods

3.1 Preparation of the Reagents

1. Prepare Puck's saline A: For 1 l, 8.0 g NaCl, 0.4 g KCl, 1.0 g glucose, 0.15 g $MgCl_2 \cdot 6H_2O$, 0.35 g $NaHCO_3$, 5 mg phenol red. Adjust pH to 7.4.

2. Prepare fatty-acid-free BSA solution: Dissolve 50 g of fatty-acid-free BSA in 250 ml of DMEM/F12, filter, split into 40 ml aliquots, and store at –20 °C.

3. Prepare linoleic acid (LA) complex: Mix 25 ml of a 20 g/l LA solution in sterile 0.1 M Na_2CO_3 with 135 ml of Puck's saline A and with 40 ml of fatty-acid-free BSA solution. Bubble nitrogen through the solution in a screw-capped glass tube. Heat at 50 °C for 1 h. Filter. Divide into 2 ml aliquots and store at –20 °C protected from light.

4. Prepare solution A: For 1 l, 1.80 g glucose, 7.60 g NaCl, 0.22 g KCl, 0.14 g Na_2HPO_4, 7.15 g HEPES, 1.24 mg phenol red. Adjust pH to 7.6 with NaOH.

5. Prepare a 4× stock solution of dispase grade II: Dissolve 9.6 mg of dispase per ml in solution A. Divide into 2 ml aliquots and store at –20 °C. The working solution of dispase can be obtained by thawing the 4× stock solution and diluting it to a 1× solution with Puck's saline A.

6. Prepare growth medium: DMEM/F12 buffered with 15 mM HEPES, 10 µg/ml insulin, 5 ng/ml epidermal growth factor, 2 % calf serum, 0.5 mg/ml fraction V BSA, 5 µg/ml LA complex, 100 U/ml penicillin, 100 U/ml streptomycin, 50 µg/ml gentamycin, and 20 U/ml nystatin (*see* **Note 2**).

7. Prepare mincing medium: DMEM/F12 buffered with 15 mM HEPES, 100 U/ml penicillin, 100 U/ml streptomycin, 100 µg/ml gentamycin and 60 U/ml nystatin.

3.2 Preparation of Mouse Mammary Cell Suspension

1. Dissect no. 4 and 5 glands aseptically and remove the lymph nodes (*see* **Note 3** and **Note 4**). Place the glands in a 100 mm Petri dish with 0.5–1 ml of mincing medium (*see* **Note 5**).

2. Mince the tissue with two sterile scalpels, one in each hand, until pieces of about 0.5 mm^3 are obtained (*see* **Note 6**).

3. Place the minced tissue in a 50 ml conical tube containing an appropriate volume (10 ml/g of tissue) of digesting medium, consisting of mincing medium supplemented with 2 mg/ml collagenase and 100 U/ml hyaluronidase. Incubate for 3 h at 37 °C, with shaking at 125 rpm, with the tube placed at a 45° angle (*see* **Note 7**).

4. Transfer the samples to 15 ml tubes and centrifuge at $200 \times g$ for 5 min.

5. Discard the supernatant. Wash the cell pellet in 10 ml of PBS containing 5 % calf serum and centrifuge the tubes at $200 \times g$ for 5 min. Repeat four times, to give a total of five washes.

6. Resuspend the cells in an appropriate volume of growth medium supplemented with 2 % Matrigel (2×10^5 cells/ml). The resulting suspension will contain mostly single cells, with some very small aggregates (*see* **Note 8**).

3.3 Primary Culture of Mouse Mammary Cells

1. Prepare the appropriate number of six-well plates, by adding a thin layer of Matrigel. To do this, place the plates on ice and spread 200 μl of Matrigel over the surface of each well with the bottom end of a sterile micropipette tip. Incubate the plates for 30 min at 37 °C (*see* **Note 9**).

2. Add 1.5 ml of mammary cell suspension per well.

3. Incubate for 4 days and then add 1 ml of fresh growth medium supplemented with 2 % Matrigel. Under these conditions, most cells will be organized in spheres, surrounded by a few spindle-shaped fibroblastic cells.

4. Every three days, carefully remove 1 ml of medium and replace it with 1 ml of fresh growth medium supplemented with 2 % Matrigel.

3.4 Establishment of Mouse Mammary Myoepithelial Cell Lines

1. After 10–14 days, recover the cells from the Matrigel by treatment for 15–30 min with dispase (*see* **Note 10**). Centrifuge the cells ($200 \times g$, 5 min) and wash the cell pellet five times with PBS containing 5 % calf serum.

2. Prepare collagen I-coated dishes. To do this, spread 100 μl/cm² of 50 μg/ml collagen I solution in sterile PBS over 100 mm Petri dishes and incubate for 1 h at 37 °C. Wash the dishes twice with PBS.

3. Add the cell suspension maintained in growth medium at a density of 2×10^6 cells per dish. There may be some very small aggregates, but most of the suspended cells will be single cells (*see* **Notes 8** and **11**). The growth medium should be replenished every 3 days.

4. When required, brief dispase treatment (5 min) should be performed, to eliminate spindle-shaped fibroblastic cells. The dish should then be washed five times with PBS containing 5 % calf serum and the epithelial cells should be maintained in growth medium (*see* **Note 12**).

5. Split the cells 1:2, at about 80 % confluence. Mammary epithelial cells grow slowly and may need to be maintained in culture for about 4 weeks to obtain a sufficiently high density of cells for the splitting of the culture.

6. To detach the cells from the dish, treat them for 15–30 min with dispase. Wash five times with PBS containing 5 % calf serum and plate the cells on collagen-coated dishes (*see* **Note 13**).

7. Once the epithelial cells reach the crisis and massive cell death is observed (between passages 3 and 5), a few clones of immortalized cells will emerge.

8. Rapidly subclone the cell population by conventional approaches (either by ring cloning or by cloning at limiting dilutions). To favor the development of myoepithelial cell lines, isolate

myoepithelial cell clones by flow cytometry cell sorting with anti-β4 integrin antibody before subcloning [4] (*see* **Note 14**).

9. Once established, myoepithelial cell lines usually grow faster than primary cells. Passage them by conventional trypsin treatment, with splitting in a 1:5 ratio, possibly allowing the cells to grow directly on the plastic.

3.5 Characterization of the Mouse Mammary Myoepithelial Cell Lines

Once stable cell lines have been established, perform conventional immunofluorescence staining and immunoblotting to check for the expression of myoepithelial markers, such as keratin 5 and 14 and α-SMA (*see* **Note 15**).

3.6 In Vitro Analysis of Myoepithelial Cell Contractility by Time-Lapse Video Microscopy

1. Seed glass-bottomed dishes with a low density of cells and place the dishes in the incubator for 16–20 h (*see* **Note 16**).

2. One hour before starting the experiment, rinse the dish with PBS to eliminate debris and add 1 ml of fresh growth medium.

3. Start the time-lapse imaging experiment by following the spontaneous surface area fluctuations of the cells present in the selected fields over a period of 15 min (one image captured per minute). Planar surface area can be measured with ImageJ software. In our hands, spontaneous fluctuations did not exceed 10 % of the initial surface area, defined as a mean value over 15 min of observation (*see* **Note 17**).

4. Test the capacity of the cell line to respond to OT by carefully adding 1 ml of growth medium supplemented with 2 nM OT (final concentration, 1 nM) to the dish. Capture one image per minute and follow the response for 1 h. A contractile response will lead to a significant decrease in cell surface area (contraction) followed by recovery of the initial cell surface area (relaxation) (Fig. 3) (*see* **Note 18**).

5. Determine the cell contraction parameters. Determine the primary contractile response (time course and degree of contraction, i.e., changes in planar surface area) and the capacity for relaxation (time and stability of recovery) of the myoepithelial cell line investigated, and compare these parameters with those of control cells (*see* **Note 19**).

6. If the cell contraction parameters of genetically modified myoepithelial cells differ from those of control cells, the molecular mechanisms underlying the abnormal contractile phenotype can be analyzed by treating the cells with pharmacological inhibitors or by the forced expression of modified (constitutively active or inactivated) intracellular signaling intermediates. To determine whether MLCK or ROCK is involved, use specific inhibitors of these kinases (ML7 and Y27632, respectively). Treat the control myoepithelial cell line for 1 h with various doses of the specific inhibitors before OT treatment, to

+ OT

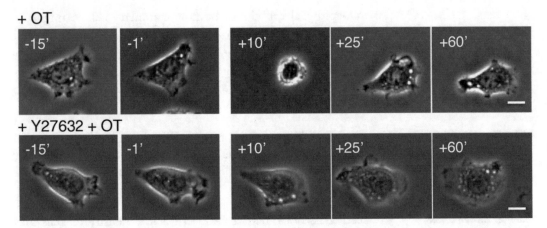

Fig. 3 In vitro myoepithelial cell contraction assay. Time-lapse video recording of the changes in myoepithelial cell shape accompanying contraction in response to OT. Microphotographs of cultured myoepithelial cells treated with OT, taken before or after treatment at the time points indicated in the *top left corner*. *Lower panels* show cells treated with the ROCK inhibitor Y27632 before OT. Note that Y27632 completely inhibited contraction. Bars: 10 µm

define the optimal concentration (the minimum dose of inhibitor completely abolishing cell contraction) and the suboptimal concentration (maximum dose of inhibitor with no significant effect on the initial response to OT). Investigate the change in optimal and suboptimal concentrations in the genetically modified myoepithelial cell line.

4   Notes

1. Follow the manufacturer's instructions for the preparation of a sterile 1 mg/ml collagen stock solution in 0.1 M acetic acid. Store at 4 °C.

2. The use of low serum concentration has been reported to favor epithelial cell growth, as fibroblasts require higher serum concentrations [9].

3. We recommend establishing cell lines from mammary glands isolated from mice in late pregnancy, at 16–18 days of gestation. At this stage, the gland is functionally similar to the fully differentiated lactating gland and provides the highest yield of epithelial cells (about 2.5×10^6 cells/0.1 g of tissue).

4. In animals in the late stages of pregnancy, no. 4 and 5 glands are easy to dissect, whereas no. 1–3 glands, when dissected, may be contaminated by salivary gland and/or muscle tissue. Only the inguinal lymph node and the lymph node located at the top of the hindlimb on no. 4 glands must be removed to avoid contamination of the culture.

5. The minimal volume of mincing medium preventing the tissue from drying out should be used. The use of a large amount of medium leads to the dispersion of mammary fragments, impeding efficient mincing.

6. The use of fragments of uniform size is important, to ensure a homogeneous digestion.

7. The incubation time should be adjusted slightly and digestion should be stopped when a cloudy homogeneous solution is obtained.

8. Viable cells are counted with a standard hemocytometer after trypan blue staining.

9. Culture in collagen gels rather than Matrigel has been reported [9, 11]. Our highest yields of myoepithelial cell clones were obtained with the protocol described here.

10. The use of dispase rather than trypsin results in a greater viability of the recovered cells [9].

11. The cell suspension can be plated directly on uncoated dishes, although we have found this approach to be less efficient.

12. Note that the culture contains several cell types, including fibroblasts, myoepithelial and luminal epithelial cells, with luminal epithelial cells 5–10 times more abundant than myoepithelial cells. Myoepithelial cells are less cuboidal than luminal cells. However, they are less elongated than fibroblasts and form many more cell/cell contacts.

13. If the culture still contains fibroblasts, start with a short dispase treatment, rinse twice with PBS and then continue treatment with fresh dispase until the epithelial cells are detached.

14. Myoepithelial and luminal cells can usually be distinguished on the basis of their level of β4-integrin expression, with a much stronger signal being obtained for myoepithelial cells.

15. Antibodies against other smooth muscle-specific proteins, such as smooth muscle-calponin or caldesmon can also be used.

16. If the cells do not spread out correctly on glass coat the dish with collagen.

17. Cells with an unstable surface area (fluctuation of more than 10%) and cells that do not spread out well (with refringent borders), or are abnormally large (usually polynucleated) or are dividing are excluded from the experiment.

18. In our hands, surface area reached a minimum approximately 6 min after OT stimulation, and initial surface area was recovered about 20 min after OT treatment. We recommend testing different doses of OT and choosing the minimal concentration leading to a contraction of approximately 80% of the cells.

19. In our experiments, mammary myoepithelial cells from the established cell line, carrying conditional alleles of the *itga3* gene (control cells), were compared with the same cells after infection with Cre-adenoviruses resulting in deletion of α3β1 integrin (mutant cells).

Acknowledgments

The work in the team of M.A.G. was supported by *La Ligue Nationale Contre le Cancer (Equipe Labellisée 2013)* and grants from *Agence Nationale de la Recherche* ANR-08-BLAN-0078-01 and ANR-13-BSV2-0001. M.A.G. is *Directeur de Recherche* and K.R. is *Chargé de Recherche* at the *Institut National de la Santé et de la Recherche Médicale (INSERM)*.

References

1. Reversi A, Cassoni P, Chini B (2005) Oxytocin receptor signaling in myoepithelial and cancer cells. J Mammary Gland Biol Neoplasia 10(3):221–229. doi:10.1007/s10911-005-9583-7

2. Hartshorne DJ, Ito M, Ikebe M (1989) Myosin and contractile activity in smooth muscle. Adv Exp Med Biol 255:269–277

3. Somlyo AP, Somlyo AV (2003) Ca2+ sensitivity of smooth muscle and nonmuscle myosin II: modulated by G proteins, kinases, and myosin phosphatase. Physiol Rev 83(4):1325–1358, doi:10.1152/physrev.00023.2003 83/4/1325 [pii]

4. Raymond K, Cagnet S, Kreft M, Janssen H, Sonnenberg A, Glukhova MA (2011) Control of mammary myoepithelial cell contractile function by alpha3beta1 integrin signalling. EMBO J 30(10):1896–1906. doi:10.1038/emboj.2011.113, emboj2011113 [pii]

5. Haaksma CJ, Schwartz RJ, Tomasek JJ (2011) Myoepithelial cell contraction and milk ejection are impaired in mammary glands of mice lacking smooth muscle alpha-actin. Biol Reprod 85(1):13–21. doi:10.1095/biolreprod.110.090639, biolreprod.110.090639 [pii]

6. Li S, Chang S, Qi X, Richardson JA, Olson EN (2006) Requirement of a myocardin-related transcription factor for development of mammary myoepithelial cells. Mol Cell Biol 26(15):5797–5808, doi:10.1128/MCB.00211-06 26/15/5797 [pii]

7. Sun Y, Boyd K, Xu W, Ma J, Jackson CW, Fu A, Shillingford JM, Robinson GW, Hennighausen L, Hitzler JK, Ma Z, Morris SW (2006) Acute myeloid leukemia-associated Mkl1 (Mrtf-a) is a key regulator of mammary gland function. Mol Cell Biol 26(15):5809–5826, doi:10.1128/MCB.00024-06 26/15/5809 [pii]

8. Taddei I, Faraldo MM, Teulière J, Deugnier MA, Thiery JP, Glukhova MA (2003) Integrins in mammary gland development and differentiation of mammary epithelium. J Mammary Gland Biol Neoplasia 8(4):383–394, doi:10.1023/B:JOMG.0000017426.74915.b9

9. Medina D (2000) The preneoplastic phenotype in murine mammary tumorigenesis. J Mammary Gland Biol Neoplasia 5(4):393–407

10. Raymond K, Kreft M, Janssen H, Calafat J, Sonnenberg A (2005) Keratinocytes display normal proliferation, survival and differentiation in conditional beta4-integrin knockout mice. J Cell Sci 118(Pt 5):1045–1060, doi:10.1242/jcs.01689 118/5/1045 [pii]

11. Streuli M (1996) Protein tyrosine phosphatases in signaling. Curr Opin Cell Biol 8(2):182–188

Chapter 9

Using 3D Culture of Primary Mammary Epithelial Cells to Define Molecular Entities Required for Acinus Formation: Analyzing MAP Kinase Phosphatases

Malgorzata Gajewska and Sara McNally

Abstract

Three-dimensional (3D) cell cultures on reconstituted basement membrane (rBM) enable the study of complex interactions between extracellular matrix (ECM) components and epithelial cells, which are crucial for the establishment of cell polarity and functional development of epithelia. 3D cultures of mammary epithelial cells (MECs) on Matrigel (a laminin-rich ECM derived from the Engelbreth-Holm-Swarm (EHS) murine tumor) promote interactions of MECs with the matrix via integrins, leading to formation of spherical monolayers of polarized cells surrounding a hollow lumen (acini). Acini closely resemble mammary alveoli found in the mammary gland. Thus, it is possible to study ECM-cell interactions and signalling pathways that regulate formation and maintenance of tissue-specific shape and functional differentiation of MECs in 3D under in vitro conditions. Here we present experimental protocols used to investigate the role of mitogen-activated protein kinase phosphatases (MKPs) during development of the alveoli-like structures by primary mouse mammary epithelial cells (PMMEC) cultured on Matrigel. We present detailed protocols for PMMEC isolation, and establishment of 3D cultures using an "on top" method, use of specific kinase and phosphatases inhibitors (PD98059 and pervanadate, respectively) administered at different stages of acinus development, and give examples of analyses carried out post-culture (Western blot, immunofluorescence staining, and confocal imaging).

Key words Primary mouse mammary epithelial cells, Extracellular matrix, 3D cultures, Mitogen-activated protein kinase phosphatases

1 Introduction

Mechanisms controlling the process of development and differentiation of the mammary epithelium have been a subject of a wide range of studies. As recorded in other chapters in this book, in vivo mouse models have provided us with valuable information in this regard, however, these models are expensive to maintain, and each experiment has to be well motivated from the ethical point of view before it can be executed. Thus, a significant place has been found for in vitro studies in investigating specific cellular processes and signalling pathways important in mammary epithelial cell function.

Finian Martin et al. (eds.), *Mammary Gland Development: Methods and Protocols*, Methods in Molecular Biology, vol. 1501,
DOI 10.1007/978-1-4939-6475-8_9, © Springer Science+Business Media New York 2017

In particular, in order to recapitulate various aspects of cell organization within the mammary epithelium, 3D culture systems have been developed, in which cells are cultured, supported on extracellular matrix components resembling the tissue specific conditions seen in the mammary gland. 3D cultures of MECs on a reconstituted laminin-rich basement membrane successfully reconstitute the organisation of normal mammary epithelial cells as cysts (or acini). Proper formation of tubular structures by the epithelial cells can be achieved by varying the ECM composition (reviewed in [1–4]). These model systems have proved to be most useful in defining the molecular supports necessary to trigger and maintain the organisation of the mammary epithelium.

Basement membrane (BM), which is a thin layer of ECM directly enveloping epithelial cells, is principally composed of a lattice-type network of proteins, including collagen type IV, laminins, heparin sulphate proteoglycan, nidogen, and dystroglycan [5]. The nature and composition of ECM were recognized as key determinants of normal growth, polarization and differentiation of mammary epithelial cells. In the late 1970s Emerman and Pitelka [6] showed that mouse MECs polarize, resume their native shape, and assemble a BM when cultured on floating collagen gels. Moreover, addition of lactogenic hormones in these conditions resulted in production of caseins by the MECs [6]. The induction of milk protein synthesis was later shown to depend on the assembly of a BM in the presence of collagen [7]. Further studies identified the significant role of another ECM molecule, laminin1, in the differentiation of MECs, as it could induce the expression of β-casein and drive cell polarization [8, 9]. The necessary molecular signals from BM to the epithelial cells were shown to be mediated by integrins that reside on the basal surface of polarized epithelial cells [10]. A commercially available rBM, Matrigel, is nowadays commonly used in 3D cultures of mammary epithelial cells. Matrigel is a laminin-rich BM/ECM derived from the EHS murine sarcoma. It successfully supports the growth, differentiation, and survival of a wide range of cells that depend on a BM including MECs. Matrigel is stored as a frozen solution, and thawed at 4 °C prior to use, and it gels at 37 °C within 15–30 min, forming a layer of rBM [11]. When normal MECs are cultured on Matrigel they proliferate and organize into 3D spheroids. The cells on the outside of the spheroids, which are in direct contact with the matrix, develop an axis of apicobasal polarity, illustrated by the basal secretion of matrix components (laminin 5), the apical orientation of the Golgi, and the appropriate localization of junctional complexes (lateral localization of desmosomes and gap junctions, and apicolateral localization of tight junction proteins, such as zona occludens, i.e., ZO-1). Next, two distinct populations of cells can be distinguished within each spheroid—an outer layer of polarized cells remaining in direct contact with the matrix, and the inner subset of cells lacking matrix contact, which subsequently undergo apoptotic cells death, and a hollow lumen of each acini is formed [12–14].

Establishing a 3D culture system, in which MECs form alveoli-like structures, enabled scientist to study signalling pathways that control interactions between cells and matrix, the polarization process and functional differentiation of these secretory cells. Streuli and co-workers have shown that β1-integrins, which mediate the ECM-derived signals, regulate the capacity of prolactin to drive MEC differentiation [15], and that the BM is involved in the regulation of cytoskeleton organization, which provides a permissive environment for mammary differentiation. Furthermore, MECs lacking β1-integrin receptors fail to polarize properly when cultured on Matrigel, and do not form typical multicellular acini [16]. This is connected with dramatic reduction of focal adhesion kinase (FAK) activation and decreased phosphorylation of adaptor protein paxillin [16]. Our research group has also been using 3D cultures to study signalling pathways involved in regulation of cell polarization, establishment of cell-cell contacts, and lumen clearance in response to the contact of MECs with rBM. We use a model of primary mid-pregnant mouse mammary epithelial cells (PMMEC), as well as the human mammary epithelial cell line MCF-10A cultured on growth factor-reduced (GFR) Matrigel. We and others have demonstrated that formation of acini by MECs requires supplementation of growth medium with EGF, insulin or IGF-I and glucocorticoid hydrocortisone for proper stimulation of the correct number of rounds of cell replication needed to form a spherical structure, that will undergo processes resulting in lumen clearance [17, 18]. When a cell line is used it is possible to initiate formation of spheroids by a single cell that proliferates after seeding on Matrigel, because it is easy to obtain and handle a suspension of single cells not forming aggregates [14]. In the case of PMMEC the cells are directly isolated from the mammary gland, and a dispersed suspension of cells seeded on rBM rapidly associates to initiate acinus formation; thus the rounds of replication are limited and the acini develop faster, usually within 4 days of culture [18, 19].

Our studies on PMMEC cultured on Matrigel enabled us to show that organization of cells into 3D alveoli-like structures requires hydrocortisone, which activates downstream signalling pathways mediated via mitogen activated protein kinases (MAPK) and c-Jun N-terminal kinase (JNK). Glucocorticoids regulated the expression of tight junction proteins ZO1 and occludin on mRNA and protein levels. Cells forming acini in the presence of glucocorticoid receptor antagonist RU486 failed to polarize, and the spheroids did not develop hollow lumen [18]. A similar effect was obtained when PMMEC were treated with a specific JNK inhibitor, SP600125. Further studies have revealed that pharmacological inhibition of JNK or JNK1 knockdown using small interfering RNA (siRNA) results in the generation of cell assemblies that do not polarize or exit the cells cycle, and fail to undergo apoptosis due to prolonged activation of MAPK pathway observed as sustained phosphorylation of ERK1/2 kinases in comparison to control conditions [19]. MAPK pathway is responsible among others for cell survival and our results indicated an important relationship between signalling pathways mediated by JNK and MAPK. In fact, we showed that inhibition of ERK1/2 phosphorylation with kinase inhibitor PD98059 was sufficient to reverse the phenotype caused by JNK blockage allowing formation of small acini composed of cells showing normal apico-basal polarization, and having lumen in the centre. Furthermore, we showed that ERK1/2 activation, which under normal conditions occurs at the beginning of acini development and subsequently decreases, is additionally controlled by mitogen-activated protein kinase phosphatases (MKP2 and MKP3), whose expression was also abrogated in spheroids cultured in the presence of JNK inhibitor [19]. MKPs belong to a family of dual specificity phosphatases (DUSPs) which control ERK1/2 activation via an inhibitory feedback loop [20, 21]. Expression of MKPs is induced by increased phosphorylation of MAP kinases, causing attenuation of ERK1/2 activation by dephosphorylation of these kinases and inactivation of MAPK pathway. By regulating the extent of MAPK activation MKPs are able to regulate the processes of proliferation and differentiation of developing cells [20]. It is worth noting that MKP2 and MKP3 differ in subcellular localization, regulating the activity of ERK kinases in different cellular compartments. Following activation, ERK1/2 dissociate from the cytoplasmic anchoring complex and enter the nucleus, where they accumulate during mid-G1 phase [22]. The nuclear localization of ERK1/2 was shown to be essential for growth factor-induced DNA replication and cell transformation [23, 24]. It has been demonstrated that MKP1 and MKP2 are located exclusively in the nucleus, whereas MKP3 act within the cytosol [22]. Since in our model system addition of JNK-inhibitor, SP600125 caused failure of acinus formation and sustained high levels of phosphorylated ERK1/2, we were interested to see whether inhibition of MKPs will result in a similar effect. To test this hypothesis we used pervanadate, a potent inhibitor of protein tyrosine phosphatases [25, 26]. Addition of pervanadate to the 3D culture of PMMECs 17 h after seeding, which is the time when cells already had generated spherical structures, and when in normal conditions the polarization process begins, resulted in disrupted polarization and development of cell assemblies similar to those formed in the presence of JNK inhibitor (Fig. 3). On the other hand, pharmacological inhibition of ERK1/2 by PD98059 from the beginning of 3D culture caused formation of small acini with normal phenotype, whereas when the inhibitor was added 17 h post-seeding the acini developed a hollow lumen after 48 h of culture, sooner than in control conditions (Fig. 2). However, when cells were treated with PD98059 and pervanadate (pervanadate being administered 4 or 17 h after ERK1/2 inhibitor) the formation of acinar structures was abrogated, which might have resulted from restoration of MAPK pathway activity, disruption of polarization and inhibition of cell death processes. These data indicate that JNK pathway is crucial for normal development of alveoli by MECs, whereas activation of MAPK pathway needs to be tightly regulated and attenuated among others by MKPs, at the later stages of acini formation.

In vitro 3D culture systems enable researchers to perform relatively simple experiments using inhibitors of specific kinases, as well as siRNA/shRNA approach to knockdown the expression of specific genes in order to study cell-ECM interactions in detail, which consequently initiate processes of cell polarization, development of cell-cell contacts and functional differentiation. Such experiments not only allow for a tight control of culture conditions, but also enable a longer period of observation of cells forming the 3D structures, giving an opportunity to study very transient processes, as well as those occurring within a few days or weeks. This chapter describes an example of experiments performed on PMMECs cultured on Matrigel that aimed to elucidate the role of protein tyrosine phosphatases in the development of mammary alveoli. Presented protocols have been routinely used in our laboratory in the studies on signalling pathways involved in the regulation of acini formation by mammary epithelial cells.

2   Materials

2.1   Equipment for PMMECs Isolation and Culture

1. Laminar flow hood.

2. Orbital incubator shaker.

3. Water bath.

4. Centrifuge.

5. Autoclaved 250 ml conical flask and 500 ml glass bottles.

6. Autoclaved tips for automatic pipettes (p10, p200, p1000).

7. Sterile 50 ml polypropylene tubes.

8. Sterile scalpel blades.

9. Sterile vacuum filtration flasks.

10. Sterile syringe filters (with 0.22 μm membranes).

11. Sterile 60 mm culture plates and chamber slides.

12. Sterile polystyrene pipettes (5, 10, 25 ml).

13. Pipette aid.

14. Humidified incubator (37 °C, 5 % CO_2).

15. Inverted microscope with phase-contrast optics.

2.2 PMMEC Isolation and Culture

1. Collagenase digestion mixture: 480 mg Ham's F-10 powder medium, 70 mg trypsin, 150 mg collagenase A, 2 ml of foetal bovine serum; prepared in a final volume of 50 ml (with deionized water—dH$_2$O). The collagenase digestion mixture is subsequently adjusted to pH 7.4 and filtered through a 0.22 μm sterile syringe filter.

2. DNase mixture used to achieve isolation of single epithelial cells: 480 mg Ham's F-10 powder medium, 250 μl of 10 mg/ml DNase, 250 μl of 1 M MgCl$_2$; brought up to a final volume of 50 ml (with dH$_2$O) and passed through a 0.22 μm syringe sterile filter.

3. Ham's F12 culture medium routinely supplemented with penicillin/streptomycin (50 units/ml of penicillin, 50 μg/ml streptomycin), hydrocortisone 1 μg/ml (stock solution—1000×: 1 mg/ml in 100% ethanol), insulin 5 μg/ml (stock solution—1000×: 5 mg/ml in 5 mM HCl), epidermal growth factor (EGF) 5 ng/ml (stock solution—1000×: 5 μg/ml prepared in Ham's F12 medium).

4. GFR Matrigel.

5. PD98059 stock solution of 10 mM in DMSO (*see* **Note 1**).

6. 0.2 M Na$_3$VO$_4$.

7. 0.2 M H$_2$O$_2$ solutions.

8. Catalase (10,000–40,000 units of enzyme/mg of protein) for preparation of pervanadate solution used as inhibitor of protein-tyrosine phosphatases (PTP) (*see* **Note 2**).

2.3 Protein Extraction

2.3.1 Whole-Cell Extracts Preparation from PMMEC

1. Radio-immunoprecipitation (RIPA) buffer: 0.79 g Tris base (final conc. 50 mM), pH 7.4, 0.90 g NaCl (final conc. 150 mM) (*see* **Note 3**); dissolve both compounds in 75 ml of deionized water (dH$_2$O) and adjust to pH 7.4 with HCl; then add 10 ml of 10% NP-40 (final conc. 1%), 2.5 ml of 10% Na-deoxycholate (final conc. 0.25%), 0.2 ml of 0.5 M EDTA (final conc. 1 mM); adjust volume to 100 ml with dH$_2$O and store at 4 °C. Prior to use add to 1 ml of full RIPA buffer: 10 μl of 100× protease inhibitor cocktail (containing: 4-(2-aminoethyl)benzenesulfonyl fluoride hydrochloride against serine proteases; aprotinin against serine proteases; bestatin hydrochloride against aminopeptidases; E-64—[N-(*trans*-epoxysuccinyl)-1-leucine 4-guanidinobutylamide] against cysteine proteases; leupeptin hemisulfate salt against both serine and cysteine proteases; pepstatin A against acid proteases), 10 μl of 100× phosphatase inhibitor cocktail (containing: Na$_3$VO$_4$ inhibiting a number of ATPases and protein tyrosine phosphatases, Na$_2$MoO$_4$ inhibiting acid and phosphoprotein phosphatases, Na$_2$C$_4$H$_4$O$_6$ inhibiting acid phosphatases, imidazole inhibiting alkaline phosphatases), and 10 μl of 100 mM PMSF (final conc. 1 mM).

2.3.2 Isolation of Nuclear/Cytosolic Fractions from PMMEC

1. Lysis buffer for cytosolic fraction (20 ml volume): 0.2 ml (10 mM) 1 M HEPES, pH 7.9, 0.2 ml (10 mM) 1 M KCl, 30 μl (1.5 mM) 1 M MgCl$_2$; fill up to 20 ml with dH$_2$O and store at 4 °C. Prior to use add to 1 ml of lysis buffer: 0.5 μl (0.5 mM) 1 M DTT, 0.2 μl (0.2 mM) of 100 mM PMSF, 10 μl of 100× protease inhibitor cocktail, 10 μl of 100× phosphatase inhibitor cocktail.

2. Nuclear extraction buffer (20 ml volume): 0.2 ml (10 mM) 1 M Tris–HCl, pH 7.4, 0.2 ml (10 mM) 1 M NaCl, 0.4 ml (10 mM) 0.5 M EDTA, 0.2 ml (1%) Triton X-100, 60 μl (3 mM) 1 M MgCl$_2$; fill up to 20 ml with dH$_2$O and store at 4 °C. Prior to use add (amounts for 1 ml of lysis buffer): 40 μl (4 mM) 100 mM PMSF, 10 μl of 100× protease inhibitor cocktail, 10 μl of 100× phosphatase inhibitor.

2.4 Immuno-fluorescent Staining of PMMEC Grown in 3D Cultures

1. Phosphate-buffered saline (PBS).

2. Fixation buffer: 3.7% paraformaldehyde solution prepared in PBS, pH 7.5.

3. Permeabilization buffer: 0.5% Triton-X100 prepared in PBS.

4. Blocking solution: 5% Normal goat serum (NGS)—diluted with PBS to 5% NGS solution.

5. Washing solution: 0.1% Tween 20 in PBS.

6. 4′,6-Diamidino-2phenylindole (DAPI)—for nuclei staining: Stock solution of 5 μg/ml in water.

7. Mounting medium with antifade reagent.

8. Cover slips.

3 Methods

3.1 Mammary Gland Collection

Mammary glands from 14- to 16-day pregnant mice are used as source of PMMEC. We use mice of the albino CD-1 strain which are maintained as an outbred strain. In order to collect the mammary glands we usually use 10–15 mid-pregnant mice, which are euthanized humanely. The mammary glands are dissected out using the following procedure:

1. Pin down the euthanized mice with the ventral surface facing up.

2. Spray the skin with 70% ethanol to disinfect the surface.

3. Grab the skin anteriorly using forceps near the urethral opening, and it is cut along the ventral midline using scissors, from the groin to the chin.

4. Make 10–15 mm incisions from the center of the torso to each side of the animals.

5. Peel the subcutaneous skin away from the peritoneal cavity using forceps, exposing the mammary glands on each side.

6. Dissect the mammary glands out using forceps and sterile scalpel blades, and put aseptically into a sterile 50 ml polypropylene tube placed on ice.

3.2 Isolation of PMMEC from the Mammary Glands

1. Prepare a fresh collagenase digestion mixture (*see* Subheading 2.2) prior to isolation (*see* **Note 4**).

2. Take the mammary glands out of the tube, place it in a 100 mm tissue culture dish and mince criss-crossing two sterile scalpel blades (*see* **Note 5**).

3. Place the minced glands in a sterile 250 ml glass conical flask with the collagenase digestion mixture (~4 ml of digestion mixture per gram of tissue). Swirl to allow the minced tissue to separate.

4. Carry out the digestion for 90 min on a shaking table set at 250 rpm, at 37 °C.

5. Use the following stringent washing protocol to isolate epithelial cells from fibroblasts. All of the washes and centrifugation steps are carried out in 50 ml sterile polypropylene tubes. Carry out the selective centrifugation as follows:

(a) Move the digested cell suspension to a 50 ml tube and centrifuge for 30 s at 100 rpm to pellet undigested alveolar structures.

(b) Transfer the supernatant to a fresh tube and centrifuge at 800 rpm for 3 min.

6. Prepare the DNase mixture according to Subheading 2.2 of the protocol, resuspend the pellet in the DNase mixture, and incubate at 37 °C for 30 min on a shaking platform at 150 rpm (*see* **Note 6**).

7. Transfer the cell suspension to a fresh 50 ml tube and centrifuge at 800 rpm for 3 min.

8. Discard the supernatant and resuspend the pellet in 30–50 ml of preheated (37 °C) Ham's F12 culture medium (*see* Subheading 2.2), depending on the size of the pellet.

9. Add the cell suspension to Ham's F12 medium containing pen/strep with the following hormones and growth factors: hydrocortisone 1 μg/ml, insulin 5 μg/ml, and EGF 5 ng/ml.

3.3 3D Culture of PMMEC on Laminin-Rich ECM, Using the "On-Top" Assay

In our experiments we use the cell culture protocol in which MECs are cultured on top of a thin layer of Matrigel [18, 19, 26] (*see* Note 7).

1. Prior to seeding of cells coat the culture dishes and/or chamber slides with a thin layer of GFR Matrigel, which is spread evenly on the culture surface, containing at least 9 mg/ml protein concentration (*see* **Note 8**).

2. Use 200 µl of liquid GFR Matrigel to coat a 60 mm tissue culture plastic dish; or a 4-well chamber slide using 50 µl GFR Matrigel per well on ice (*see* **Note 9**).

3. Leave the culture dishes and chamber slides for 15 min in a humidified incubator at 37 °C, with 5 % CO_2 for Matrigel to solidify.

4. In order to generate 3D spherical structures, seed PMMEC at a density of 2.4×10^6 cells/ml onto the prepared 60 mm tissue culture plastic dishes (5 ml per dish), or at a density of 5×10^3 cells/ml on 4-well chamber slides.

5. Grow the cells in a fetal bovine serum-free proliferation medium consisting of Ham's F12 (Pen/Strep) with the following hormones and growth factors: hydrocortisone 1 µg/ml, insulin 5 µg/ml and EGF 5 ng/ml.

6. Culture the cells at 37 °C in a humidified atmosphere of 5 % CO_2 for up to 96 h, refreshing the medium after 48 h of culture.

3.3.1 Blocking Protein-Tyrosine Phosphatases (PTP) Activity in PMMEC Using Pervanadate

Our studies have shown that development of normal acini by PMMEC cultured on Matrigel requires time restriction of MAPK/ERK pathway activation. We demonstrated that sustained ERK phosphorylation leads to formation of aberrant structures with filled lumen [19]. Furthermore, we showed a correlation between the levels of phosphorylated ERK kinases, and the levels of MKP-2, and -3 in the course of acini development. Under normal (control) conditions the highest degree of ERK phosphorylation was observed during the first 16 h after seeding cells on Matrigel, and then the activity of these kinases gradually diminished; whereas the levels of MKP-2, and -3 were increasing until the 24th hour of culture, and then slightly decreased in the 48th hour [19]. Thus, in order to investigate the role of both MKPs in regulation of MAPK/ERK pathway and mammary acini formation we performed experiments with the use of pervanadate (VO_3, 25 mM), a known potent protein tyrosine phosphatase (PTP) inhibitor [25] (Fig. 1).

1. Prepare 0.2 M sodium pervanadate (Na_3VO_4) stock solution by weighing 0.736 g of Na_3VO_4 and dissolving it in 20 ml of dH_2O (*see* **Note 10**). Adjust the pH to 10 by adding 2–3 drops of concentrated HCl (*see* **Note 11**) and boil the solution until colourless. Cool down to room temperature and store at room temperature (*see* **Note 12**).

2. Prepare 1 ml fresh 20 mM VO_3 stock solution by mixing equal volumes of 20 mM Na_3VO_4 and 20 mM H_2O_2 (2.3 µl of 30 % H_2O_2 and 100 µl of 0.2 M Na_3VO_4; fill up with 897,7 µl dH_2O).

3. Incubate the mixture for 30 min at room temperature. Stop the reaction by adding 1 mg of catalase (*see* **Note 13**) and keep on ice until use.

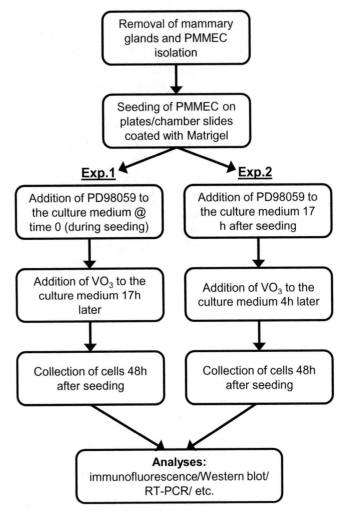

Fig. 1 Scheme of in vitro experiments used to investigate the effect of pharmacological inhibition of ERK1/2 by PD98059 added to 3D cultures of PMMEC, followed by restoration of kinase activity using pervanadate (VO_3)—potent inhibitor of protein tyrosine phosphatases

Experiment 1: Inhibition of MAPK/ERK pathway by PD98059 added to the medium at the time of seeding cells, and inhibition of MKPs 17 h later—role of MAPK/ERK signalling pathway in the first hours of mammary acini development (Fig. 2).

1. Prepare PMMEC, plates and chamber slides according to the protocols presented in Subheadings 3.2 and 3.3.

2. Seed 2.4×10^6 cells/ml into 60 mm tissue culture plastic dishes coated with GFR Matrigel, or 5×10^3 cells/ml on 4-well chamber slides, and grow them in 5 and 1 ml, respectively, of a fetal bovine serum-free proliferation medium, consisting of Hams F12 (Pen/Strep) supplemented with hormones: (hydrocortisone 1 μg/ml, insulin 5 μg/ml, EGF 5 ng/ml).

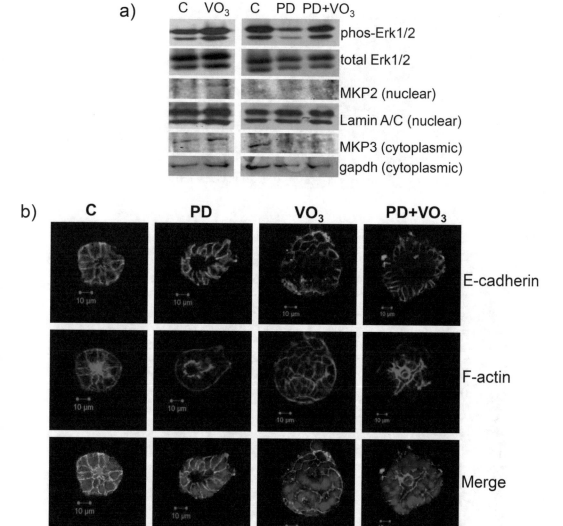

Fig. 2 Role of MAP kinases ERK1/2 and mitogen-activated protein kinase phosphatases (MKPs) at early stages of acini formation by PMMEC cultured on Matrigel. The activity of ERK1/2 was inhibited by PD98059 (PD) added from the beginning of 3D culture (time 0), followed by restoration of kinases' activity using protein tyrosine phosphatase inhibitor: pervanadate (VO₃) added after subsequent 17 h

3. Add 5 μl of 10 mM PD98059 stock solution per 60 mm dish, or 1 μl of 10 mM PD98059 stock solution per well in order to obtain a final concentration of 10 μm PD98059.

4. Culture the cells at 37 °C in a humidified atmosphere of 5 % CO_2.

5. Replace the medium after 17 h and add 25 μm VO₃ to appropriate dishes (1.25 μl of 20 mM VO₃ per 1 ml of medium) or wells in a chamber slide (VO₃, and PD+ VO₃). Add 1.25 μl of catalase C30 per 1 ml of medium to the controls.

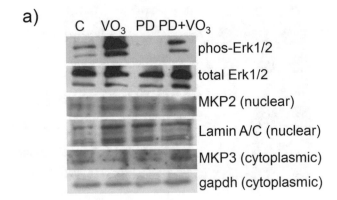

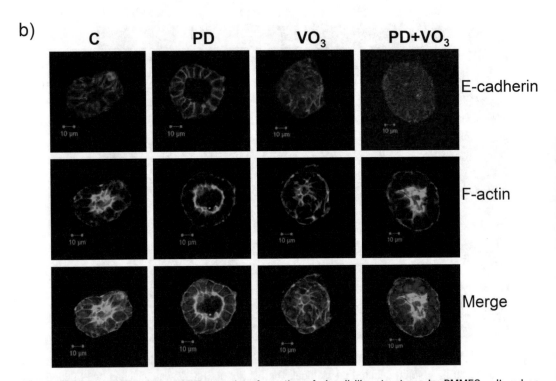

Fig. 3 MAP kinases ERK1/2 and MKPs regulate formation of alveoli-like structures by PMMEC cultured on Matrigel influencing the processes of cell polarisation and formation of hollow lumen of the acini. The activity of ERK1/2 was inhibited by PD98059 (PD) added to the culture medium 17 h post-seeding. Protein tyrosine phosphatase inhibitor: pervanadate (VO$_3$) was added after subsequent 4 h (at 21 h of cell culture)

6. 48 h after seeding harvest the cells from the culture dishes for Western blot analysis, or fix cells in the in the chamber slides for immunofluorescent staining (Fig. 2).

Experiment 2: Inhibition of MAPK/ERK pathway by PD98059 added to the medium 17 h after seeding the cells, and inhibition of PTPs 4 h later (21 h after seeding PMMEC)—role of MAPK/ERK signalling pathway in the later stages of mammary acini formation (Fig. 3).

1. Prepare PMMEC, plates and chamber slides are prepared according with the protocols presented above.

2. Seed 2.4×10^6 cells/ml into 60 mm tissue culture plastic dishes coated with GFR Matrigel, or 5×10^3 cells/ml on 4-well chamber slides, and grow them in 5 and 1 ml, respectively, of a fetal bovine serum-free proliferation medium, consisting of Hams F12 (Pen/Strep) supplemented with hormones: (hydrocortisone 1 μg/ml, insulin 5 μg/ml, EGF 5 ng/ml).

3. Culture the cells at 37 °C in a humidified atmosphere of 5 % CO_2.

4. Replace the medium after 17 h and add 5 μl of 10 mM PD98059 stock solution per dish, or 1 μl of 10 mM PD98059 stock solution per well in order to obtain a final concentration of 10 μm.

5. After another 4 h (21st hour of culture) add 25 μm VO_3 to appropriate dishes (1.25 μl of 20 mM VO_3 per 1 ml of medium) or wells in a chamber slide (VO_3, and PD+ VO_3). Add 1.25 μl of catalase per 1 ml of medium to the controls.

6. 48 h after seeding harvest the cells from the culture dishes for Western blot analysis, or fix cells in the in the chamber slides for immunofluorescent staining (Fig. 3).

3.4 Protein Extraction

3.4.1 Protein Extraction for Measuring MAPK Activity

When analyzing the levels of kinases we usually use whole-cell extracts isolated using RIPA buffer, as it is one of the most reliable buffers used to lyse cultured mammalian cells from both plated cells and cells pelleted from suspension cultures. This buffer enables extraction of cytoplasmic, membrane and nuclear proteins and is compatible with many applications, including reporter assays, protein assays, immunoassays, and protein purification.

1. Carefully remove the medium and wash the cells in ice-cold PBS (on ice) (see **Note 14**).

2. Scrape the cells into 1 ml of PBS and centrifuge for 3 min at 14,000 rpm (20,000 g) to pellet the cells, remove the PBS, and store the pellet at –80 °C prior to protein extract preparation (see **Note 15**).

3. Resuspend the pelleted acini in 60–100 μl of RIPA buffer, depending on the size of the pellet.

4. Incubate the samples on ice for 30 min.

5. Centrifuge the sample at 14,000 rpm (20,000 g) at 4 °C for 20 min to remove all insoluble particulate matter.

6. Retain the resulting supernatant, snap freeze it in liquid nitrogen, store it at –80 °C, and discard the pellet.

3.4.2 Cell Fractionation for Measuring MKP-2 and 3 Levels

When analyzing the levels of MKP-2 we use nuclear extracts, as this PTP is localized and acts mainly in the nucleus of cells; whereas the level of MKP-3 is determined in the cytosolic fraction (see **Note 16**).

In order to analyze the levels of the abovementioned proteins cell fractioning protocol is performed, using two different lysis buffers.

1. Resuspend the pelleted acini in 60–100 μl of lysis buffer (*see* Subheading 2.3.2), depending on the size of the pellet (*see* **Note 14**).

2. Incubate the samples on ice for 30 min.

3. Centrifuge the resuspended sample at 2500 rpm (1125*g*) at 4 °C for 30 min.

4. Snap freeze the supernatant (crude cytosolic fraction) in liquid nitrogen and store it at –80 °C until further use.

5. The remaining pellet containing nuclear proteins and membranes is resuspended (nuclear and membrane/insoluble fraction) in 50 μl of nuclear extraction buffer (*see* Subheading 2.3.2).

6. Incubate the samples on ice for 30 min.

7. Centrifuge the samples at 14,000 rpm (20,000 *g*) at 4 °C for 10 min.

8. Snap freeze the supernatant (nuclear extract) in liquid nitrogen and store it at –80 °C until further analysis by Western blotting.

3.5 Immuno-fluorescence Staining of PMMEC

1. Remove the medium from the cells and rinse the chamber slides once with ice-cold PBS.

2. Fix the cells with 3.7 % paraformaldehyde, pH 7.5 for 20 min at RT.

3. Remove the paraformaldehyde and wash the cells once in PBS.

4. Permeabilize the cells by incubation in 0.5 % Triton-X-100/PBS solution for 15–20 min with gentle agitation on an oscillating table.

5. Wash the cells twice with PBS.

6. Incubate cells in blocking buffer (5 % Normal Goat Serum in PBS) for 1 h at room temperature, with gentle agitation (*see* **Note 17**).

7. Remove the outer chambers of the slides and blot excess fluid blotted from the periphery of the fixed cells.

8. Dilute the antibodies in blocking buffer according to the manufacturer's instructions. For a four-chambered slide, drop 25 μl of diluted antibody in the center of each well and lay a cover slip prepared from parafilm gently over the slide.

9. Incubate the primary antibody overnight at 4 °C in a humidity chamber, to prevent dehydration.

10. Remove the antibody solution by immersing slides in PBS (*see* **Note 18**).

11. Dry the slides and remove excess moisture.

12. Dilute the fluorescent secondary antibodies (FITC goat anti-mouse IgG or TRITC goat anti-mouse IgG) in 500 µl of blocking buffer and place 25 µl onto slides as described in **step 8** of protocol. Additionally, stain F-actin directly using phalloidin conjugated to Alexa Fluor 594 dye (red fluorescence) or to FITC (green fluorescence). Dilute the fluorescent conjugated phalloidin 250× in 500 µl of blocking buffer and place 25 µl on the slides.

13. Incubate the slides for 40–50 min at RT in the dark.

14. Wash the slides briefly in PBS.

15. Incubate the slides in a 1 µg/ml solution of DAPI (stock solution 10 mg/ml) in PBS for 30 s, rinse off the excess DAPI solution in PBS, and blot excess liquid with tissue paper.

16. Cover the slides with one drop of antifade mounting medium per well before applying a coverslip.

17. To insure that the cover slip remains in place, apply clear nail varnish around its perimeter.

18. Keep the slides in the dark at 4 °C until visualisation.

19. Visualize the fluorescence with a laser scanning confocal microscope (*see* **Note 19**). Repeat all immunofluorescence analyses at least three times on independently generated cell cultures.

4 Notes

1. PD98059 is a potent and selective inhibitor of MAPK/ERK kinase, which mediates its inhibitory properties by binding to the ERK-specific MAP kinase MEK, therefore preventing phosphorylation of ERK1/2 (p44/p42 MAPK) by MEK1/2.

2. Always prepare fresh prior to use.

3. Tris base prevents protein denaturation, while NaCl prevents nonspecific protein aggregation.

4. All subsequent steps in this preparation should be carried out in the laminar flow hood.

5. It is essential to mince the tissue well in order to get effective collagenase/trypsin digestion.

6. DNase treatment is necessary to separate single epithelial cells by stopping cell-cell aggregation.

7. The "on-top" assay requires relatively small amounts of Matrigel used to coat the culture surface of plates and chamber slides in comparison to the 3D embedded assay. Furthermore, it allows using simple fixation/immunofluorescence staining protocols for confocal microscopy, which do not involve preparation of frozen or paraffin sections of spheroids prior to staining.

8. Matrigel is stored at –20 °C, and it has to be thawed at 0–4 °C temperature (in ice/water bucket) prior to plating in order to obtain a liquid state. Throughout the entire procedure Matrigel has to be kept on ice to prevent premature gelling of the matrix!

9. The coating is done on ice, to keep the matrix in the liquid state during the procedure. We use sterile, disposable inoculation loops to spread Matrigel evenly on the surface of culture dishes or chamber slides. Depending on the size of the culture dish we use 100 µl of Matrigel on 30 mm dish, 200 µl (can be reduced to 180 µl) on 60 mm dish, and 400 µl on 100 mm dish. It is important to start coating the dishes when Matrigel is completely liquidized (thawed on ice/water mixture), otherwise a lot of bubbles may be formed during spreading the Matrigel, and there may be a need to use bigger volumes of the matrix.

10. The pervanadate solution has to be prepared freshly before each experiment.

11. The solution may turn yellow.

12. If the pH of the cooled solution is not 10, the procedure of pH adjustment and boiling has to be repeated.

13. Catalase will destroy any unreacted H_2O_2 in the solution.

14. We usually extract proteins from a single 60 mm culture dish/ experimental condition.

15. Cell scrapers are used to harvest cells for further analysis. During this procedure Matrigel is scraped along with the cells, thus ECM proteins are present in the cells' suspension in PBS. That is why it is important to centrifuge the cells at high speed (14,000 rpm, 20,000 g). After centrifugation PMMEC will be gathered on the bottom of the Eppendorf, and a thin layer of gel (Matrigel) is visible above. We recommend aspirating the gel-layer along with PBS in order to minimize contamination of the sample with ECM components during protein extraction procedures. Alternatively, cells can be recovered from Matrigel using enzymatic or nonenzymatic procedure with a cell recovery solution, which depolymerizes Matrigel matrix gels without enzymatic digests and lengthy incubation periods at high temperatures. Cells are released without damage thereby avoiding biochemical changes during incubation and digestion of extracellular portions of cell-surface receptors and adhesion molecules. However, some cells may be lost following this protocol, so it is recommended to use bigger culture dishes (100 mm) or a few replicates for each experimental variant. We use this recovery solution before RNA extraction, i.e., before suspending cells in TRIzol or QIAzol reagent.

16. When analyzing protein expression in different cell fractions (nuclear and cytosolic) it is important to use reference proteins that are characteristic for a specific fraction. We detect the levels of lamin A/C (nuclear membrane structural components) as the reference protein of the nuclear fraction, and gapdh levels (glyceraldehyde 3-phosphate dehydrogenase) as the reference in cytosolic extracts. Remember that commonly used beta-actin can be detected in both fractions, so it is impossible to determine potential contamination of nuclear extracts with cytosolic proteins using beta-actin as the reference protein.

17. The blocking solution will reduce nonspecific antibody binding reactions.

18. This procedure will also remove the parafilm cover slip without damaging the cells.

19. We use an Enterprise Blue diode (405 nm), krypton/argon (458, 477, 488, 504 nm), helium-neon (He-Ne) 543 nm, and He-Ne 633 nm lasers. Single focal plane images containing single- or multiple-channel colors (fluorescence outputs) are acquired. All confocal images are taken at the midpoint section of acini (lasers emitting a number of lines in the UV and visible ranges 351, 364, 458, 477, 488, 514, 543, and 633 nm).

References

1. Johnson KR, Leight JL, Weaver VM (2007) Demystifying the effects of a three-dimensional microenvironment in tissue morphogenesis. Methods Cell Biol 83:54–583

2. Swamydas M, Eddy JM, Burg KJ, Dréau D (2010) Matrix compositions and the development of breast acini and ducts in 3D cultures. In Vitro Cell Dev Biol Anim 46:673–684

3. Inman JL, Bissell MJ (2010) Apical polarity in three-dimensional culture systems: where to now? J Biol 9:2

4. Lo AT, Mori H, Mott J, Bissell MJ (2012) Constructing three-dimensional models to study mammary gland branching morphogenesis and functional differentiation. J Mammary Gland Biol Neoplasia 17:103–110

5. Campbell JJ, Watson CJ (2009) Three-dimensional culture models of mammary gland. Organogenesis 5:43–49

6. Emerman JT, Pitelka DR (1977) Maintenance and induction of morphological differentiation in dissociated mammary epithelium on floating collagen membranes. In Vitro 13:16–68

7. Streuli CH, Bissel MJ (1990) Expression of extracellular matrix components is regulated by substratum. J Cell Biol 110:1405–1415

8. Roskelly CD, Desprez PY, Bissel MJ (1994) Extracellular matrix-dependent tissue-specific gene expression in mammary epithelial cells requires both physical and biochemical signal transduction. Proc Natl Acad Sci 91:12378–12382

9. Li ML, Aggeler J, Farson DA, Hatier C, Hassell J, Bissell MJ (1987) Influence of a reconstituted basement membrane and its components on casein gene expression and secretion in mouse mammary epithelial cells. Proc Natl Acad Sci 84:136–140

10. Miranti CK, Brugge JS (2002) Sensing the environment: a historical perspective on integrin signal transduction. Nat Cell Biol 4:83-E90

11. Kleinman HK, Martin GR (2005) Matrigel: basement membrane matrix with biological activity. Semin Cancer Biol 15:378–386

12. Barcellos-Hoff MH, Aggeler J, Ram TG, Bissell MJ (1989) Functional differentiation and alveolar morphogenesis of primary mammary cultures on reconstituted basement membrane. Development 105:223–235

13. Debnath J, Mills KR, Collins NL, Reginato MJ, Muthuswamy SK, Brugge JS (2002) The role of apoptosis in creating and maintaining luminal space within normal and oncogene-expressing mammary acini. Cell 111:29–40

14. Debnath J, Muthuswamy SK, Brugge JS (2003) Morphogenesis and oncogenesis of

MCF-10A mammary epithelial acini grown in three-dimensional basement membrane cultures. Methods 30:256–268

15. Streuli CH, Bailey N, Bissell MJ (1991) Control of mammary epithelial differentiation: basement membrane induces tissue-specific gene expression in the absence of cell-cell interaction and morphological polarity. J Cell Biol 115:1383–1395

16. Naylor MJ, Li N, Cheung J, Lowe ET, Lambert E, Marlow R, Wang P, Schatzmann F, Wintermantel T, Schüetz G, Clarke AR, Mueller U, Hynes NE, Streuli CH (2005) Ablation of beta1 integrin in mammary epithelium reveals a key role for integrin in glandular morphogenesis and differentiation. J Cell Biol 171:717–728

17. Marshman E, Streuli CH (2002) Insulin-like growth factors and insulin-like growth factor binding proteins in mammary gland function. Breast Cancer Res 4:231–239

18. Murtagh J, McArdle E, Gilligan E, Thornton L, Furlong F, Martin F (2004) Organization of mammary epithelial cells into 3D acinar structures requires glucocorticoid and JNK signaling. J Cell Biol 166:133–143

19. McNally S, McArdle E, Gilligan E, Napoletano S, Gajewska M, Bergin O, McCarthy S, Whyte J, Bianchi A, Stack J, Martin F (2011) c-Jun N-terminal kinase activity supports multiple phases of 3D-mammary epithelial acinus formation. Int J Dev Biol 55:731–744

20. Misra-Press A, Rim CS, Yao H, Roberson MS, Stork PJ (1995) A novel mitogen-activated pro-tein kinase phosphatase. Structure, expression, and regulation. J Biol Chem 270:14587–14596

21. Brondello J-M, Brunet A, Poueysségur J, McKenzie FR (1997) The dual specificity mitogen-activated protein kinase phospha-tase-1 and -2 are induced by the p42/p44MAPK cascade. J Biol Chem 272:1368–1376

22. Volmat V, Camps M, Arkinstall S, Poueysségur J, Lenormand P (2001) The nucleus, a site for signal termination by sequestration and inacti-vation of p42/p44 MAP kinases. J Cell Sci 114:3433–3443

23. Kim-Kaneyama J, Nose K, Shibanuma M (2000) Significance of nu clear relocaliza-tion of ERK1/2 in reactivation of c-fos transcription and DNA synthesis in senes cent fibroblasts. J Biol Chem 275:20685–20692

24. Robinson MJ, Stippec SA, Goldsmith E, White MA, Cobb MH (1998) A constitutively active and nuclear form of the MAP kinase ERK2 is sufficient for neurite outgrowth and cell trans-formation. Curr Biol 8:1141–1150

25. Zhao Z, Tan Z, Diltz CD, You M, Fischer EH (1996) Activation of mitogen-activated pro-tein (MAP) kinase pathway by pervanadate, a potent inhibitor of tyrosine phosphatases. J Biol Chem 271:22251–22255

26. Furlong F, Finlay D, Martin F (2005) PTPase inhibition restores ERK1/2 phosphorylation and protects mammary epithelial cells from apoptosis. Biochem Biophys Res Commun 336:1292–1299

Chapter 10

A 3D Fibroblast-Epithelium Co-culture Model for Understanding Microenvironmental Role in Branching Morphogenesis of the Mammary Gland

Zuzana Koledova and Pengfei Lu

Abstract

The mammary gland consists of numerous tissue compartments, including mammary epithelium, an array of stromal cells, and the extracellular matrix (ECM). Bidirectional interactions between the epithelium and its surrounding stroma are essential for proper mammary gland development and homeostasis, whereas their deregulation leads to developmental abnormalities and cancer. To study the relationships between the epithelium and the stroma, development of models that could recapitulate essential aspects of these interacting systems in vitro has become necessary. Here we describe a three-dimensional (3D) co-culture assay and show that the addition of fibroblasts to mammary organoid cultures promotes the epithelium to undergo branching morphogenesis, thus allowing the role of the stromal microenvironment to be examined in this essential developmental process.

Key words 3D culture, Branching morphogenesis, Extracellular matrix, Fibroblasts, Mammary epithelium, Matrigel, Organoids, Organotypic assay, Paracrine signaling

1 Introduction

The mouse mammary gland is a complex organ. It is comprised of an epithelial ductal tree that is embedded in a rather complex stromal microenvironment, consisting of the ECM and multiple stromal cell types, including fibroblasts, endothelial cells, and infiltrating leukocytes. In addition to providing a scaffold for the organ, the stroma regulates the function, proliferation, differentiation, and invasion of mammary epithelial cells via an intricate network of chemical signaling and physical controls. These interactions between the epithelial cells and stroma are tightly regulated to ensure proper mammary gland development and physiological function [1]. Despite remarkable progress in the past decade, including that based on genetic, transplantation, and tissue recombination studies, the molecular mechanisms underlying the epithelial-stromal cross talk remain poorly understood. As a result, there has been a

Finian Martin et al. (eds.), *Mammary Gland Development: Methods and Protocols*, Methods in Molecular Biology, vol. 1501, DOI 10.1007/978-1-4939-6475-8_10, © Springer Science+Business Media New York 2017

growing demand for the development of ex vivo culture methods that could recapitulate various aspects of mammary gland development, especially those concerning epithelial-stromal interactions.

Conventional, two-dimensional (2D) culture of mammary epithelial cells does not resemble the structure or the function of the mammary epithelium in vivo [2]. To preserve the 3D architecture of mammary epithelium, the inclusion of ECM matrices, an essential component of the stromal microenvironment, within the culture system becomes necessary. The commonly used matrices include collagen type I, Matrigel (a reconstituted basement membrane product), or their combinations [2–4]. Importantly, by manipulating matrix composition, biophysical properties of the ECM (such as stiffness) can be altered and their effects on mammary epithelial morphogenesis can be studied. Several studies demonstrated that fundamental processes of mammary epithelial growth and morphogenesis, including cell shape, polarity, functional differentiation, or invasiveness, are regulated by matrix stiffness [5] or distinct patterns of fiber organization [6].

Further development of 3D epithelial cultures with the addition of stromal cells has enabled complex epithelial-stromal interactions to be examined in a defined ECM context. Compared with whole-mount samples or explant cultures, these surrogate models of mammary gland are amenable to experimental manipulations and are readily available to high-throughput imaging and chemical screening [7]. Together, the 3D co-cultures allow distinct components of the stroma and the epithelium to be perturbed so that their contributions to normal tissue morphogenesis or tumor formation could be assessed.

Several 3D co-culture models of epithelial cells with fibroblasts have been reported [2, 8–11]. However, these models use either mammary epithelial cell lines or primary breast epithelial cells that have been cultured on plastic dishes, whose extreme stiffness have profound impact on cell behaviour [12]. To develop a co-culture model that resembles the in vivo system, we have developed a protocol that uses primary mammary organoids. We found that the physiological bi-layered epithelial organization is an excellent tool for studying processes of branching morphogenesis and their regulation by mammary fibroblasts, especially by paracrine signaling. Here we describe the protocol for isolation of mammary epithelial organoids and mammary fibroblasts from mouse mammary glands using enzymatic digestion and differential centrifugation and for their embedding in Matrigel to form a 3D co-culture (Fig. 1).

2   Materials

2.1   Mice

Female mice of various strains and ages (*see* **Note 1**) can be used to prepare mammary organoids and fibroblasts using this protocol. We recommend, however, the use of donor mice of a pubertal age

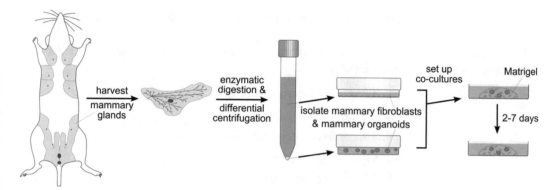

Fig. 1 A schematic drawing of mammary organoid and fibroblast isolation and 3D co-culture setup

(5–10 weeks), when mammary branching morphogenesis is actively ongoing (*see* **Note 2**).

2.2 Reagents and Solutions

All reagents for tissue culture should be sterilized by autoclaving or filtration through a 0.22 μm filter.

1. Phosphate-buffered saline (PBS) without Ca^{2+} and Mg^{2+}.

2. Insulin (powder): Dissolve 50 mg insulin in 50 ml of 0.1 M HCl, pH 2–3. Filter sterilize and store at –20 °C as 250 μl aliquots.

3. Collagenase solution: 2 mg/ml Collagenase A from *Clostridium histolyticum*, 2 mg/ml trypsin, 5% (v/v) fetal bovine serum (FBS), 5 μg/ml insulin, 50 μg/ml gentamicin in Dulbecco's modified Eagle medium (DMEM)/F12. In a 50 ml tube or 100 ml flask, combine 47.5 ml of DMEM/F12 with 2.5 ml of FBS, 50 μl of gentamicin, and 250 μl of insulin. Warm the solution in a 37 °C water bath. Add 0.1 g of collagenase and 0.1 g of trypsin to the warmed solution and shake at 37 °C until dissolved (10–15 min). Filter-sterilize and keep warm until used. Prepare a fresh solution for each primary cell preparation.

4. Deoxyribonuclease I (DNase I): Dissolve 2000 U in 1 ml of 0.15 M NaCl. Filter sterilize and make 40 μl aliquots. Store at –20 °C.

5. Bovine serum albumin (BSA) solution: 2.5% BSA in PBS.

6. Basal organoid co-culture medium: 1× Insulin-transferrin-selenium (ITS), 1× penicillin/streptomycin (100 U/ml penicillin, 100 μg/ml streptomycin) in DMEM/F12.

7. Fibroblast medium: 10% FBS, 1× ITS, 1× penicillin/streptomycin in DMEM.

8. Trypsin-EDTA (0.05% trypsin, 0.2% EDTA).

9. Growth factor-reduced Matrigel.

10. Fixation solution: 4 % paraformaldehyde (PFA) in PBS.

2.3 Instruments, Equipment, and Other Materials

All metal instruments for tissue dissection should be sterilized by autoclaving.

1. Dissection board and pins.

2. Standard forceps, straight.

3. Graefe iris forceps, straight.

4. Operating scissors.

5. Iris/eye scissors, straight.

6. Sterile cotton buds.

7. Sterile polystyrene Petri dish.

8. Sterile disposable scalpels no. 20 or 23.

9. Benchtop incubator orbital shaker.

10. Centrifuge with swing-bucket rotor.

11. CO_2 incubator, set to 37 °C and 5 % CO_2.

12. Centrifuge tubes, 15 and 50 ml.

13. Disposable plastic serological pipettes, 5, 10, 25 ml.

14. Electronic pipettor.

15. Adjustable volume pipettes and pipette tips.

16. Ice bucket.

17. Inverted microscope.

18. Hemacytometer.

19. 24-well cell culture plate, flat bottom.

20. 100 mm cell culture dishes.

21. Heat block.

3 Methods

3.1 Harvesting Mouse Mammary Glands

Euthanize donor mice according to the approved protocol (such as by cervical dislocation). Proceed with dissection immediately. Tissue could be stored at 4 °C overnight but doing so reduces cell

Fig. 2 (continued) Minced mammary glands in collagenase solution at the beginning of digestion (**e**) and after 30 min of digestion (**f**). (**g**) After centrifugation, digested mammary gland tissue separates into a fatty top layer, a middle aqueous layer, and cell pellet at the bottom of the tube. (**h**) Differential centrifugation separates mammary cells according to their weight. The heavier epithelial organoids sink to the bottom of the tube, while the stromal cells remain in the supernatant. After four rounds of differential centrifugation, the epithelial cell pellet is *white*. (**i**) The supernatant from differentially centrifuged samples was collected, pooled, and centrifuged to collect stromal cells. (**j**) Mammary organoids. (**k–m**) Stromal fraction. The stromal fraction was plated on a cell culture dish (**k**) and incubated for 30 min in an incubator; during this time, fibroblasts have attached to the dish, while other cell types have remained in suspension (**l**). The dish was washed two times with PBS to remove unattached cells, leaving predominantly fibroblasts (*arrowheads*) attached to the dish, with a minor contamination by epithelial cells (*arrow*). (**n**) Mammary fibroblasts, cultured for 24 h (passage number 1). (**o**) Mammary fibroblasts, passage number 4

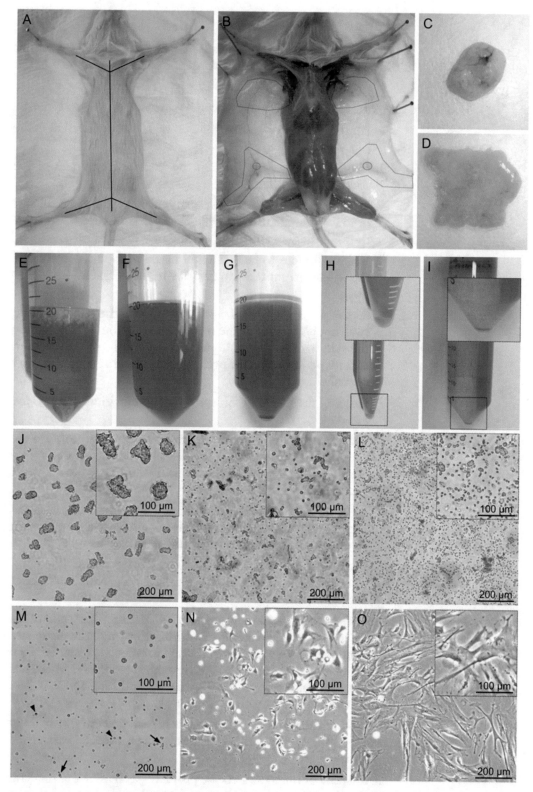

Fig. 2 Isolation of mammary organoids and fibroblasts. (**a**) A schematic drawing of mammary gland surgical access. *Black lines* indicate suggested cuts. (**b**) Exposed mammary glands. *Dotted lines* indicate the approximate region of mammary glands to be collected. *Red lines* indicate the lymph nodes which are to be excised prior to harvesting the glands. (**c**, **d**) Dissected mammary glands collected in a dish (**c**) and minced (**d**). (**e**, **f**)

survival after harvest (*see* **Note 3**). Tissue dissection is carried out in a laminar flow hood to ensure aseptic conditions.

1. Wash the thorax, abdomen, and sides of the mouse thoroughly with 70% ethanol to disinfect the skin incision area (*see* **Note 4**).

2. Pin the mouse to the dissection board facing up, stretching the legs wide (Fig. 2a).

3. Using standard forceps, grab the skin in the middle of the abdomen and pull it up.

4. Using operating scissors, make an initial cut in the medial part of lower abdomen, below the forceps. Take care to cut only through the skin and avoid cutting peritoneum.

5. Use the forceps to hold the skin in the initial incision and use the scissors to cut the skin cranially to the neck and caudally to the groin.

6. Separate the skin from the peritoneum with the use of a cotton swab on both sides of the medial incision line: While holding the skin with forceps and pulling it up, push the cotton swab against peritoneum and peel the skin of peritoneum, moving laterally.

7. Holding the skin with forceps, make four incisions towards the limbs using scissors.

8. Using forceps to hold the skin and a cotton swab to push the peritoneum back, peel the skin off (one side at a time), opening the abdomen and exposing the mammary glands. Pin the skin to the dissection board.

9. Identify the mammary glands. Using a cotton swab, push the abdominal and thoracic mass and leg muscles (yellow to brown colour) back to allow access to the mammary tissue.

10. Identify the lymph nodes in #4 mammary glands (Fig. 2b). Using iris scissors and iris forceps, dissect the lymph nodes out and discard them.

11. Collect the mammary glands #3, #4, and #5: Use a pair of iris forceps to hold the mammary gland; gently pull the mammary gland away from the skin and using iris scissors, dissect it from the skin. Take care not to cut through the skin. When collecting the mammary glands #3, take a good care not to collect the muscle (brownish color; *see* **Note 5**).

12. Collect the mammary gland tissue in a sterile Petri dish in a few (1–3) ml of sterile PBS (*see* **Note 6**) and proceed with tissue processing.

13. Dispose of the mouse corpse properly (such as by freezing and disposing in biological hazard/clinical waste bins), clean the dissection tools (such as by submerging them in cleaning and decontaminating solution, e.g., 10% Trigene/Distel, for

30 min, followed by washing in water and sterilization in auto-clave) and dissection board (spray with the cleaning and decon-taminating solution and wash with water).

3.2 Mammary Gland Digestion and Differential Centrifugation to Obtain Mammary Organoids and Fibroblasts

1. In a cell culture hood, transfer the freshly dissected mammary gland tissue into a fresh, sterile Petri dish (Fig. 2c).

2. Use 2–3 scalpels to cut the mammary gland tissue finely to a homogenous mince of approximately 1 mm³ pieces (Fig. 2d).

3. Transfer the minced tissue to a 50 ml tube with collagenase solution (*see* **Note 7**) (Fig. 2e).

4. Incubate the tissue in collagenase solution at 37 °C for 30 min while shaking it at 100 rpm (*see* **Note 8**) (Fig. 2f).

5. Centrifuge the tube at $450 \times g$ for 10 min at room temperature (RT).

6. For all subsequent steps, use BSA-coated pipettes and tubes (*see* **Note 9**). To coat a pipette, aspirate a 2.5% BSA solution into the pipette to cover its whole working surface, then return the used 2.5% BSA solution back into the stock tube/bottle containing the rest of the BSA solution (*see* **Note 10**). To coat a tube, carefully pipette the 2.5% BSA solution on the walls of the tube or fill the tube, then return the BSA solution back into the stock tube/bottle.

7. After centrifugation, there will be three layers: a top fatty layer (white), a middle aqueous phase, and epithelial pellet (red) on the bottom of the tube (Fig. 2g). Using a BSA-coated pipette, transfer the fatty layer (~8 ml) into a 15 ml tube. Then carefully aspirate the remaining supernatant from the 50 ml tube, resuspend the pellet in 5 ml of DMEM/F12 and set it aside.

8. For the 15 ml tube: Add 5 ml of DMEM/F12 to the fatty suspension and mix it well with the pipette. Centrifuge the tube at $450 \times g$ for 10 min at RT.

9. The centrifugation will again differentiate the solution into three layers. Aspirate and discard the top fatty and middle aqueous layers. To the pellet left at the bottom of the 15 ml tube, add the resuspended pellet from the 50 ml tube (from **step 7**). Wash the inside of the 50 ml tube with 5 ml of DMEM/F12 to collect all the cells and transfer them to the 15 ml tube. Mix the cells with a pipette.

10. Centrifuge the 15 ml tube at $450 \times g$ for 10 min at RT.

11. Aspirate and discard the supernatant. Resuspend the cell pellet in 4 ml of DMEM/F12 and add 40 μl of DNase I. Shake the tube by hand or on an orbital shaker for 3–5 min.

12. Add 6 ml of DMEM/F12, and mix the suspension well by pipetting.

13. Centrifuge the tube at $450 \times g$ for 10 min at RT.

14. Aspirate and discard the supernatant. Resuspend the pellet in 9 ml of DMEM/F12.

15. Perform differential centrifugation (*see* **Note 11**): Centrifuge the tube at $450 \times g$ for 10 s at RT (*see* **Note 12**). Transfer the supernatant into a fresh 50 ml tube. Resuspend the pellet in 9 ml of DMEM/F12.

16. Repeat **step 15** four more times (five times in total), collecting the supernatant in the 50 ml tube and finishing with transferring the supernatant into the 50 ml tube and resuspending the pellet in a small volume (200 µl to 1 ml) of basal organoid medium.

17. The pellet fraction contains epithelial organoids. The organoids can be used in 3D Matrigel cultures straight away or can be cultured in suspension on non-adherent dishes in basal organoid co-culture medium overnight (*see* **Note 13**).

18. Centrifuge the 50 ml tube with the collected supernatant at $600 \times g$ for 3 min at RT.

19. Aspirate the supernatant and resuspend the pellet in 10 ml of fibroblast medium.

20. Transfer the suspension into a 100 mm cell culture dish (*see* **Note 14**).

21. Incubate the cell culture dish at 37 °C for 30 min.

22. After 30 min, the fibroblasts will have attached to the cell culture dish while most other stromal cell types as well as epithelial organoids have remained in the suspension. Remove the unattached cells by aspirating the medium from the dish.

23. Wash the cell culture dish twice with PBS (5–10 ml) to get rid of any leftover unattached cells.

24. Add 10 ml of fresh fibroblast medium and incubate the plate at 37 °C.

3.3 Culture of Primary Mammary Fibroblasts

After isolation of primary mammary fibroblasts, some (small) contamination of epithelial cells (and possibly other stromal cell types) will be unavoidably present. However, these cells will not efficiently proliferate in fibroblast culture conditions, as they will die and/or be rapidly diluted out from the culture within a few days. If there is a considerable contamination of fibroblast culture by other cell types, it is possible to enrich the fibroblast population using the same principle as during fibroblast isolation: Trypsinize the cells, seed them onto a new cell culture dish, and let them set for 15 min. Then remove unattached cells by aspirating the medium and washing two times with PBS, and add a fresh fibroblast medium.

Check fibroblast culture daily under microscope. When the culture reaches 80% confluency, split the culture into new cell culture dishes; alternatively, primary mammary fibroblasts can be cryopreserved for later use.

1. Aspirate the culture medium from cell culture dishes.
2. Wash the dish twice with 5 ml of PBS.
3. Add 1.5 ml of trypsin-EDTA to the dish.
4. Incubate the dish at 37 °C for 2–3 min.
5. Monitor the cell dissociation process by watching carefully under a microscope; the dish can be gently tapped on a side to help release the cells.
6. Add 5 ml of fibroblast medium to the dish and pipette the suspension several times to detach remaining cells from the bottom and to resuspend cells.
7. Transfer the cell suspension to a 15 ml tube. Centrifuge at $600 \times g$ for 2 min.
8. Aspirate the supernatant and resuspend the pellet in fibroblast medium and split (1:3–1:6 according to the area) into new dishes with fresh fibroblast medium. Alternatively, resuspend the pellet in freezing medium (*see* **Note 15**) and transfer to a cryotube. Freeze the cells at –80 °C (*see* **Note 16**).

3.4 3D Co-culture of Mammary Organoids and Fibroblasts

For 3D co-culture, we recommend using freshly prepared mammary organoids and short-term cultured (i.e., for 1–4 passages) mammary fibroblasts (*see* **Note 17**). The method described here refers to cell numbers and medium volumes optimized for 24-well format of co-cultures. If you wish to set up bigger or smaller-scale co-cultures, adjust the cell numbers and medium volumes accordingly.

1. To prepare mammary organoids, collect freshly isolated epithelial organoids from mammary tissue or collect the organoids cultured overnight in non-adherent plates.
2. Centrifuge the organoids for 3 min at $450 \times g$ and resuspend the pellet in 1–2 ml of basal organoid co-culture medium.
3. Keep the suspension on ice and count the number of organoids in suspension (*see* **Note 18**).
4. To prepare mammary fibroblasts, trypsinize mammary fibroblasts off the cell culture dish as described above and resuspend the fibroblast pellet in 3 ml of PBS.
5. Centrifuge for 2 min at $600 \times g$.
6. Resuspend the fibroblast pellet in 3 ml of PBS.
7. Repeat **steps 5** and **6** one more time (for total three washes with PBS) (*see* **Note 19**). After final centrifugation for 2 min at $600 \times g$, resuspend the pellet in 1–2 ml of basal organoid medium.

8. Keep the fibroblast suspension on ice and count the number of fibroblasts in suspension using a hemocytometer.

9. To prepare the cell culture plate, use a small volume of Matrigel (20 µl per well) to cover the central part of the bottom of wells of a 24-well cell culture plate, creating a round patch of 0.5–1 cm diameter (*see* **Note 20**). Adjust the number of Matrigel-coated wells as required for the experiment.

10. Incubate the plate at 37 °C for 15 min in an incubator (5% CO_2) to solidify Matrigel (*see* **Note 21**).

11. Calculate the volumes of organoid suspension, fibroblast suspension, and Matrigel needed to make up the co-cultures (*see* Table 1 for suggested cell numbers and volumes of media). For one well of a 24-well plate, we recommend to plate 100 organoids combined with 100–1000 fibroblasts per organoid in 70 µl of Matrigel (*see* **Note 22**).

12. In a small tube, mix the required volumes of organoid suspension and fibroblast suspension together. Adjust the volume of suspension to 10 µl per well: If the combined volume of cells in the medium is less than 10 µl per well, add the required volume of basal organoid medium and mix well but gently with a pipette. If the combined volume of cells in medium is larger than 10 µl per well, centrifuge the tube at $450 \times g$ for 10 min, carefully remove the excessive volume of medium, and resuspend the cell pellet in the remaining medium.

13. Place the cell suspension on ice to cool before adding any Matrigel.

14. Add the appropriate volume of Matrigel (70 µl per well) to the cells and mix it carefully, avoiding bubble formation. Keep the cell-Matrigel mixture on ice at all times.

15. Place the Matrigel-covered cell culture plate on a heating plate (37 °C).

16. Carefully plate 80 µl of the cell-Matrigel mixture onto each Matrigel-covered patch in the well, forming a dome-shaped structure.

17. Incubate the plate for 30–60 min in an incubator (37 °C, 5% CO_2) to solidify the Matrigel.

18. Add 1 ml of pre-warmed (37 °C) basal organoid medium into each well (**Notes 23** and **24**).

19. Incubate the co-cultures in the cell incubator, changing medium every 2–3 days. Within a few days, epithelial branching can be observed (Fig. 3).

20. When the experiment is finished, it is possible to fix the co-cultures with 4% PFA. First, carefully aspirate the cell culture medium, then add 1 ml of 4% PFA per well and refrigerate (4 °C) overnight. Fixed co-cultures are stable for several weeks.

Table 1
Suggested numbers of cells and volumes of media for 3D co-culture in a 24-well format

# of fibroblasts per organoid	# of organoids	# of fibroblasts	Matrigel (µl)	Volume of cells in medium (µl)
100	50	5,000	70	10
100	100	10,000	70	10
100	200	20,000	70	10
100	300	30,000	70	10
250	50	12,500	70	10
250	100	25,000	70	10
250	200	50,000	70	10
250	300	75,000	70	10
500	50	25,000	70	10
500	100	50,000	70	10
500	200	100,000	70	10
500	300	150,000	70	10
1000	50	50,000	70	10
1000	100	100,000	70	10
1000	200	200,000	70	10
1000	300	300,000	70	10

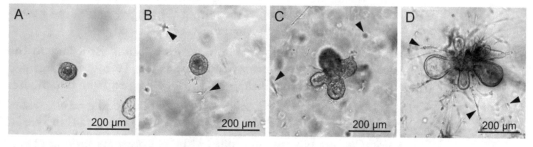

Fig. 3 The presence of fibroblasts induces mammary epithelial branching. (**a–d**) Photographs of mammary organoids in 3D co-cultures with fibroblasts, day 4. (**a**) 3D culture of mammary organoids only. (**b–d**) 3D co-culture of mammary organoids with 10 (**b**), 100 (**c**), or 1000 (**d**) fibroblasts per organoid. Note that with increasing number of fibroblasts in co-cultures, more epithelial branching is induced. *Arrowheads* indicate fibroblasts

4   Notes

1. There is considerable amount of variability in body weight and thus mammary gland size between strains. Also, the yield of tissue will slightly increase with the age of donor mouse from puberty to adulthood: the older the mouse, the more mammary gland tissue and hence more epithelium and fibroblasts it will yield. Moreover, progressive epithelial proliferation and invasion of the fat pad during puberty increases the epithelial content in mammary gland.

2. The developmental stage and parity of donor mice need to be considered because they have a major impact on epithelial morphogenesis, due to, e.g., hormonal status.

3. To ensure a high viability and quality of cells, do not delay mouse dissection and processing the tissue unless unavoidable. Moreover, within 1 h postmortem rigor mortis sets in and the mouse body becomes stiff and more difficult to handle. Also, the mammary gland tissue loses its elasticity, and becomes tougher and difficult to separate from the skin.

4. A thorough soaking with 70 % ethanol is needed to minimize bacterial and fungal contamination of the excised tissue and the subsequent cell cultures. Preferably, the dissection hood should be specifically designated for dissection and should not be used for work with cell cultures unless it is thoroughly cleaned and disinfected.

5. It is extremely important not to collect the muscle; the following procedure will not be able to exclude the muscle tissue and it will contaminate the final epithelial preparation.

6. It is not necessary to use PBS when collecting mammary gland tissue. Its use, however, is recommended because it prevents tissues from drying out during prolonged harvesting procedures, e.g., when several mice are processed.

7. Adjust the volume of collagenase solution according to the amount of mammary gland tissue: Use 10 ml of collagenase solution for mammary gland tissue obtained from one mouse.

8. The digestion time might slightly vary depending on the tissue-to-collagenase solution ratio or how finely chopped the tissue is.

9. Coating pipettes and tubes with BSA prevents sticking of epithelium to the walls and avoids loss of tissue during the extraction procedure.

10. The 2.5 % BSA solution can be reused many times, provided that the solution is kept sterile and regularly checked for contamination.

11. Differential centrifugation separates fibroblasts and epithelial organoids based on their size and density. Epithelial organoids

are larger and heavier; therefore they sink to the bottom of the tube faster than fibroblasts. Fibroblasts (and other types of stromal cells) will remain in the supernatant. The speed and time of centrifugation have been optimized for mammary gland tissue. However, this technique is not absolute; after each differential centrifugation some organoids (smaller fraction) will remain in the supernatant and some stromal cells (small fraction) will sink with organoids to the bottom. Therefore, several rounds of differential centrifugation are needed. We recommend five rounds of differential centrifugation for optimal yield of epithelium and fibroblasts. It is important to keep the pulses of $450 \times g$ short (10 s); increasing the time of centrifugation will increase the amount of stromal cells in the epithelial fraction, whereas decreasing the time of centrifugation will increase the amount of epithelial organoids in the stromal fraction and decrease the total yield of epithelium.

12. For most centrifuges, the total time of centrifugation from the very start of centrifugation is measured, not the time of centrifugation at the selected speed. The centrifugation time in this protocol applies to the time of centrifugation at $450 \times g$. Therefore, to perform differential centrifugation we recommend setting the time of centrifugation to 1 min and watching carefully, when the rotor reaches $450 \times g$, measuring the 10 s from this moment and then stopping the centrifugation manually by pressing the stop button.

13. Using non-adherent dishes to culture mammary organoids in suspension is essential; if adherent dishes are used, organoids will adhere to the dish and lose their 3D spherical structure.

14. Mammary glands from 1 to 2 pubertal mice will usually yield enough stromal cells suitable for one 100 mm cell culture dish; if more mammary gland tissue is processed, split the stromal cell suspension into more cell culture dishes accordingly.

15. Freezing medium for fibroblasts: heat-inactivated FBS with 10% dimethyl sulfoxide. Use 1 ml of freezing medium for cells from one 100 mm plate.

16. Freeze the cells slowly (reducing the temperature at approximately 1 °C/min) using a controlled rate cryo-freezer or a cryo-freezing container.

17. Freshly prepared fibroblasts might be unsuitable for immediate use in co-cultures because of the relatively harsh treatment needed to harvest them—after the initial collagenase/trypsin treatment during bulk tissue digestion, another treatment with trypsin is needed to retrieve fibroblasts attached to the cell culture dish at the end of the fibroblast isolation procedure. The fibroblasts might need some time to recover from the stress; therefore, we recommend using fibroblasts cultured for

a few days/passages instead. On the other hand, we do not recommend using fibroblasts cultured for more than five passages because by this time fibroblasts cease to proliferate. Also, prolonged culture might lead to cell culture adaptation as well as other changes in cell behaviour.

18. Counting mammary organoids: A hemocytometer is not useful for counting large structures as organoids. To count organoids, apply 20 μl of well-mixed organoid suspension into a 35 mm cell culture dish, spreading the drop of suspension wide with the pipette tip. Count the number of organoids under microscope. If the sample is too dense, dilute it by adding PBS (e.g. 20 μl) to the sample on 35 mm dish as needed.

19. It is critical to wash the fibroblast pellet with PBS to get rid of any FBS left from the fibroblast medium because it interferes with ductal morphogenesis.

20. Keep the Matrigel on ice or in a cool block at all times. Use ice-cold sterile PBS to cool pipette tips before using them to pipette the Matrigel.

21. Prevent prolonged incubation of the Matrigel-covered plate at 37 °C as it could lead to Matrigel drying, thus hindering cell-Matrigel interaction.

22. It is possible to increase the number of organoids and the number of fibroblasts in co-cultures; however, we recommend to adjust the volume of Matrigel accordingly, too. Otherwise the 3D co-cultures get over-crowded and will not be suitable for microscopy assessment.

23. The temperature of the medium and that of the solidified Matrigel must be the same to avoid formation of bubbles in the Matrigel. Bubbles interfere with the assessment of the co-cultures under a microscope.

24. It is possible to add growth factors, inhibitors or other agents to the basal organoid co-culture medium as required for the experiment.

Acknowledgements

This work was supported by grants from the National Institutes of Health (R03 HD060807), Breast Cancer Now, and ShanghaiTech University (to P.L.), and by the project "Employment of Best Young Scientists for International Cooperation Empowerment" (CZ.1.07/2.3.00/30.0037) co-financed from European Social Fund and the state budget of the Czech Republic and by the grant "Junior investigator 2015" from Faculty of Medicine, Masaryk University (to Z.K.). The Wellcome Trust Centre for Cell-Matrix Research, University of Manchester, was supported by core funding from the Wellcome Trust (088785/Z/09/Z).

References

1. Polyak K, Kalluri R (2010) The role of the microenvironment in mammary gland development and cancer. Cold Spring Harb Perspect Biol 2:a003244. doi:10.1101/cshperspect.a003244

2. Krause S, Maffini MV, Soto AM et al (2008) A novel 3D in vitro culture model to study stromal-epithelial interactions in the mammary gland. Tissue Eng Part C Methods 14:261–271

3. Wang F, Weaver VM, Petersen OW et al (1998) Reciprocal interactions between beta1-integrin and epidermal growth factor receptor in three-dimensional basement membrane breast cultures: a different perspective in epithelial biology. Proc Natl Acad Sci USA 95:14821–14826

4. Lee GY, Kenny PA, Lee EH et al (2007) Three-dimensional culture models of normal and malignant breast epithelial cells. Nat Methods 4:359–365

5. Kass L, Erler JT, Dembo M et al (2007) Mammary epithelial cell: influence of extracellular matrix composition and organization during development and tumorigenesis. Int J Biochem Cell Biol 39:1987–1994

6. Barnes C, Speroni L, Quinn KP et al (2014) From single cells to tissues: interactions between the matrix and human breast cells in real time. PLoS One 9, e93325. doi:10.1371/journal.pone.0093325

7. Basham KJ, Kieffer C, Shelton DN et al (2013) Chemical genetic screen reveals a role for desmosomal adhesion in mammary branching morphogenesis. J Biol Chem 288:2261–2270

8. Heneweer M, Muusse M, Dingemans M et al (2005) Co-culture of primary human mammary fibroblasts and MCF-7 cells as an in vitro breast cancer model. Toxicol Sci 83:257–263

9. Chhetri RK, Phillips ZF, Troester MA et al (2012) Longitudinal study of mammary epithelial and fibroblast co-cultures using optical coherence tomography reveals morphological hallmarks of pre-malignancy. PLoS One 7, e49148. doi:10.1371/journal.pone.0049148

10. Lühr I, Friedl A, Overath T et al (2012) Mammary fibroblasts regulate morphogenesis of normal and tumorigenic breast epithelial cells by mechanical and paracrine signals. Cancer Lett 325:175–188

11. Zhong A, Wang G, Yang J et al (2014) Stromal-epithelial cell interactions and alteration of branching morphogenesis in macromastic mammary glands. J Cell Mol Med. doi:10.1111/jcmm.12275

12. Schedin P, Keely PJ (2011) Mammary gland ECM remodeling, stiffness, and mechano-signaling in normal development and tumor progression. Cold Spring Harb Perspect Biol 3:a003228. doi:10.1101/cshperspect.a003228

Chapter 11

Next Generation RNA Sequencing Analysis Reveals Expression of a Transient EMT Profile During Early Organization of MCF10A Cells in 3D

Benedikt Minke, Drieke Vandamme, Thomas Schwarzl, Walter Kolch, and Finian Martin

Abstract

RNA sequencing is a technique widely used to identify and characterize gene expression patterns. We demonstrate that this method can be applied to screen expression profiles in mammary epithelial cells cultured in 3D, supported by a natural laminin-rich extracellular matrix, but requires several specific steps in the preparation of the RNA samples. Here we describe the use of RNA sequencing to analyze mRNA patterns in MCF10A human mammary epithelial cells cultured under 3D conditions in a laminin-rich extracellular matrix. We focus on our methods for total RNA extraction at early time points during the formation and maturation of 3D acinus structures in these cultures and provide examples of our results and downstream analysis.

Key words MCF10A cells, Mammary epithelial cells, Acinus, Acinus formation, EMT, Epithelial–mesenchymal transition, RNA sequencing, Next generation sequencing, Id-1

1 Introduction

MCF10A cells are an immortalized human epithelial cell line derived from mammary tissue from a patient with fibrocystic disease [1]. The cell line is non-tumorigenic and is widely used in breast cancer studies as a nonmalignant control cell line [2, 3]. MCF10A cells are characterized as "normal" and nonmalignant as they do not form tumors in xenograft studies. The unique quality of these cells is that they exhibit three-dimensional organization when cultured on a laminin rich extracellular matrix (ECM), such as Matrigel (secreted by Engelbreth–Holm–Swarm (EHS) mouse sarcoma cells). In this environment, MCF10A cells form acini over a period of about 8 days.

Acinus formation can be divided, both temporally and functionally, into a series of separable events: (1) cell cluster/spheroid

Finian Martin et al. (eds.), *Mammary Gland Development: Methods and Protocols*, Methods in Molecular Biology, vol. 1501, DOI 10.1007/978-1-4939-6475-8_11, © Springer Science+Business Media New York 2017

formation, (2) cell proliferation, (3) cell polarization, and finally (4) lumen clearance. The first 4 days of this development are dominated by cell proliferation and cell cluster/spheroid formation. The former one is driven by the EGF which is present in the culture medium; the latter, when dispersed cells have been cultured in the ECM (Matrigel), is initiated by cell–matrix and driven by cell–cell interaction. When the spheroids have attained an optimum size, there follows a phase of cell polarization and reorganization, during which the cells on the surface of the spheroids, which are in contact with the ECM, polarize and generate a structured epithelial sheet while the luminal cells remain unorganized. These luminal cells will eventually die by processes that may include apoptosis to clear the lumen. This final mature acinus is maintained as an organized structure [4].

Since the model is an in vitro system that reiterates important aspects of normal mammary epithelial cell behavior, the final phase of acinus formation [cell polarization and lumen clearance] has been very actively studied, in particular, the effect of oncogene expression on acinus formation [5–7]. Early stages of acinus formation by dispersed MCF10A cells, however, have not been investigated. Such studies provide an important insight into changes undergone by the MCF10A cells prior to acinus maturation. We have recently shown that in the very early stages of acinus formation from dispersed cells, MCF10A cells exhibit a distinct behavior: ECM engagement, migration-associated single cell polarization, migration, and cell cluster formation is seen [8].

We hypothesized that the phases of pre-acinus development are defined by specific gene expression patterns. Our main aim was to define the pattern of gene expression that accompanies the initial clustering of cells into unpolarized spheres. We performed total RNA sequencing analysis at five time points between 24 and 48 h after seeding dispersed MCF10A cells in Matrigel under conditions that favor acinus development. We identified different gene expression trends with time, clustered them into groups, and subjected them to "function" analysis. We could detect a transient increased expression of a significant group of genes associated with the process of epithelial–mesenchymal transition (EMT) in the period prior to acinus maturation. This EMT expression event was driven by an early BMP-signaling event and interestingly, blocked the final maturing of the acini.

2 Materials

2.1 Tissue Culture

1. Media for normal growth and passage is DMEM-F12 supplemented with 1% L-glutamine and 5% horse serum, 20 ng/ml EGF, 0.5 μg/ml hydrocortisone, 100 ng/ml cholera toxin, 10 μg/ml insulin, 1% penicillin/streptomycin.

2. Assay media is growth media, but with a reduced horse serum content (2 % horse serum).

3. Trypsin 0.05 %, in medium.

4. Cell culture dishes or flasks.

2.2 3D Cultures

1. Extracellular matrix: Growth Factor Reduced Matrigel.

2. Cell Culture dishes, 6 cm.

3. Inoculation loop.

2.3 Total RNA Extraction Components

1. TRI reagent.

2. 1-bromo-3-chloropropane.

3. Isopropanol.

4. RNAse-free filter tips.

5. Cell scraper.

6. Clean, not-autoclaved tubes.

2.4 Optional DNAse Treatment

1. Recombinant DNase (rDNase).

2. DNase inactivation reagent.

2.5 Reverse Transcription Components

1 Reverse transcriptase kit.

3 Methods

3.1 Tissue Culture

All media should be prepared in 500 ml batches with all ingredients except EGF added. This is then filter-sterilized through a 0.2 μm filter and can be stored at 4 °C for up to 2 months.

3.2 3D Culture

1. Prior to seeding, thaw Matrigel on ice (*see* **Note 1**), and keep the thawed Matrigel in a tissue culture hood on an ice tray.

2. Cool the cell culture dishes on ice (*see* **Note 2**).

3. Add 150 μl Matrigel to a cooled 6 cm cell culture dish and carefully create an even film using the inoculation loop (*see* **Notes 3** and **4**).

4. Place the dish in the incubator at 37 °C for at least 20 min.

5. Meanwhile, prepare the MCF10A cells using trypsin to detach them from the plastic flask surface. Shake the flask thoroughly if the cells do not detach after 5–10 min (*see* **Note 5**).

6. Count cells, for example using an automated cell counter.

7. Mix between 150,000 and 200,000 cells per 6 cm dish with the required media (5 ml for a 6 cm dish) and add 20 ng/ml

fresh EGF (*see* **Note 6**) and 20 μl/ml Matrigel for the overlaying matrix.

8. Add small molecule inhibitors immediately if required.

9. Leave the cells for up to 4 days, before changing the media and adding fresh EGF (*see* **Note 7**).

3.3 Total RNA Extraction

All the following steps should be carried out in a fume hood.

1. Bring 10 ml pipettes, cell scraper, filter pipette tips (200 and 1 μl), clean not-autoclaved tubes (two per sample) (*see* **Notes 8** and **9**), BCP, TRI Reagent into the hood.

2. Remove media, but do not wash cells (*see* **Note 10**).

3. Add 1 ml TRI Reagent per 6 cm dish.

4. Use a cell scraper to generate the lysate by thoroughly scraping and mixing, and transfer the solution to another tube.

5. Incubate the lysed cells at room temperature for 5 min.

6. Add 100 μl of BCP per 1 ml TRI Reagent, shake vigorously for 15 s, and store at room temperature for 15 min.

7. Centrifuge at $12,000 \times g$, 4 °C, for 15 min.

8. Transfer aqueous phase to clean tube (this contains the RNA). If required, store the other phases for further work at 4 °C.

The following steps can be carried out outside the fume hood.

1. Precipitate the RNA by adding 500 μl of Isopropanol per 1 ml of TRI reagent used.

2. Incubate the samples at room temperature for 10 min.

3. Centrifuge at $12,000 \times g$, 4 °C, for 8 min.

4. Remove the supernatant and wash the pellet with 1 ml 75% ethanol per original 1 ml of TRI reagent used by vortexing.

5. Centrifuge at $7500 \times g$, 4 °C for 5 min.

6. Remove the supernatant and air-dry the pellet (*see* **Note 11**).

7. Dissolve the pellet thoroughly in RNAse-free water (usually 30 μl).

8. Measure the RNA concentration and quality immediately.

9. Adjust the concentration to 100–200 ng/μl (depending on downstream protocol).

3.4 (Optional) DNAse Treatment

1. Use 1 μl rDNase for up to 10 μg of RNA in a 50 μl reaction. These reaction conditions will remove up to 2 μg of genomic DNA from total RNA.

2. Incubate at 37 °C for 30 min.

3. Add DNase inactivation reagent (0.1 volume) and mix well.

4. Incubate for 2 min at RT.

5. Centrifuge at $10,000 \times g$ for 1.5 min, then transfer RNA to a fresh tube.

3.5 Storage

Aliquot the RNA into amounts [by volume] needed for various downstream steps, to avoid freeze-thawing. For example:

(a) 2 μl for RNA analyzer.

(b) 10 μl for cDNA for qPCR.

(c) At least 10 μl for RNA sequencing.

Store all RNA at −80 °C until needed (*see* **Note 12**).

3.6 Quality Control

1. RNA analyzer.

(a) Test RNA quality using a RNA Nano Chip for quality control of the RNA sample.

2. qPCR.

(a) Perform Reverse Transcription using the High-Capacity cDNA Reverse Transcription Kit or similar [for protocol *see*: www3.appliedbiosystems.com/cms/groups/mcb_support/documents/generaldocuments/cms_042557.pdf].

(b) Measure multiple known target transcripts by qPCR before running the much more costly RNA sequencing to be sure the sample quality meets the standards needed.

3.7 Library Generation and RNA Sequencing

1. Sequence 2 μg of total RNA per sample with TruSeq RNA sample preparation Kit v2 (Illumina) according to manufacturer's protocol [*see*: support.illumina.com/content/dam/illumina-support/documents/documentation/chemistry_documentation/samplepreps-truseq/truseqrna/truseq-rna-sample-prep-v2-guide-15026495-f.pdf]. Size and purity of the libraries should be analyzed on a Bioanalyzer High Sensitivity DNA chip.

2. Libraries are clustered using TruSeq Single-Read Cluster Kit v5-CS-GA (Illumina) and sequenced on an Illumina Genome Analyzer IIx with a TruSeq SBS Kit v5-GA (Illumina).

3.8 Align the Read Mapping and Statistical Analysis

1. Align the sequence reads to the human reference genome GRCh37 (hg19) using Burrows-Wheeler Aligner (BWA) [9]. Gene counts are summarized using the program htseq-count from the HTseq package (www.huber.embl.de/users/anders/HTSeq/).

2. Make multidimensional scaling (MDS) plots using the R/Bioconductor packages edgeR [10, 11]. Differentially expressed genes were called using general linear models in edgeR [11, 12]. *p*-Values were adjusted for multiple testing with the Benjamini–Hochberg correction and a corrected

p-value cutoff of 0.05 was used. The mRNA-seq data was deposited in ArrayExpress (http://www.ebi.ac.uk/arrayexpress) under accession number E-MTAB-2969.

3. Analyze the retrieved gene lists for overrepresented pathways, biological functions, and upstream regulators using Ingenuity Pathway Analysis (IPA) Systems.

4. P-values reported for IPA results are generated by IPA using a right sided Fisher exact test for overrepresentation analysis, Benjamini–Hochberg correction for multiple hypothesis testing correction, and a z-score algorithm for upstream regulator analysis, p-values <0.05 were considered significant.

5. USe FPKMs created with cufflinks for cluster analysis with Short Time Series Expression Miner (STEM) [13]. We used STEM by MIT to identify any evident patterns in our Data (*see* Fig. 1).

6. Figures 1 and 2 provide an example of the output of our analysis of the changing gene expression patterns in the early stages (0–48 h) of acinus formation by dispersed cultured MCF10A cells supported on a laminin-rich matrix (Matrigel): Fig. 1a shows the number of unique transcripts that are significantly different (increased or decreased) from control (0 h) at the various time points studied; Fig. 1b shows shared [and unique] expressed genes at the time points 24, 34, 36, 38, and 48 h analyzed using a Venn diagram. Genes from each time point have an oval of different color (blue, yellow, green, red) assigned to them. Numbers of shared transcripts are shown as the area where these ovals overlap. While every time point has a number of associated unique transcripts, this decreases with time. Overlap is higher between early time points; the 48 h time point is less similar to them. Over 4600 genes are represented in all samples (grey center).

7. After quality control, compare all RNA datasets to the control (c.vs.#) and to each other (#.vs.#), to obtain LogFC, p-values and false discovery rates (FDR). These are then clustered hierarchically and visualized in a heat map Fig. 1c. Each sample is represented in a column, each gene in a row (cutoff >1.5 log OR < −1.5 log and FDR < 0.05). Similarity is shown by branch length. Genes are color-coded with red = positive logFC, blue = negative logFC, white = low logFC and cutoff). Interestingly, at 24 h two relatively sizable groups of uniquely (and significantly) overexpressed transcripts are detected. Finally, Fig. 1d shows differences in gene expression and hierarchical clustering. We hypothesized that at this early phase of acinus development cell behavior might be more mesenchymal-like than epithelial. STEM analysis (Fig. 2a) identifies occurrence of expression of EMT-linked transcripts in our

a

	Up	Down	Total
24h	3511	5185	8696
34h	3838	5602	9440
36h	3831	5348	9179
38h	3496	5347	8843
48h	2291	3665	5956

b

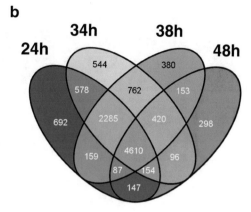

c

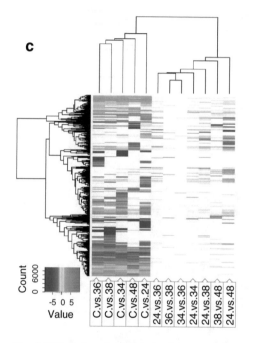

d

Molecular and CellularFunctions	24h	48h
Cellular Movement	251	157
Cellular Development	328	184
Cell Morphology	221	136
Cellular Growth and Proliferation	344	198
Cell-To-Cell Signalling and Interaction	348	109

Fig. 1 RNA sequencing analysis of gene expression in the early stages of acinus formation by MCF10A cells. (**a**) Table showing the total number of up and down regulated genes present in the samples, without cutoff; shown are the total number of differentially expressed genes, the number of upregulated and downregulated genes. (**b**) Overlap in gene expression: Venn diagram made using "Venny." The cohort of transcripts for the time points, 24–48 h, were compared to show overlap between samples. Transcripts from each time point have been assigned an oval of different color. Number of shared transcripts is shown as/in the area where ovals overlap. (**c**) Differences in gene expression; hierarchical clustering: All RNA datasets were compared to the control (c.vs.#) and to each other (#.vs.#), to obtain LogFC, p-values, and false discovery rates (FDR). LogFC values were clustered hierarchically and visualized in a heat map. Each sample is represented in a column, each gene in a row (cutoff >1.5 log OR < -1.5 log and FDR < 0.05). Similarity is shown by branch length. Genes are color-coded by intensity of their expression (*see* color key, *red* = positive logFC, *blue* = negative logFC, *white* = low logFC and cutoff). And, (**d**) Gene ontology analysis of expressed transcripts at time-points 24 and 48 h: Datasets were subjected to Ingenuity Pathway Analysis and Gene Ontology (GO) terms were identified for individual genes. Shown are the most common GO terms identified in the category "Molecular and Cellular Functions" for transcripts expressed at 24 and 48 h and the number of genes identified with each GO term

a

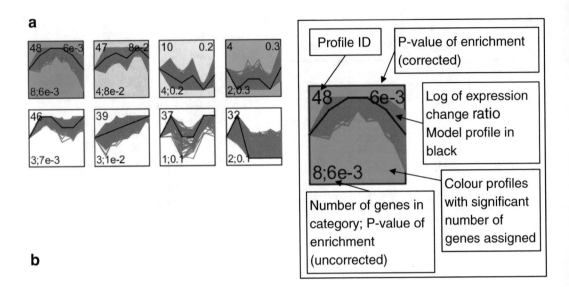

b

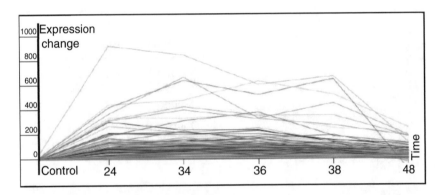

ID2	ID1	FOSL1	SNAI2	MMP3	NOTCH3	HRAS
ID3	KRT16	MMP25	LOXL2	TRIM28	TGFB1	SNAI1
LCN2	JUNB	SDC1	MMP14	ATL1	SOX9	AKT1

Fig. 2 RNA sequencing: differences in gene expression in the early stages of acinus formation by MCF10A cells; Short Time Series Expression Miner [STEM] analysis. (**a**) Clustering was performed using STEM. Genes were first clustered according to their expression profile and then cut off by significance (*colored squares* = significant). Profiles were then sorted by number of genes in the cluster falling into the GO term category "Epithelial to Mesenchymal Transition" (EMT). Examples of significant and nonsignificant profiles are shown [an explanatory "key" is provided, left-hand cartoon]. (**b**) STEM analysis profile 48: Expression profile shows an upregulation from 0 to 24 h, a plateau, and/or subsequent reversion towards base level by 48 h; 1591 genes fell into this category (*p*-value < 0.001). The "most" upregulated genes from profile 48 that are known to be linked to EMT are listed (*bottom*). (**c**) STEM analysis profile 48: Top ten "most" up- regulated EMT genes. Line graph showing the expression profile of the ten most strongly upregulated EMT genes, sorted by their overexpression levels at 24 h; the highest expression genes, ID1 and ID3, are more than tenfold overexpressed. And, (**d**) qPCR validation

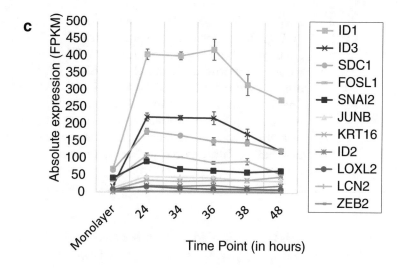

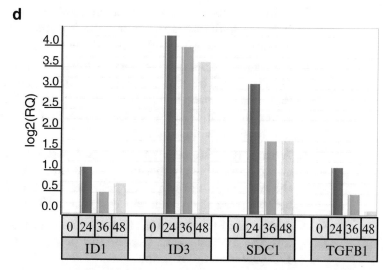

Fig. 2 (continued) for RNA sequencing analysis. Expression profiling of EMT markers ID1, ID3, SDC1, and TGFB1: Q-RT-PCR was performed on total RNA isolated from MCF10A cells cultured under normal assay conditions on Matrigel. Cells were harvested and total RNA was isolated at 24, 36, and 48 h. Values were normalized against time 0 h control and 18s ribosomal RNA as endogenous control (Key: *blue* = 24 h, *green* = 36 h, *yellow* = 48 h. From *left* to *right*: ID1, ID3, SDC1, TGFB1)

dataset. Genes are first clustered according to their expression profile and then cut off by significance (colored squares = significant). Profiles are then sorted by number of genes in the cluster falling into the GO term category "Epithelial to Mesenchymal Transition." We considered profile 48 the most potentially interesting: Here the expression of the EMT transcripts rose from time zero, peaking at $t = 24$ h and fell away by $t = 48$ h. Figure 2b, c confirm that this does in fact represent a significant transient expression of a cohort of classical EMT-associated transcripts; with the data in Fig. 2d confirmed by quantitative RT-PCR analysis the gene expression results obtained by RNA sequencing. Our studies therefore associate the transient expression of a significant cohort of EMT-associated mRNAs with the early phase of acinus formation by dispersed MCF10A cells, a time when the cells are mobilized to form unpolarized cell clusters, the "pre-acini" which in turn mature in size by proliferation before "differentiating" to form acini.

4   Notes

1. It is recommended to store Matrigel frozen as 1 ml aliquots.

2. Keeping the dishes ice-cold will help to disperse the Matrigel.

3. Other protocols use significantly more Matrigel per area; however, we find this is the minimum required for normal 3D organization of MCF10A cells and any more is unnecessary.

4. Matrigel solidifies quickly when above 4 °C. The inside of the pipette tip will be coated with quite a lot of it and the amounts that actually make it into the dish may vary. If tip boxes are stored in the fridge at 4 °C, the tips are cold and less Matrigel gets retained within the tips.

5. When MCF10A cells are of higher passage numbers, they attach more strongly to the plastic. At this stage, even shaking the flask might not help. The best way to solve this problem is by increasing the trypsin concentration to 0.25 %.

6. Always use EGF from fresh frozen aliquots instead of mixing it into the 500 ml media bottle—this prolongs the shelf life as EGF deteriorates over time.

7. This medium should not contain Matrigel.

8. Autoclaving often involves a lot of manual handling and can introduce RNases and should be avoided for RNA work.

9. Using 2 ml tubes is recommended, as the total working volume will be quite high due to the extra Matrigel.

10. Washing at this stage could introduce RNases.

11. Cover the tubes while air-drying in order to prevent anything from settling inside.

12. Store the RNA frozen even for short term as to avoid degradation.

Acknowledgements

We wish to thank Stephen Madden, Tapesh Santra, and Markus Schroeder for help with various aspects of the in silico analysis. We would also like to thank members of the UCD-Conway Core Facilities for their assistance. Science Foundation Ireland funded the activities of Systems Biology Ireland.

References

1. Soule HD, Maloney TM, Wolman SR, Peterson WD, Brenz R, McGrath CM et al (1990) Isolation and characterization of a spontaneously immortalized human breast epithelial cell line, MCF-10. Cancer Res 50(18):6075–6086

2. Spink BC, Cole RW, Katz BH, Gierthy JF, Bradley LM, Spink DC (2006) Inhibition of MCF-7 breast cancer cell proliferation by MCF-10A breast epithelial cells in coculture. Cell Biol Int 30(3):227–238

3. Ciftci K, Su J, Trovitch PB (2003) Growth factors and chemotherapeutic modulation of breast cancer cells. J Pharm Pharmacol 55(8):1135–1141

4. Debnath J, Brugge JS (2005) Modelling glandular epithelial cancers in three-dimensional cultures. Nat Rev Cancer 5(9):675–688

5. Barcellos-Hoff MH, Aggeler J, Ram TG, Bissell MJ (1989) Functional differentiation and alveolar morphogenesis of primary mammary cultures on reconstituted basement membrane. Dev Camb Engl 105(2): 223–235

6. Debnath J, Muthuswamy SK, Brugge JS (2003) Morphogenesis and oncogenesis of MCF-10A mammary epithelial acini grown in three-dimensional basement membrane cultures. Methods (San Diego, CA) 30(3):256–268

7. Debnath J, Mills KR, Collins NL, Reginato MJ, Muthuswamy SK, Brugge JS (2002) The role of apoptosis in creating and maintaining luminal space within normal and oncogene-expressing mammary acini. Cell 111(1):29–40

8. McNally S, Mcardle E, Gilligan E, Napoletano S, Gajewska M, Bergin O et al (2011) c-Jun N-terminal kinase activity supports multiple phases of 3D-mammary epithelial acinus formation. Int J Dev Biol 55(7-9):731–744

9. Li H, Durbin R (2009) Fast and accurate short read alignment with Burrows-Wheeler transform. Bioinformatics 25(14):1754–1760

10. Gentleman RC, Carey VJ, Bates DM, Bolstad B, Dettling M, Dudoit S et al (2004) Bioconductor: open software development for computational biology and bioinformatics. Genome Biol 5(10):80

11. Robinson MD, McCarthy DJ, Smyth GK (2010) edgeR: a Bioconductor package for differential expression analysis of digital gene expression data. Bioinformatics 26(1):139–140

12. McCarthy DJ, Chen Y, Smyth GK (2012) Differential expression analysis of multifactor RNA-Seq experiments with respect to biological variation. Nucleic Acids Res 40(10):4288–4297

13. Ernst J, Bar-Joseph Z (2006) STEM: a tool for the analysis of short time series gene expression data. BMC Bioinformatics 7(1):191

Chapter 12

A 3D Culture Model to Study How Fluid Pressure and Flow Affect the Behavior of Aggregates of Epithelial Cells

Alexandra S. Piotrowski-Daspit, Allison K. Simi, Mei-Fong Pang, Joe Tien, and Celeste M. Nelson

Abstract

Cells are surrounded by mechanical stimuli in their microenvironment. It is important to determine how cells respond to the mechanical information that surrounds them in order to understand both development and disease progression, as well as to be able to predict cell behavior in response to physical stimuli. Here we describe a protocol to determine the effects of interstitial fluid flow on the migratory behavior of an aggregate of epithelial cells in a three-dimensional (3D) culture model. This protocol includes detailed methods for the fabrication of a 3D cell culture chamber with hydrostatic pressure control, the culture of epithelial cells as an aggregate in a collagen gel, and the analysis of collective cell behavior in response to pressure-driven flow.

Key words Mechanical stress, Fluid flow, Micropatterning, 3D culture, Fluid pressure

1 Introduction

1.1 Studying the Physical Microenvironment

In addition to the biochemical signals in the microenvironment, many physical aspects of the microenvironment can affect cell behavior [1, 2]. Physical factors including stiffness, pressure, flow, shear stress, and stretch can cause changes in cell behavior that help maintain tissue homeostasis during development, and alterations in these physical factors are frequently associated with initiation and progression of disease [3–5]. For example, increased matrix stiffness is a telltale sign of cancer, and can promote tumorigenesis [6]. A stiffer microenvironment can alter epithelial plasticity by promoting epithelial-to-mesenchymal transition (EMT) in mammary epithelial cells [7], a process linked to cancer invasion. Furthermore, most malignant solid tumors have elevated interstitial fluid pressure (IFP) compared to normal tissue [8]. High IFP has been associated with poor prognosis and metastasis to lymph nodes [9, 10].

Finian Martin et al. (eds.), *Mammary Gland Development: Methods and Protocols*, Methods in Molecular Biology, vol. 1501, DOI 10.1007/978-1-4939-6475-8_12, © Springer Science+Business Media New York 2017

Because of the connection between physical factors and disease, it is necessary to develop experimental models that recapitulate the various physical properties of the microenvironment [11]. Cells exist primarily in 3D within living tissues (though epithelial and endothelial cells can exist as quasi-2D monolayers). Interactions between cells, their neighbors, and the surrounding extracellular matrix (ECM) are crucial for maintaining normal tissue function and homeostasis [12]. Therefore, 3D culture models are often used to mimic the structure and function of the tissue microenvironment [13, 14].

1.2 Interstitial Fluid Pressure (IFP)

Solid tumors have a high IFP that results from abnormal, leaky blood vessels and impaired lymphatic drainage [15, 16]. This feature has poor implications in cancer, as elevated IFP can lead to therapeutic resistance by hindering the delivery of drugs into solid tumors [10, 17–19]. IFP has also been shown to influence the migratory and invasive behaviors of single-cell suspensions of MDA-MB-231 breast cancer cells in collagen gels. In one study, a hydrostatic pressure differential of culture medium was established across the suspensions. Single-cell tracking showed that IFP increased the percentage of migratory cells and the speed at which they moved, and cells were observed to migrate primarily in the direction of flow via autologous chemotaxis, a phenomenon that has been previously reported by the same group using modified 3D Boyden chambers [20, 21]. In similar studies, IFP was also found to affect the migratory behavior of MDA-MB-231 cells, except that cells migrated against the flow direction, particularly when seeded densely [22, 23]. The response of cancer cells to physical cues such as elevated IFP seems to depend on the context in which the signals are presented, and neither of the previous models captures the behavior of an intact aggregate of tumor cells, suggesting the need for a new model. Here we describe a microfluidic approach to model the effects of IFP on an aggregate of tumor cells in 3D.

We have developed a culture model in which a defined hydrostatic pressure differential of culture medium may be applied across an aggregate of epithelial cells surrounded by a gel of type I collagen [24]. Our technique allows for the generation and control of the fluid pressure profile experienced by the aggregate of epithelial cells, and enables us to examine the effects of IFP on collective migration of those cells. An ideal model system is as close to in vivo conditions as possible; here, the 3D model mimics the physiological conditions of a dense tumor tissue. This approach allows one to visualize directly cell migration and tumor invasion in 3D from a preexisting aggregate, as well as to analyze changes in gene expression using in situ assays including immunostaining, and bulk analyses including immunoblotting and real-time RT-PCR. The model can also be used as a platform to screen for therapeutics that inhibit cancer cell invasion under different pressure conditions. Although we use this model in the context of cancer, it could also be used to study the effects of pressure and/or fluid flow in other physiological or pathological contexts.

2 Materials

Prepare collagen mixture on ice. Keep reagents at 4 °C.

2.1 Preparation of PDMS Chamber and Cavity Surrounded by Collagen

1. Polydimethylsiloxane (PDMS).
2. PDMS curing agent.
3. Lithographically patterned silicon master.
4. ¼″ hole punch.
5. 150-mm petri dishes.
6. 100-mm petri dishes.
7. 100-mm tissue-grade polystyrene culture dishes.
8. 24 mm × 50 mm #1½ glass coverslips.
9. 18 mm × 18 mm #2 glass coverslips.
10. 70 % (v:v) ethanol.
11. 0.12 mm × 30 mm acupuncture needles.
12. 1.5 ml microcentrifuge tubes.
13. 10× Hank's balanced salt solution (HBSS).
14. 0.1 N NaOH.
15. Bovine dermal type I collagen (non-pepsinized).
16. Cell culture medium. For example, 1:1 Dulbecco's Modified Eagle's Medium : Ham's F12 Nutrient Mixture (DMEM/F12 (1:1)) supplemented with: 10 % fetal bovine serum (FBS), and 50 µg/ml gentamicin.
17. Sterile phosphate-buffered saline (PBS).
18. 1 % (w:v) bovine serum albumin (BSA) in PBS. Store at 4 °C.
19. Handheld drill.

2.2 Formation of 3D Epithelial Cell Aggregates

1. Culture medium (*see* Subheading 2.1, **item 16**).
2. 0.05 % 1× trypsin–EDTA.
3. Collagen gels with channels, assembled between PDMS chambers and glass coverslips.
4. PDMS.
5. PDMS curing agent.
6. ¼″ hole punch.

2.3 Immuno-fluorescence Imaging

1. 16 % paraformaldehyde.
2. PBS.

3. 1.5 ml microcentrifuge tubes.

4. PBS with 0.3% (v:v) Triton X-100 (0.3% PBST).

5. Normal goat serum.

6. Rabbit anti-E-cadherin antibody.

7. Alexa 488 goat anti-rabbit antibody.

8. Nuclear counterstain, such as Hoechst 33342.

9. Aluminum foil.

10. Inverted microscope with phase-contrast and fluorescence capabilities and a 10×/0.30 NA objective.

2.4 Image Analysis: Measuring the Extent of Collective Migration

1. ImageJ (National Institutes of Health, Bethesda, MD).

3 Methods

Here we describe an engineered 3D culture model that can be used to study the effects of pressure gradients and fluid flow on the migratory/invasive behavior and gene expression profile of an aggregate of epithelial cells.

3.1 Preparation of PDMS Chamber and Cavity Surrounded by Collagen

1. Mix the PDMS prepolymer and curing agent at a 10:1 (w:w) ratio. Aim for a total weight of approximately 50 g. Remove the entrapped air bubbles by degassing in a vacuum chamber (~15 min). Pour the bubble-free mixture onto a lithographically patterned silicon master in a 150-mm petri dish. The silicon master should have features that produce channels that are approximately 20 mm long, 1 mm wide, and 1 mm tall spaced approximately 1.5 cm apart. Cure the PDMS in an oven at 60 °C for at least 2 h.

2. Once the PDMS is cured, carefully peel the PDMS from the silicon wafer, removing any PDMS from the bottom of the master. Using a clean razor blade, cut off the excess PDMS from around the molded features.

3. Using a clean razor blade, cut the polymerized PDMS into chambers containing individual channels ~1.25 cm wide and ~2.5 cm long (Fig. 1a). Use a ¼″ hole punch to bore holes on either side of each channel well. Bore one of the holes (hole B in Fig. 1b) in the middle of the channel such that the distance between the holes is 8–10 mm. Sterilize the chambers carefully in a biosafety cabinet (cell culture hood) by sonicating the chambers in 70% ethanol, washing briefly with 100% ethanol, and aspirating the excess liquid (*see* **Note 1**).

4. Drill a rectangular hole (~15 mm × ~40 mm) in the middle of a 100-mm tissue culture dish.

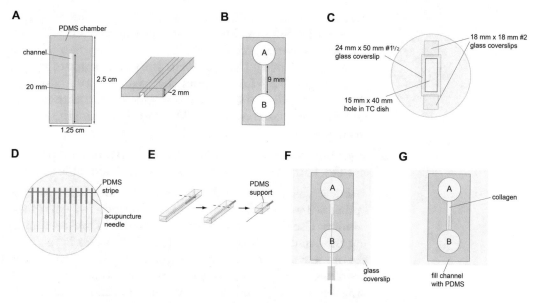

Fig. 1 Schematic diagrams detailing preparation of PDMS chamber and channel surrounded by collagen. (**a**) Top and perspective views of a PDMS chamber with a single channel, including dimensions. (**b**) Top view of a PDMS chamber with a single channel showing locations of holes A and B. (**c**) Top view of 100 mm tissue culture dish showing the placement of glass coverslips over a rectangular hole drilled in the middle of the dish. (**d**) Top view of a 100 mm petri dish containing acupuncture needles held in place by a stripe of PDMS. (**e**) Schematic detailing the preparation of an acupuncture needle with a PDMS support block. (**f**) Top view of a PDMS chamber conformally adhered to a glass surface with acupuncture needle setup prior to the addition of collagen into hole A. (**g**) Top view of a PDMS chamber containing a collagen channel with a cavity in the shape of the acupuncture needle from (**f**), with the channel adjacent to well B filled with PDMS

5. In a biosafety cabinet, wash the modified tissue culture dish containing the rectangular hole with 100% ethanol to sterilize.

6. In a biosafety cabinet, carefully sterilize one 24 mm×50 mm #1½ glass coverslip and two 18 mm×18 mm #2 glass coverslips by sonicating in 70% ethanol, washing briefly with 100% ethanol, and aspirating the excess liquid.

7. Lay down the 18 mm×18 mm #2 glass coverslips parallel to the ~15 mm sides of the hole in the 100-mm tissue culture dish, leaving a few mm of space between the coverslips and the hole (Fig. 1c).

8. Mix the PDMS prepolymer and curing agent and remove air bubbles as described previously. Aim for a total weight of approximately 5 g. Using a 200 μl pipet tip, add a layer of PDMS on top of the modified 100 mm tissue culture dish around the hole and between the hole and the coverslips. Then, carefully lay the 24 mm×50 mm #1½ coverslip on top of the two 18 mm×18 mm #2 coverslips such that it covers the hole forming a seal with the PDMS. Cure the PDMS in an oven at 60 °C for at least 2 h.

9. Mix the PDMS prepolymer and curing agent as described previously. Aim for a total weight of approximately 20 g. Again, remove the entrapped air bubbles by degassing in a vacuum chamber (~15 min). Using a pipette tip, paint a stripe of PDMS in a straight line ~1.5 cm from the edge of a 100-mm petri dish. Lay down 12–14 120 μm diameter acupuncture needles with handles placed on the PDMS (Fig. 1d). Cure the PDMS in an oven at 60 °C for at least 1 h.

10. Pour the remainder of the 20 g PDMS mixture on top of the needles. Cure the PDMS in an oven at 60 °C for at least 2 h.

11. Using a clean razor blade, separate the individual needles embedded within PDMS and remove a portion (~1 cm) of the handles. Use the razor blade to cut away the PDMS surrounding the needle, leaving a small rectangular portion in the middle of the needle to be used for support (Fig. 1e). Sterilize the needles carefully in a biosafety cabinet by sonicating in 70% ethanol, washing briefly with 100% ethanol, and aspirating the excess liquid.

12. Coat the needles with 1% BSA in PBS for at least 4 h at 4 °C and then wash with PBS and ddH$_2$O.

13. In a biosafety cabinet, place three of the PDMS chambers in the middle of one modified glass-bottom 100-mm tissue culture dish such that the channels are parallel to one another (and perpendicular to the 50 mm sides of the 24 mm × 50 mm #1½ glass coverslip) with the open face of the channels against the glass (*see* **Note 2**). Carefully thread the cleaned needles into the side of the chamber next to hole B such that the tip of the needle is in between the two wells formed by holes A and B and in the middle of the channel (Fig. 1f). Secure the needle in place by conformally adhering the PDMS support to the tissue culture dish next to the PDMS chamber and the 24 mm × 50 mm #1½ glass coverslip. Cool the tissue culture dishes to 4 °C for at least 2 h (*see* **Note 3**).

14. In a cold (4 °C) 1.5 ml microcentrifuge tube, prepare a neutralized solution of collagen. Add 30.6 μl 10× HBSS, 18.4 μl 0.1 N NaOH, 244.8 μl collagen, and 12.2 μl cell culture medium for a final concentration of collagen of approximately 4 mg/ml (*see* **Note 4**). Mix slowly by pipetting up and down; try not to introduce bubbles. If bubbles are induced, centrifuge the mixture briefly at 15,700 rcf.

15. Add 15 μl of neutralized collagen solution to the well formed by hole A (well A) of the PDMS chamber, opposite to the side containing the needle (Fig. 1f). Tilt the culture dish on its side to allow the collagen to flow down and fill the channel, tapping the dish as necessary (*see* **Note 5**). Gently aspirate the excess collagen from well A. Incubate at 37 °C for 20 min.

16. Add 20 μl of cell culture medium to both wells at the gel surface in order to wet the gel. Gently remove the needle from the chamber: bend it 90° where it exits the PDMS support, and gently pull it straight out while holding down the PDMS support.

17. Add a small amount of PDMS (prepared as above and incubated at 60 °C for 15 min) to plug the channel next to the well formed by hole B (well B) that contained the needle (Fig. 1g). Add 20–50 μl of cell culture medium to both wells once the PDMS is cured. Incubate the channels overnight at 37 °C.

3.2 Formation of 3D Epithelial Cell Aggregates

1. Aspirate the cell culture medium from the wells in the PDMS chamber.

2. Trypsinize epithelial cells (in this case MDA-MB-231 human breast carcinoma cells) and resuspend in cell culture medium at a final concentration of approximately 10^7 cells per ml.

3. Add 50 μl of the concentrated suspension of cells to well B and allow the cells to fill the cavity by convection (Fig. 2a, b).

4. Resuspend the cells in well B after 2–5 min. Then, once the cavity is completely filled, aspirate the cell suspension from the well.

5. Wash well B twice with 50 μl of medium.

6. Add fresh medium to the rim of well A and slightly over the rim of well B. Incubate the seeded channels at 37 °C for 48 h, changing the medium every 24 h.

7. After 48 h, discard any samples in which cells have migrated outside of the original shape of the aggregate.

8. Mix the PDMS prepolymer and curing agent as described above. Aim for a total weight of approximately 50 g. Pour the bubble-free mixture onto a 150-mm petri dish. Cure the PDMS in an oven at 60 °C for at least 1 h.

9. Using a clean razor blade, cut the PDMS into blocks that are approximately 1 cm by 1 cm. Use a ¼″ hole punch to bore a hole in the center of the blocks, creating PDMS gaskets (Fig. 2c).

10. To set up a hydrostatic pressure differential across the cell aggregates in the collagen channels within the PDMS chambers, add up to four PDMS gaskets on top of one of the wells on one side of the channel. Ensure that the holes in the PDMS gaskets align with the wells in the chamber (Fig. 2d). Seal the PDMS gaskets conformally to the chamber by gently pressing down.

11. Add culture medium to the rim of both wells. Maintain the pressure differential by replenishing the medium on the higher pressure side every 12 h (*see* **Note 6**).

12. Culture the cell aggregates for up to 9 days at 37 °C.

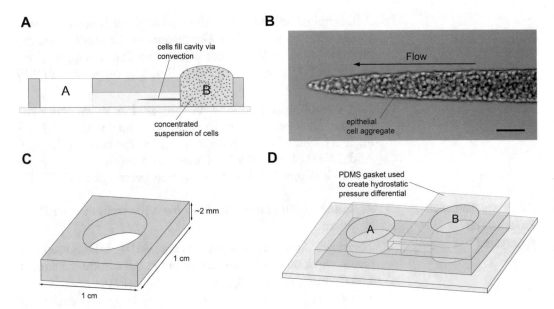

Fig. 2 Schematic diagrams detailing the formation of 3D mammary epithelial cell aggregates and the control of hydrostatic pressure profiles across the channels. (**a**) Side view of a cavity surrounded by collagen within a PDMS chamber being seeded with mammary epithelial cells via convection from a concentrated cell suspension in well B. (**b**) Phase-contrast image of a seeded cavity depicting the direction of convective flow during seeding. Scale bar 100 μm. (**c**) Perspective view of a PDMS gasket to be used to control the profile of hydrostatic pressure. (**d**) Perspective view of the culture model showing the creation of a hydrostatic pressure differential across the channel

3.3 Immuno-fluorescence Analysis

1. Prepare a fixative by diluting 16% paraformaldehyde 1:4 (v:v) in PBS.

2. Aspirate the medium from the PDMS chamber and add fixative to well B until the well is slightly overfilled. Incubate at room temperature for ~18 min.

3. Aspirate the fixative and fill well B to the rim with PBS. Repeat twice, each after 20 min, for three total washes.

4. *To stain for nuclei*: prepare 4 ml of a 1:1000 (v:v) solution of Hoechst 33342 in PBS. To stain for E-cadherin or another marker that is detected with antibodies, skip to **step 8**.

5. Aspirate the PBS from well B and fill it to the rim with the Hoechst solution. Incubate at room temperature for 15–20 min.

6. Aspirate the Hoechst solution and wash well B three times with PBS as in **step 3**. Stained samples can be stored in PBS at 4 °C until further use.

7. If desired, visualize the stained nuclei as described in Subheading 3.4.

8. *To stain for E-cadherin or other protein marker* (Fig. 3): aspirate PBS. Peel the PDMS chamber from the tissue culture dish.

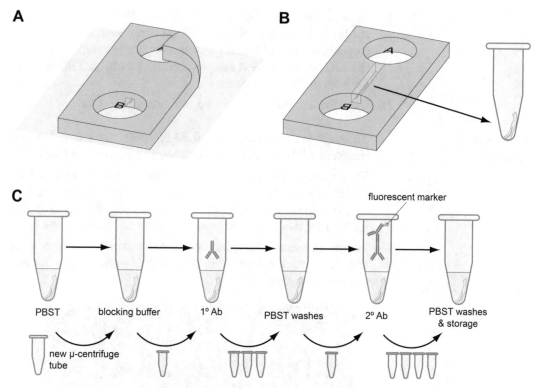

Fig. 3 Schematic of antibody staining procedure. (**a**) The PDMS chamber is peeled off of the tissue culture dish. (**b**) The collagen channel with embedded cell aggregate is carefully removed from the underside of the chamber and placed in a 1.5 ml microcentrifuge tube. (**c**) Washes and staining all take place inside new microcentrifuge tubes

Carefully remove the collagen channel and embedded cell aggregate from the underside of the PDMS chamber and place in a 1.5 ml microcentrifuge tube (*see* **Note 7**).

9. Add 300 μl of PBST to the tube and incubate the sample at room temperature for 15 min.

10. Prepare blocking buffer by diluting goat serum 1:10 (v:v) in PBST.

11. Remove the sample and place in a new 1.5 ml microcentrifuge tube. Add 300 μl of blocking buffer and incubate on a shaker at room temperature at least 4 hours.

12. Prepare a 1:200 (v:v) solution of primary rabbit anti-mouse E-cadherin antibody in blocking buffer (*see* **Note 8**).

13. Remove the sample and place in a new 1.5 ml microcentrifuge tube. Add 300 μl of the primary antibody solution and incubate on a shaker overnight at 4 °C.

14. Remove the sample and place in a new 1.5 ml microcentrifuge tube. Add 300 μl of PBST and incubate on a shaker at room temperature for 30 min. Repeat with a fresh microcentrifuge tube and PBST aliquot every 30 min for 3–4 h.

15. Repeat **steps 12–14** with the Alexa 488 goat anti-rabbit antibody. Wrap the microcentrifuge tubes with aluminum foil to prevent photobleaching of the secondary antibody. After the final wash, stained samples can be stored in PBS at 4 °C until further use (*see* **Note 9**).

16. Visualize samples as described in Subheading 3.4.

3.4 Imaging Techniques

1. To image samples, use a 10×/0.30 NA objective focused on the midplane of the tip of the epithelial cell aggregate (*see* **Note 10**).

2. To monitor cell migration over time in the culture model under various flow conditions, capture phase-contrast images of the epithelial cell aggregates (Figs. 4a and 5a) on each day for up to 9 days using an inverted phase-contrast microscope.

3. To image stained samples, transfer the samples onto a glass slide, and place a drop of PBS on top of each sample to keep the sample hydrated.

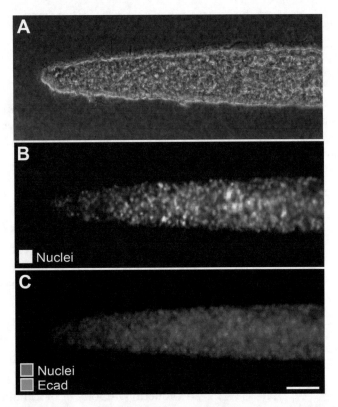

Fig. 4 Visualization of cell aggregates using phase-contrast, small molecule dyes, and fluorescent antibodies. (**a**) An epithelial cell aggregate under a control pressure profile (no hydrostatic pressure differential across the channel) on Day 6 visualized using phase-contrast imaging. (**b**) Image of the sample from (**a**) stained using the nuclear marker Hoechst 33342. Nuclei are shown in *white*. (**c**) Sample from (**a**) stained for E-cadherin (shown in *green*). Nuclei are shown in *red*. Scale bar 100 μm

4. To visualize cell nuclei, image fixed Hoechst 33342-stained samples under UV illumination (Figs. 4b and 5b).

5. To visualize samples stained for E-cadherin, image using an inverted fluorescence microscope (Fig. 4c).

3.5 Image Analysis: Measuring the Extent of Cell Invasion from the Initial Aggregate

1. Open the phase-contrast image files in ImageJ.

2. Set the scale of the image according to microscope calibrations by selecting "Set Scale…" under the "Analyze" menu.

3. Using the line tool on the main menu, draw a line along the length of a collectively migrating cohort protruding from the aggregate (Fig. 5c) (*see* **Note 11**).

4. Measure the length of the line by clicking "Measure" under the "Analyze" menu. This will output the length of the line in the units specified.

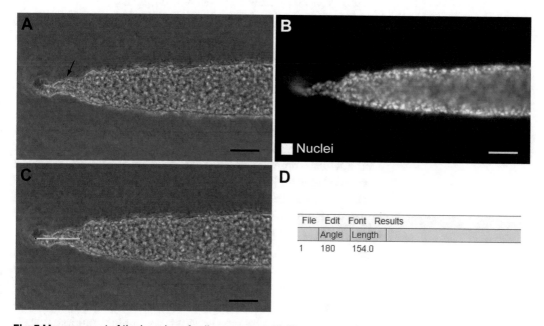

Fig. 5 Measurement of the invasion of cell aggregates. (**a**) Phase-contrast image of an epithelial cell aggregate on Day 6 under a pressure profile, obtained by holding well A at a higher hydrostatic pressure than well B. Invasion from the tip of the cell aggregate is shown with an *arrow*. (**b**) Image of the sample from (**a**) stained using Hoechst 33342. Nuclei are shown in *white*. (**c**) Example of using the line tool in ImageJ to measure the length of the invasive protrusion from the epithelial cell aggregate tip in the sample from (**a**). (**d**) Measurement output from ImageJ. Scale bars 100 μm

4 Notes

1. Cleaning should eliminate all leftover debris and dust particles.

2. Press down on the PDMS chambers to ensure that they are conformally adherent to the tissue culture dish.

3. The PDMS chambers must be chilled prior to the addition of collagen to prevent premature gelation.

4. When using a new bottle of collagen, check the pH of the final mixture. We used a pH of 8.5–9. To alter the pH, adjust the volume of NaOH accordingly.

5. To help the collagen mixture flow into the channels, turn the tissue culture dish on its side (with the channels perpendicular to the work surface) while tapping.

6. The hydrostatic pressure differential across the aggregate can range from 0.4 to 1.6 cm H_2O depending on how many PDMS blocks are used. Interstitial flow velocities are on the order of 1 μm/s (average flow rates of 20–100 μl/day).

7. Remove the collagen channel containing the epithelial cell aggregate from the PDMS chamber with forceps. Be sure to grip the channel from the end adjacent to well A.

8. Here, we stained for E-cadherin, but the same protocol may be used for other primary antibodies.

9. In addition to staining, one can also use real time RT-PCR analysis to quantify changes in gene expression. For this analysis, it is necessary to combine at least six samples per condition to obtain an adequate amount of RNA. Samples must also be incubated with collagenase (we recommend a 2 mg/ml solution of collagenase from *Clostridium histolyticum* (Sigma) in culture medium) prior to RNA extraction using an RNeasy Mini Kit (Qiagen).

10. To visualize 3D features of the mammary epithelial cell aggregates (live or fixed), capture confocal stacks of the samples using an inverted spinning disk confocal microscope (200 images, 1 μm apart).

11. We defined collective migration as protrusions from the primary aggregate, still attached to the latter, containing multiple nuclei.

Acknowledgements

This work was supported in part by grants from the NIH (GM083997, HL110335, and HL118532), pilot project funding from the NIH/NCI Physical Sciences-Oncology Center at Princeton University (U54CA143803), Concept Award W81XWH-09-1-0565 from the Breast Cancer Research Program of the Department of Defense (to J.T.), the David & Lucile Packard Foundation, the Alfred P. Sloan Foundation, the Camille & Henry Dreyfus Foundation, and the Burroughs Wellcome Fund.

References

1. Discher DE, Janmey P, Wang YL (2005) Tissue cells feel and respond to the stiffness of their substrate. Science 310(5751):1139–1143

2. Schwarz US, Bischofs IB (2005) Physical determinants of cell organization in soft media. Med Eng Phys 27(9):763–772

3. Jaalouk DE, Lammerding J (2009) Mechanotransduction gone awry. Nat Rev Mol Cell Biol 10(1):63–73

4. Kolahi KS, Mofrad MR (2010) Mechanotransduction: a major regulator of homeostasis and development. Wiley Interdiscip Rev Syst Biol Med 2(6):625–639

5. Nelson CM, Gleghorn JP (2012) Sculpting organs: mechanical regulation of tissue development. Annu Rev Biomed Eng 14:129–154

6. Paszek MJ, Zahir N, Johnson KR, Lakins JN, Rozenberg GI, Gefen A, Reinhart-King CA, Margulies SS, Dembo M, Boettiger D, Hammer DA, Weaver VM (2005) Tensional homeostasis and the malignant phenotype. Cancer Cell 8(3):241–254

7. Lee K, Chen QK, Lui C, Cichon MA, Radisky DC, Nelson CM (2012) Matrix compliance regulates rac1b localization, nadph oxidase assembly, and epithelial-mesenchymal transition. Mol Biol Cell 23(20):4097–4108

8. Baxter LT, Jain RK (1989) Transport of fluid and macromolecules in tumors. I. Role of interstitial pressure and convection. Microvasc Res 37(1):77–104

9. Hompland T, Ellingsen C, Ovrebo KM, Rofstad EK (2012) Interstitial fluid pressure and associated lymph node metastasis revealed in tumors by dynamic contrast-enhanced mri. Cancer Res 72(19):4899–4908

10. Milosevic M, Fyles A, Hedley D, Pintilie M, Levin W, Manchul L, Hill R (2001) Interstitial fluid pressure predicts survival in patients with cervix cancer independent of clinical prognostic factors and tumor oxygen measurements. Cancer Res 61(17):6400–6405

11. Baker BM, Chen CS (2012) Deconstructing the third dimension: how 3d culture microenvironments alter cellular cues. J Cell Sci 125(Pt 13):3015–3024

12. Kleinman HK, Philp D, Hoffman MP (2003) Role of the extracellular matrix in morphogenesis. Curr Opin Biotechnol 14(5):526–532

13. Vidi PA, Bissell MJ, Lelievre SA (2013) Three-dimensional culture of human breast epithelial cells: the how and the why. Methods Mol Biol 945:193–219

14. Bissell MJ, Radisky DC, Rizki A, Weaver VM, Petersen OW (2002) The organizing principle: microenvironmental influences in the normal and malignant breast. Differentiation 70(9-10):537–546

15. Heldin CH, Rubin K, Pietras K, Ostman A (2004) High interstitial fluid pressure – an obstacle in cancer therapy. Nat Rev Cancer 4(10):806–813

16. Boucher Y, Jain RK (1992) Microvascular pressure is the principal driving force for interstitial hypertension in solid tumors: implications for vascular collapse. Cancer Res 52(18):5110–5114

17. Chauhan VP, Stylianopoulos T, Boucher Y, Jain RK (2011) Delivery of molecular and nanoscale medicine to tumors: transport barriers and strategies. Annu Rev Chem Biomol Eng 2:281–298

18. Lunt SJ, Fyles A, Hill RP, Milosevic M (2008) Interstitial fluid pressure in tumors: therapeutic barrier and biomarker of angiogenesis. Future Oncol 4(6):793–802

19. Netti PA, Baxter LT, Boucher Y, Skalak R, Jain RK (1995) Time-dependent behavior of interstitial fluid pressure in solid tumors: implications for drug delivery. Cancer Res 55(22):5451–5458

20. Haessler U, Teo JC, Foretay D, Renaud P, Swartz MA (2012) Migration dynamics of breast cancer cells in a tunable 3d interstitial flow chamber. Integr Biol (Camb) 4(4):401–409

21. Shields JD, Fleury ME, Yong C, Tomei AA, Randolph GJ, Swartz MA (2007) Autologous chemotaxis as a mechanism of tumor cell homing to lymphatics via interstitial flow and autocrine ccr7 signaling. Cancer Cell 11(6):526–538

22. Polacheck WJ, German AE, Mammoto A, Ingber DE, Kamm RD (2014) Mechanotransduction of fluid stresses governs 3d cell migration. Proc Natl Acad Sci U S A 111(7):2447–2452

23. Polacheck WJ, Charest JL, Kamm RD (2011) Interstitial flow influences direction of tumor cell migration through competing mechanisms. Proc Natl Acad Sci U S A 108(27):11115–11120

24. Tien J, Truslow JG, Nelson CM (2012) Modulation of invasive phenotype by interstitial pressure-driven convection in aggregates of human breast cancer cells. PLoS One 7(9), e45191

Part III

Mammary Stem Cells

Chapter 13

Purification of Distinct Subsets of Epithelial Cells from Normal Human Breast Tissue

Mona Shehata and John Stingl

Abstract

The mammary epithelium is composed of a variety of specialized cell types that function in a coordinated fashion to produce and eject milk through multiple cycles of pregnancy. The ability to identify and purify these subsets of cells in order to interrogate their growth and differentiation capacities, as well as to characterize the molecular mechanisms that regulate their behavior, is essential in identifying the processes associated with breast cancer initiation and progression. This methods chapter outlines the step-by-step methods for dissociating human breast reduction specimens to a single cell suspension of viable cells. As well, strategies for purifying four distinct subsets of epithelial cells by using fluorescence-activated cell sorting and protocols for interrogating the growth and differentiation properties of these purified cells at clonal densities in adherent culture are also described.

Key words Human breast tissue, Flow cytometry, Cell culture, Stem cells, Colony-forming cells

1 Introduction

The human mammary epithelium is composed of a series of branched ducts that drain the distally positioned terminal ductal-lobuloalveolar units (TDLUs; [1]). These TDLUs undergo considerable lobulo-alveolar expansion during pregnancy, with the newly formed alveolar cells secreting milk during lactation. The mammary epithelium itself is composed of two general lineages of cells, the luminal cells and the myoepithelial cells. The luminal cell layer can be further subdivided into two subtypes of cells, hormone-sensing estrogen receptor[+] (ER[+]) and ER[-] [2–10]. The ER[-] luminal cells are characterized by high expression of the ELF5 transcription factor, which has been reported to specify alveolar cell fate [11]; thus these cells have been hypothesized to represent progenitor cells that will generate alveolar daughter cells during pregnancy.

A number of reports have been published over the years that have used flow cytometry to identify and characterize the cells that make up the human mammary epithelium [2–4, 6–8, 10, 12–15].

Finian Martin et al. (eds.), *Mammary Gland Development: Methods and Protocols*, Methods in Molecular Biology, vol. 1501,
DOI 10.1007/978-1-4939-6475-8_13, © Springer Science+Business Media New York 2017

Although a myriad of cell surface and intracellular markers have been identified as being useful for flow-sorting mammary cells, we have found that a combination of just three markers (EpCAM, CD49f, and aldehyde dehydrogenase 1 (ALDH1) activity) used in conjunction with markers (CD45 and CD31) to deplete unwanted contaminating hematopoietic and endothelial cells can be used to resolve and isolate four types of breast epithelial cells [2]. These include the basal cells, the luminal progenitors, of which undifferentiated and differentiated subsets can be identified, and the non-clonogenic luminal cells [2]. The basal cell population, which has an EpCAM^{low}CD49f^{high} phenotype, is composed of predominantly myoepithelial cells, but also contains a subpopulation of bipotent progenitors that generate colonies composed of both luminal and basal cells in colony-forming cell (CFC) assays. The luminal progenitor cell population, which has an EpCAM^{high}CD49f$^+$ phenotype, is highly enriched for cells that generate pure luminal cell colonies in vitro. This luminal progenitor population is composed of a spectrum of cells that span from those that exhibit low levels of luminal cell differentiation, but express ALDH1 and ELF5 and have high cloning efficiencies in vitro, to cells that are ALDH$^-$, but express higher levels of luminal cell differentiation and have slightly lower cloning efficiencies in vitro. The non-clonogenic luminal (NCL) cell population, which has an EpCAM^{high}CD49f$^-$ phenotype, is composed primarily of cells exhibiting high levels of luminal cell differentiation, including ER expression. As their name implies, these cells are largely deficient in CFC potential, and thus have been historically perceived as being composed of terminally differentiated cells. However, it is important to keep in mind that this lack of detectable cloning potential in vitro may just indicate sub-optimal culture conditions rather than an intrinsic lack of proliferation potential.

The methods below describe the step-by-step procedure for collecting and dissociating normal human breast tissue derived from reduction mammoplasty surgeries, and how these samples can be cryopreserved for later use. The methods also describe the procedure for preparing the cells for fluorescence-activated cell sorting (FACS) and how to resolve the different subpopulations on the flow cytometry dot plots. Finally, the methods also describe an in vitro CFC assay that permits the quantitation of the different types of progenitor cells within breast cell populations.

2 Materials

2.1 Dissociation of Human Tissue

1. Scalpels.
2. Sterile forceps.
3. Rotary shaker.
4. Glass petri dish (~15 cm in diameter).

5. 250 ml dissociation flasks.

6. Collagenase/hyaluronidase solution (10× concentration stock). Stored at –20 °C in 1 ml aliquots.

7. 5 µg/ml insulin (Sigma catalog number I-1882). Stored at –20 °C in 100 µl aliquots.

8. DMEM/F12 supplemented with 10 mM Hepes (referred to as DMEM/F12/H) and 5 % fetal bovine serum (FBS). Stored at 4 °C.

9. 7.5 % BSA fraction V solution. Stored at 4 °C.

10. 50 µg/ml gentamicin solution. Stored at 4 °C.

11. 1.8 ml RNAse/DNAse free cryovials.

12. 10 % neutral buffered formalin.

13. Dimethyl sulfoxide (DMSO)-based freezing mix: 6 % DMSO, 50 % FBS in DMEM/F12/H.

14. Mr. Frosty freezing container.

15. Liquid nitrogen.

2.2 Single Cell Preparation and Antibody Staining

1. DMEM/F12/H.

2. 0.25 % trypsin–EDTA. Stored –20 °C in 2 ml aliquots.

3. Hank's Balanced salt solution liquid with calcium chloride and magnesium chloride supplemented with 10 mM Hepes and 2 % FBS. Referred to as HF. Stored at 4 °C.

4. 5 mg/ml dispase solution. Stored at –20 °C in 2 ml aliquots.

5. Deoxyribonuclease (DNase) dissolved at 1 mg/ml in DMEM/F12. Filter-sterilized with a 0.22 µm filter. Stored at –20 °C in 200 µl aliquots.

6. Ammonium chloride solution (optional). Stored at –20 °C in 50 ml aliquots.

7. 40 µm cell strainers.

8. Hemocytometer.

9. Trypan blue (0.4 %).

10. 10 % normal rat serum in HF (antibody preblocking solution).

11. Antibodies for flow cytometry: (*see* **Note 1**).

   (a) CD31-biotin (clone WM-59).

   (b) CD45-biotin (clone HI30).

   (c) EpCAM-PE (clone 9C4).

   (d) CD49f-PE/Cy7 (clone GoH3).

   (e) ALDEFLUOR™ kit.

   (f) Streptavidin APC-Cy7.

(g) 4′,6-diamidino-2-phenylindole (DAPI). Make up as a 1 mg/ml stock solution in distilled water and filter-sterilize. Store at −20 °C in 1 ml aliquots. Use at 1 μg/ml final concentration.

2.3 Human Mammary Colony Forming Assay

1. 60 mm culture dishes.

2. NIH 3T3 Swiss mouse embryo fibroblast cell line and irradiated at 50 Gy.

3. Phosphate-buffered saline (PBS).

4. Human EpiCult-B base media + supplement.

5. 10^{-6} M hydrocortisone dissolved in ethanol. Stored at −20 °C.

6. 5 % FBS. Stored at −20 °C.

7. 1 mg/ml gentamicin. Stored at 4 °C.

8. Giemsa stain.

3 Methods

3.1 Dissociation of Normal Human Mammary Tissue

1. Transport human mammary tissue from the operating room on ice in sterile specimen cups in DMEM/F12/H + 5% FBS + 50 μg/ml gentamicin.

2. Record the patient age and sample type (e.g., mastectomy, risk-reduction mastectomy, tumor, normal adjacent, and contralateral normal).

3. Estimate the size of the sample and record this in a database. Using a scalpel, remove several small non-fatty pieces of tissue for RNA, DNA and for formalin fixation/paraffin embedding.

   (a) For RNA and DNA: Place two small pieces of tissue (approximately 4 mm³) into two RNAse/DNAse free cryovials. Make sure tissue is at the bottom of the vial. Snap freeze in liquid nitrogen and immediately place in an −80 °C freezer or in a designated liquid nitrogen tank.

   (b) For formalin fixation/paraffin embedding: Place a small piece of non-fatty tissue (approximately 5 mm3 in volume) in approximately 5 ml of 10 % neutral buffered formalin and fix for 24 h. After 24 h fixation, decant the formalin and add 70 % ethanol then process the tissue for paraffin embedding.

4. Using the scalpels, roughly mince the remainder of the specimen by cutting in a crosshatch pattern. Several pairs of scalpels may be required if the blades become dull. Once the tissue is minced to small fragments (~3 mm³), transfer the tissue to a dissociation flask. Each flask should contain up to approximately two heaped tablespoons of breast tissue. If the sample is large, multiple dissociation flasks may be required.

5. Add 20 ml dissociation media per flask:

   (a) 2 ml 10× collagenase/hyaluronidase.

   (b) 13 ml DMEM/F12/H.

   (c) 5 ml BSA Fraction V solution.

   (d) 10 μl insulin (5 μg/ml final concentration).

   (e) 20 μl gentamycin stock solution (50 μg/ml final concentration).

6. Cover the opening of the flask with sterile aluminum foil. Cover the foil with a layer of Parafilm M laboratory film. Label each flask with tape and include sample name and date.

7. Gently dissociate the minced tissue on the rotary shaker in a 37 °C incubator at approximately 80 rpm until all large tissue fragments are digested. Typical digestion time is 16 h (overnight) for normal human mammary tissue (*see* **Note 2**).

3.2 Preparation of a Mammary Single Cell Suspension

1. After dissociation, transfer all the dissociated breast cell suspension to 50 ml centrifuge tubes. Multiple tubes might be necessary if multiple dissociation flasks were used. Spin the tubes at 450×g for 5 min at 4 °C. Discard the overlying fat layer and put this into a 50 ml tube to be disposed of accordingly. Discard the remaining non-fatty supernatant into Virkon or a similar decontaminating solution.

2. Add 10 ml of DMEM/F12/H to the pellet and wash once at 450×g for 5 min at 4 °C to get rid of residual collagenase and hyaluronidase.

3. After the wash, resuspend the cells in 10 ml of DMEM/F12/H and centrifuge for 4 min at 200×g at 4 °C. The pellet from this centrifugation is enriched (but is not pure) for epithelial cells. The supernatant from this slow centrifugation is enriched for human mammary fibroblasts. Carefully transfer the supernatant to a new 50 ml centrifuge tube without disturbing the epithelial enriched pellet. Centrifuge the supernatant at 450×g for 5 min at 4 °C. Once the supernatant has been centrifuged, place the tube on ice until the cryopreservation step.

4. Add 2 ml of pre-warmed trypsin–EDTA to the enriched epithelial pellet to resuspend the cells. Use more trypsin if the pellet is big. Gently pipette up and down with a P1000 pipette for 2–3 min. The sample may become stringy due to lysis of dead cells and the release of DNA.

5. Add 10 ml of cold HF and centrifuge at 450×g for 5 min at 4 °C. Remove as much of the supernatant as possible with a pipette (do NOT pour). The cells may be a large "stringy mass" floating in the HF.

6. Add 2 ml of pre-warmed 5 mg/ml dispase and 200–400 µl of 1 mg/ml DNase I. Pipette the sample for 1 min with a P1000 pipette to further dissociate cell clumps. The sample should now be cloudy, but not stringy. If still stringy, add more DNase I. Use more dispase and DNAse I solutions if the pellet is big.

7. Add 8 ml of cold HF and spin at $450 \times g$ for 5 min at 4 °C. Discard supernatant carefully.

8. Optional step (*see* **Note 3**): Resuspend the pellet in a 1:4 mixture of cold HF–ammonium chloride, centrifuge at $450 \times g$ for 5 min at 4 °C and discard the supernatant.

9. Resuspend the pellet in 10–20 ml HF and remove a 20 µl aliquot for counting. Spin the sample once more at $450 \times g$ for 5 min at 4 °C. Record the viable cell yield.

10. The different cell fractions can now be cryopreserved. It is recommended that cells are cryopreserved in DMEM/F12/H supplemented with 50 % FBS and 6 % DMSO. Make up 1.8 ml of DMSO-based freezing solution for every cryovial of cells to be frozen. Once the solution is made up, place it on ice. Resuspend the cell pellet in the appropriate volume of freezing mix (dictated by cell count—approximately $10–20 \times 10^6$ cells/vial) and aliquot to the cryotubes. Once all the cryotubes are prepared, put them into a Mr. Frosty freezing unit and place the unit into a –80 °C freezer. The next day transfer the cryovials to a liquid nitrogen tank.

11. Record all details within a database including the number of epithelial and stromal enriched vials, as well as RNA, DNA, formalin-fixed tissue samples.

3.3 Cell Staining for Flow Cytometry

1. Remove cells from liquid nitrogen and thaw in a clean water bath or a 37 °C incubator. Once thawed, spray the outside of the vials with 70 % ethanol to decontaminate and transfer to contents of the cryotube into either a 10 or 50 ml centrifuge tube, depending on the number of vials thawed. For each sample thawed, add 10 ml cold HF to resuspend the cells and centrifuge at $450 \times g$ for 5 min at 4 °C. Carefully decant supernatant and resuspend cells in 1 ml cold HF. Record the sample and number of vials used on the database.

2. Preblock the sample with HF supplemented with 10 % normal rat serum. Block on ice for 10 min.

3. During the preblock, remove 10 µl of cells and add these to 10 µl Trypan Blue and 80 µl HF for counting. Count cells using a hemocytometer in order to ensure you have a single cell suspension and to confirm the number of cells available for staining.

4. Aliquot 80–90 % of cells into the Sample tubes (Table 1 and *see* **Note 4**). Add HF to a convenient volume to distribute the

Table 1
Example of a staining table

Antibody	Single color control		FMO				Sample
Tube number	1 2 3 4 5 6 7		8 9 10			11 12	
DAPI	+		+ + +			+ +	
Lineage (CD45-biotin/CD31-biotin +streptavidin-APC-Cy7) at 1:500	+		+ + +			− +	
CD49f-AF647 at 1:100	+		+ + −			+ +	
EpCAM-PE at 1:50	+		+ − +			+ +	
ALDH	+		− + +			+ +	
ALDH+DEAB	+						

remaining cell suspension equally across the rest of the single-colored controls and fluorescence-minus-one tubes (Table 1 and *see* **Note 5**).

5. If staining for Aldefluor, follow the protocol as specified by the manufacturer. Briefly:

   (a) Label one "test (tube 6)" and one "DEAB control (tube 7)" tube for the single-colored controls for each donor sample to be tested. Place 1 ml of the cell suspension into the "test" single color control tube.

   (b) Spin the "test" and "sample (tube 12)" tubes at $450 \times g$ for 5 min at 4 °C. Discard the supernatant.

   (c) Resuspend the cells in Assay Buffer for the "test" and the "sample" tubes up to a cell concentration of 5×10^6 cells/ml.

   (d) For the single color controls: Add 5 μl of ALDEFLUOR™ DEAB Reagent to the "DEAB control" tube. Recap control tube and DEAB vial immediately.

   (e) Add 5 μl of the activated ALDEFLUOR™ Reagent per milliliter of sample to the "test" tube. Mix and immediately transfer 0.5 ml of the mixture to the DEAB "DEAB control" tube (*see* **Note 6**).

   (f) For the sample tube: Add 5 μl of the activated ALDEFLUOR™ Reagent per milliliter of sample.

   (g) Incubate "test" and "DEAB control" and "sample" tubes for 30 min at 37 °C (manufacturer's protocol does not recommend to exceed 60 min).

(h) Following incubation, centrifuge "test," "DEAB control," and "sample" tubes at $450 \times g$ for 5 min at 4 °C and discard the supernatant. Resuspend cell pellets of the "DEAB control" and "test" samples in 0.5 ml of ALDEFLUOR™ Assay Buffer and placed on ice for the remaining of the staining protocol. The cells in the sample tube can be resuspended in 3 ml of HF and sub-aliquoted to tubes 9–12 and stained with primary antibodies as summarized in Table 1.

6. Spin all the tubes except for the "DEAB control" and "test" samples at $450 \times g$ for 5 min at 4 °C.

7. Discard supernatant, being careful to remove all liquid without disturbing the pellet.

8. Stain with primary antibodies diluted in HF for a minimum of 10 min on ice (*see* **Note 7**). The primary antibodies and their appropriate dilutions (*see* **Note 8**), and the required "fluorescence minus one" controls are outlined in Table 1.

9. After primary antibody incubation, add 3 ml cold HF media to each tube. Spin all the tubes at $450 \times g$ for 5 min at 4 °C. Discard supernatant.

10. Stain with the tubes that were incubated with the "lineage" antibodies CD45-biotin and CD31-biotin with a 1:500 dilution of streptavidin-APC-Cy7 on ice for 10 min. Add 3 ml cold HF media to the remaining tubes.

11. After the streptavidin-APC-Cy7 incubation step, filter cells in all of tubes through 40 μm cell strainer into new 5 ml tubes.

12. Spin all tubes at $450 \times g$ for 5 min at 4 °C, and then discard the supernatant. Resuspend the cells at 500 μl HF/tube for controls and 1–2 ml for sample tubes ($\sim 5 \times 10^6$ cells/ml). Place tubes on ice and take to the flow analyzer/sorter. If sorting, prepare collection tubes and sort into 1.5 ml eppendorf tubes that have been pre-loaded with 0.5 ml HF.

3.4 Flow Cytometry Analysis on Single Cell Suspension

1. Place the unstained control tube onto the FACS machine and run sample. Adjust voltages of FSC and SSC as well as the desired fluorophores such that the majority of background fluorescence is within the first log decade.

2. Run single color control tubes adjusting for background spectral overlap and compensate accordingly (*see* **Note 9**).

3. To analyze the sample, collect at least 50,000 events. Gate around all events based on forward (FSC) and side (SSC) scatter, but excluding the events with the highest side scatter (Fig.1a). Then exclude doublets by gating the events in the FSC-height by FSC-area parameters (Fig. 1b). Dead and dying cells are then excluded by gating on the DAPI negative events and by avoiding debris using the FSC parameter (Fig. 1c). Also

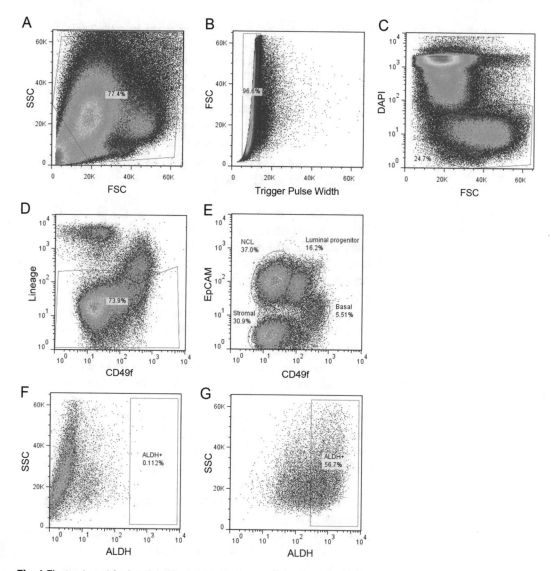

Fig. 1 Flow cytometric dot plots illustrating the gating strategy to identify viable human mammary basal, luminal progenitor and non-clonogenic luminal cells. The luminal progenitor population can be further subdivided based upon ALDEFLUOR staining. The ALDH+DEAB inhibitor control illustrates the gating strategy to identify ADLH positive cells

exclude the events expressing intermediate levels of DAPI as these have low viability. To exclude most non-epithelial (termed "lineage-negative") cells, select the events that do not express CD31 and CD45 (Fig. 1d). Draw another plot with EpCAM on the y-axis and CD49f on the x-axis (Fig. 1e), which permits visualization of the non-clonogenic luminal (EpCAM+CD49f–), luminal progenitor (EpCAM+CD49f+), basal (EpCAM^low CD49f+), and stromal (EpCAM–CD49f–) cell populations.

4. To further divide the luminal progenitor population (*see* **Note 10**) draw another plot with ALDH (AF488 parameter) on the *x*-axis and SSC on the *y*-axis. Run the DEAB+ALDH control tube. Draw a gate excluding all events of the ALDH+DEAB cells (Fig. 1f). Place the corresponding sample tube onto the cytometer. ALDH positive progenitors should be within the gate (Fig. 1g). ALDH negative progenitors can be selected by drawing another gate to the left (Fig. 1g).

5. Flow sorted samples can be sorted into 1.5–2 ml eppendorfs containing 500 μl HF media.

3.5 Human Mammary Colony Forming Assay

1. Prepare the culture media by thawing the 1 ml Human EpiCult-B supplement and add it to 100 ml Human EpiCult-B base media. In addition, supplement the media with hydrocortisone (final concentration of 0.48 μg/ml), gentamycin (final concentration of 50 μg/ml), and 5 % FBS. You will need 4 ml of media for each culture dish, and we recommend that each cell population is seeded into two replicate plates. Calculate the number of dishes required and make up enough media for an extra plate.

2. Human mammary epithelial cells grow best on a feeder layer [6]. Irradiated NIH 3T3 cells are ideal for this use. To prepare these cells, culture them in DMEM/F12/H supplemented with 5 % FBS. Once the cultures get to approximately 70 % confluence (*see* **Note 11**), harvest the cells with a brief treatment with trypsin–ETDA solution. After washing with DMEM/F12 + 5 % FBS and centrifugation at $450 \times g$ for 5 min at 4 °C, resuspend the cells at 10^6 cells/ml in DMEM/F12 + 5 % FBS in 5 ml flow cytometry tubes and irradiate with gamma ionizing irradiation to a final dose of 50 Gy. Once irradiated, 5×10^4 irradiated feeder cells can be added to every ml of Human EpiCult-B complete media (*see* **Note 12**).

3. Label the bottom of the required number of 60 mm culture dishes with a fine-tipped felt pen. Labeling only lids is not ideal as these can get easily mixed up when using a large number of dishes.

4. For every sample to be interrogated for CFC content, aliquot 8 ml of Human EpiCult-B complete media plus feeder cells into a 15 ml tube. Add the required number of mammary cells to the tube (*see* **Note 13**).

5. Mix the cells in the tube and then aliquot each 8 ml sample into two 60 mm dishes, with each dish receiving 4 ml of the cell suspension.

6. Incubate the dishes at 5 % CO_2 and at 37 °C and 10–12 days.

7. After 10–12 days, remove the media and gently wash the plates once with PBS. Completely remove the PBS and add 2 ml of acetone–methanol (1:1) per dish for 30 s. Carefully remove the

acetone–methanol (ensuring that you do not rub out the labels) and allow the dish to air-dry. Once air-dried, gently rinse the plate once with some distilled water, and then add 2 ml of Giemsa stain (diluted 1:10 with distilled water) for 2–3 min or until the color of the colonies is strong. Remove the Giemsa stain and rinse the dishes twice with distilled water and air-dry.

8. Count the number of colonies per dish using a scoring grid, or by scanning the dishes using a gel doc scanner and the image software program to manually or automatically count the colonies (*see* **Note 14**).

3.6 Characterization of Human Mammary Epithelial Colonies

Luminal and basal cells can generate colonies in culture that have morphologies that are readily distinguishable from one another [4, 6]. For example, bipotent progenitors within the basal cell population generate colonies that are composed of both luminal and basal cells. These colonies are characterized as having a nonuniform colony edge, and containing basal cells that are characterized by having a teardrop shape and having gaps between these cells (Fig. 2a). The basal cell population also contain myoepithelial cell-restricted progenitors that generate pure myoepithelial cell colonies (Fig. 2a). The luminal progenitor cell population generates pure luminal cell colonies in vitro. These colonies are characterized by having a uniform scalloped colony edge (Fig. 2b). Non-clonogenic luminal cells have limited proliferative potential and generate colonies at very low frequencies. The rare colony that is generated from these cells have a pure luminal cell colony morphology. It should be noted that the colony morphologies observed may be very dependent on the types of culture media used, and different research groups have reported different colony types being obtained from similar sorted populations [6, 12, 16].

Immunofluorescence staining for cytokeratins 14 (K14) and 18 (K18) allows basal and luminal cells, respectively, to be distinguished (Fig. 3). Colonies derived from luminal cells express K18 only (Fig. 3a, b). Basal cells usually give rise to mixed colonies that contain both K18+ luminal and K14+ basal cells. Other markers for luminal cells include K8, whereas basal cells can also be identified by expression of smooth muscle actin (Fig. 3). Stromal colonies have a dispersed and mesenchymal phenotype. These colonies should be excluded from colony counts if only epithelial progenitors are of interest.

4 Notes

1. The selected antibody conjugated fluorophores used in this protocol are most suitable for the flow cytometer that is used. End users will have to identify fluorophore combinations that are compatible with the laser configuration on their flow cytometer.

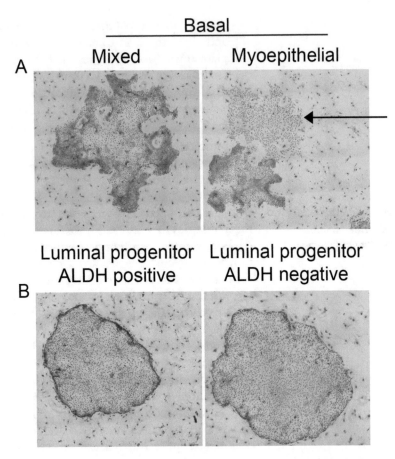

Fig. 2 Human mammary epithelial colonies grown in Human EpiCult-B™ and stained with Giemsa. (**a**) Colonies generated from basal cells. *Arrow* indicates a pure myoepithelial cell colony. (**b**) Pureluminal cell colonies generated from luminal progenitor cells

2. Some reduction mammoplasty samples may be particularly fibrous. In these cases, the dissociation time may have to be extended by a few hours, however we discourage overly long dissociation times because this can reduce viable cell yield. In some samples, there may be a large number of whitish fragments floating in the dissociation mix that appear to be resistant to further collagenase/hyaluronidase digestion. These fragments may be larger portions of ducts and TDLUs rather than undigested material. The nature of these fragments can be confirmed by removing a small aliquot of the digestion mixture, and examining it under a low power microscope.

3. If the cells liberated from the dissociated mammary tissue are going to be analyzed by flow cytometry right away, then it is recommended that the red blood cells that contaminate the

A

Luminal progenitor   Luminal progenitor
   ALDH positive   ALDH negative   Basal

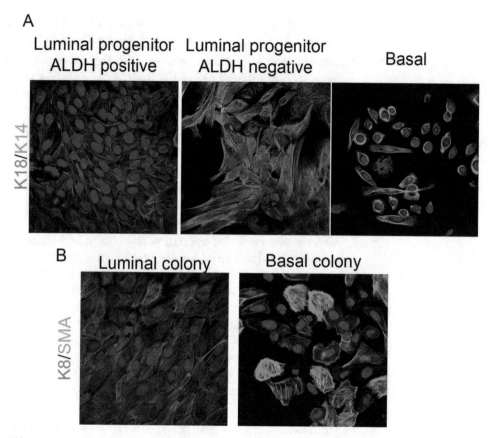

Fig. 3 Human mammary colonies stained to detect (**a**) cytokeratin (K) 18 (*red*) and K14 (green), or (**b**) K8 (*red*) and smooth muscle actin (*green*)

cell preparation be lysed by an ammonium chloride treatment. This will not be necessary if the cells are going to be cryopreserved first because this freezing and thawing process will lyse the red blood cells.

4. The type of tube that will be used will depend on the flow cytometer used for cell sorting. Some flow cytometers (e.g., BD Aria) require the cells to be in a 5 ml polystyrene tube, whereas other types (e.g., BD Influx) require the cells to be in a 5 ml polypropylene tube.

5. If sorting cells, it is advisable to aliquot the majority of the cells into the tube that will be used for sorting, rather than over-allocating cells to control tubes. If analyzing cells, then the cells can be equally distributed amongst all tubes.

6. The ALDH enzymatic reaction begins immediately upon addition of the activated substrate to the cell suspension.

It is imperative that an aliquot of the ALDEFLUOR™-reacted cells be added to the DEAB control tube without delay.

7. Antibody staining should be done with the lights dimmed to minimize photobleaching of the fluorescent dyes. As well, the cells and buffers should all be kept on ice as much as possible to inhibit cell clumping and to maximize cell viability.

8. The cell density in the diluted antibody solutions should not exceed 10^7 cells/ml, otherwise the amount of antibody for binding will be suboptimal.

9. The green fluorescence of ALDH-treated cells is very strong, it is important that the parameters are adjusted accordingly and compensation is set up accurately.

10. We have observed that approximately a third of human breast tissue samples have an additional luminal progenitor cell population that is characterized by low expression of ERBB3 [2]. This population is unusual since it has the most basal-like phenotype of all of the luminal progenitor cell subpopulations. As well, despite having a luminal progenitor gene expression profile and phenotype, it is relatively deficient at generating colonies in vitro. These cells can be resolved by including an anti-ERBB3 antibody (clone 1B4C3) conjugated to the appropriate fluorochrome to the staining protocol outlined in Table 1.

11. Multiple batches of low passage NIH 3T3 cells should be tested for their effectiveness to be used as a feeder layer. We have observed that some batches work better than others. It is imperative that cultures that are to be irradiated are derived from subconfluent cultures. In our experience, irradiated cells derived from confluent cultures do not work well as feeder cells.

12. NIH 3T3 feeder cells can be expanded in bulk in large tissue culture flasks and irradiated in bulk. Once irradiated, they can be frozen down and stored in liquid nitrogen in convenient easy to use aliquots that can be thawed on the day of the experiment.

13. Ideally, one would want to obtain between 50 and 100 colonies per dish. Colony numbers substantially higher than this run the risk of merging into one another, which will result in inaccurate colony counts. To obtain the desired number of colonies, we suggest that approximately 2000 non-sorted cells and 1000 sorted cells be seeded per dish.

14. If scoring the plates manually, use of fine-tipped felt pen to draw a grid on a 10 cm tissue culture plate such that the space between the gridlines is approximately 1 width of the field of view down a low-powered tissue culture microscope.

Acknowledgements

The authors thank Alasdair Russell for scientific discussions. They also thank Prof. Carlos Caldas and Linda Jones for facilitating and coordinating access to human tissue. They would like to acknowledge the support of The University of Cambridge, Breast Cancer Campaign, Cancer Research UK (core grant C14303/A17197), Hutchison Whampoa Limited, and NIHR Cambridge Biomedical Research Centre.

References

1. Wellings SR, Jensen HM, Marcum RG (1975) An atlas of subgross pathology of the human breast with special reference to possible precancerous lesions. J Natl Cancer Inst 55(2):231–273

2. Shehata M, Teschendorff A, Sharp G, Novcic N, Russell A, Avril S, Prater M, Eirew P, Caldas C, Watson CJ, Stingl J (2012) Phenotypic and functional characterization of the luminal cell hierarchy of the mammary gland. Breast Cancer Res 14(5):R134. doi:10.1186/bcr3334

3. Lim E, Vaillant F, Wu D, Forrest NC, Pal B, Hart AH, Asselin-Labat ML, Gyorki DE, Ward T, Partanen A, Feleppa F, Huschtscha LI, Thorne HJ, Fox SB, Yan M, French JD, Brown MA, Smyth GK, Visvader JE, Lindeman GJ (2009) Aberrant luminal progenitors as the candidate target population for basal tumor development in BRCA1 mutation carriers. Nat Med 15(8):907–913. doi:10.1038/nm.2000

4. Eirew P, Stingl J, Raouf A, Turashvili G, Aparicio S, Emerman JT, Eaves CJ (2008) A method for quantifying normal human mammary epithelial stem cells with in vivo regenerative ability. Nat Med 14(12):1384–1389. doi:10.1038/nm.1791

5. Eirew P, Stingl J, Eaves CJ (2010) Quantitation of human mammary epithelial stem cells with in vivo regenerative properties using a subrenal capsule xenotransplantation assay. Nat Protoc 5(12):1945–1956. doi:10.1038/nprot.2010.148

6. Stingl J, Eaves CJ, Zandieh I, Emerman JT (2001) Characterization of bipotent mammary epithelial progenitor cells in normal adult human breast tissue. Breast Cancer Res Treat 67(2):93–109

7. Stingl J, Eaves CJ, Kuusk U, Emerman JT (1998) Phenotypic and functional characterization in vitro of a multipotent epithelial cell present in the normal adult human breast. Differentiation 63(4):201–213. doi:10.1111/j.1432-0436.1998.00201.x

8. O'Hare MJ, Ormerod MG, Monaghan P, Lane EB, Gusterson BA (1991) Characterization in vitro of luminal and myoepithelial cells isolated from the human mammary gland by cell sorting. Differentiation 46(3):209–221

9. Petersen OW, Hoyer PE, van Deurs B (1987) Frequency and distribution of estrogen receptor-positive cells in normal, non-lactating human breast tissue. Cancer Res 47(21):5748–5751

10. Villadsen R, Fridriksdottir AJ, Ronnov-Jessen L, Gudjonsson T, Rank F, LaBarge MA, Bissell MJ, Petersen OW (2007) Evidence for a stem cell hierarchy in the adult human breast. J Cell Biol 177(1):87–101. doi:10.1083/jcb.200611114

11. Oakes SR, Naylor MJ, Asselin-Labat ML, Blazek KD, Gardiner-Garden M, Hilton HN, Kazlauskas M, Pritchard MA, Chodosh LA, Pfeffer PL, Lindeman GJ, Visvader JE, Ormandy CJ (2008) The Ets transcription factor Elf5 specifies mammary alveolar cell fate. Genes Dev 22(5):581–586. doi:10.1101/gad.1614608

12. Keller PJ, Arendt LM, Skibinski A, Logvinenko T, Klebba I, Dong S, Smith AE, Prat A, Perou CM, Gilmore H, Schnitt S, Naber SP, Garlick JA, Kuperwasser C (2012) Defining the cellular precursors to human breast cancer. Proc Natl Acad Sci U S A 109(8):2772–2777. doi:10.1073/pnas.1017626108

13. Dontu G, Al-Hajj M, Abdallah WM, Clarke MF, Wicha MS (2003) Stem cells in normal breast development and breast cancer. Cell Prolif 36(Suppl 1):59–72

14. Ginestier C, Hur MH, Charafe-Jauffret E, Monville F, Dutcher J, Brown M, Jacquemier J, Viens P, Kleer CG, Liu S, Schott A, Hayes D, Birnbaum D, Wicha MS, Dontu G (2007) ALDH1 is a marker of normal and malignant human mammary stem cells and a predictor of poor clinical outcome. Cell Stem Cell 1(5):555–567. doi:10.1016/j.stem.2007.08.014

15. Eirew P, Kannan N, Knapp DJ, Vaillant F, Emerman JT, Lindeman GJ, Visvader JE, Eaves CJ (2012) Aldehyde dehydrogenase activity is a biomarker of primitive normal human mammary luminal cells. Stem Cells 30(2):344–348. doi:10.1002/stem.1001

16. Visvader JE, Stingl J (2014) Mammary stem cells and the differentiation hierarchy: current status and perspectives. Genes Dev 28(11):1143–1158. doi:10.1101/gad.242511.114

Chapter 14

Techniques for the Reprogramming of Exogenous Stem/Progenitor Cell Populations Towards a Mammary Epithelial Cell Fate

Gilbert H. Smith and Corinne A. Boulanger

Abstract

This chapter considers the techniques necessary and required for the reprogramming of exogenous stem/progenitor cell populations towards a mammary epithelial cell fate. The protocols describe how to isolate cells from alternate mouse organs such as testicles of male mice and mix them with mammary cells to generate chimeric glands comprised of male and female epithelial cells that are fully competent. During the reformation of mammary stem cell niches by dispersed epithelial cells, in the context of the intact epithelium-free mammary stroma, non-mammary cells are sequestered and reprogrammed to perform mammary epithelial cell functions including those ascribed to mammary stem/progenitor cells. This therefore is a powerful technique for the redirection of cells from other organs/cancer cells to a normal mammary phenotype.

Key words Mammary gland, Survival surgery, Transplantation, Immunohistochemistry, Mouse mammary, Fat pad clearing, Reprogramming

1 Introduction

The capacity of any portion of the mouse mammary gland to produce a complete functional mammary outgrowth upon transplantation into an epithelium-divested fat pad is unaffected by the age or reproductive history of the donor [1, 2]. Likewise, through serial transplantations, no loss of potency is detected when compared to similar transplantations of the youngest mammary tissue tested. This demonstrates that stem cell activity is maintained intact throughout the lifetime of the animal despite aging and the repeated expansion and depletion of the mammary epithelium through multiple rounds of pregnancy, lactation and involution. These facts support our belief that mammary stem cells reside in protected tissue locales (niches), where their reproductive potency remains essentially unchanged through life. Disruption of mammary tissue removes the protection provided by the "niche" and leads to a

Finian Martin et al. (eds.), *Mammary Gland Development: Methods and Protocols*, Methods in Molecular Biology, vol. 1501,
DOI 10.1007/978-1-4939-6475-8_14, © Springer Science+Business Media New York 2017

reduced capacity of dispersed epithelial cells (in terms of the number transplanted) to produce complete functional mammary structures. We were inspired to test this niche concept in the regenerating mammary gland because we had successfully rescued mammary stem/progenitor cells from transgenic mammary tissues where regenerative capacity had been halted. We achieved this by the ectopic expression of suitable transgenes which allowed generation of two models (WAP-Notch4/Int3 × WAP-Cre/Rosa26R and WAP-TGFβ1 × WAP-Cre/Rosa26R), in which some of a population of mammary epithelial cells that were incapable of proliferation, in vivo, were marked with a lacZ-reporter reporter gene. We mixed the incompetent epithelial cells with normal wild type mammary epithelium and inoculated them into a cleared mammary fat pad and found that mixing them with normal wild type epithelial cells produced chimeric progeny during mammary gland regeneration. These results suggested that the mammary epithelial cells themselves in combination with the mammary fat pad and its stroma were essential components for a mammary stem cell niche.

The regenerative capacity of the mammary gland has been well documented [1–4]. Based on our understanding of the mammary niche, from our own work and the literature, we can hypothesize that mammary stem cells are stably maintained within specific microenvironments throughout the gland for life [5]. Mammary regeneration also occurs when dissociated epithelial cells from mammary glands are transplanted into cleared mammary fat pads, suggesting that complete mammary epithelial stem cell niches may be reconstituted de novo [6–8]. Stepwise dilution of dispersed mammary cells, or limiting dilution, results in a decrease in the percentage of inoculated fat pads that are rendered positive for mammary tree growth, implying reduction in the number of mammary epithelial stem cells [7, 9]. We hypothesize that the remaining cells encompass the epithelial signaling components and might support glandular regeneration if supplied with an extraneous source of stem/progenitor cells when injected into the mammary stroma.

Employing tissue-targeted, Cre-lox-mediated, conditional activation of a reporter gene, our laboratory has obtained evidence of the formation of a previously unrecognized mammary epithelial cell population that may originate from differentiating cells during pregnancy. This population does not undergo cell death during involution following lactation and persists throughout the lifetime of the female mouse. In transplantation studies, these cells show the capacity for self-renewal and contribute significantly to the reconstitution of resulting mammary outgrowths. In limiting dilution assays, it was found that these cells could generate both luminal and myoepithelial lineages, generate out-growths comprising both lobule and duct-limited epithelial outgrowths, and differentiate into all of the cellular subtypes recognized within the murine mammary epithelium. We have named these cells parity-induced mammary epithelial cells (PI-MEC). They undergo self-renewal, are multipotent, and contribute progeny directly to the formation of secretory acini in subsequent pregnancies. Therefore, PI-MEC represent LacZ+ lobule-limited, pluripotent epithelial progenitors, one of three distinct multipotent cell types previously identified in the mouse mammary gland [7, 9]. To determine whether LacZ+ PI-MEC could develop from uncommitted cells from another adult tissue upon interaction with a mammary microenvironment, comprised of signaling mammary epithelial cells and the mammary fat pad stroma, we mixed testicular cells (from seminiferous tubules), hematopoietic cells (from bone marrow) and thymus cells from adult whey acidic protein promoter (WAP)-Cre/Rosa26 R mice with limiting dilutions of mammary epithelial cells and inoculated them into epithelium-cleared mammary fat pads. The mammary outgrowths generated in the transplanted glands through pregnancy, lactation, and involution in the hosts were "phenotypically" normal. Therefore during the reformation of mammary stem cell niches by dispersed epithelial cells, in the context of the intact epithelium-free mammary stroma, non-mammary cells must be sequestered and reprogrammed to perform mammary epithelial cell functions including those ascribed to mammary stem/progenitor cells. This is therefore a powerful technique for the redirection of cells from other organs/tumors to a normal mammary phenotype. Figure 1 provides a summary illustration of the normal mammary niche and its disruption and incorporation of non-mammary stem/progenitor cells. Further studies have demonstrated our ability to also reprogram cells from alternate germ layers, to reprogram mouse embryonic stem cells, and both mouse and human cancer cells to a non-tumorigenic mammary fate. Below we describe our experimental methods and strategy.

2 Materials

2.1 Mice

The transgenic WAP-Cre/Rosa26 R mice were engineered and typed as described by Wagner et al. (Fig. 2) [10]. Female Nu/Nu/NCR mice were used as hosts for the transplantation studies. All mice are housed in Association for Assessment and Accreditation of Laboratory Animal Care-accredited facilities in accordance with the National Institutes of Health Guide for the Care and Use of Laboratory Animals. The National Cancer Institute Animal Care and Use Committee approved all experimental procedures.

2.2 Media and Solutions

1. Complete Media: DMEM with glutamine, PenStrep 1×, HiFBS 10%, EGF 10 ng/ml, insulin 4 µg/ml.

2. Carnoy's solution: 60% ethanol, 30% chloroform, 10% acetic acid (*see* **Note 1**).

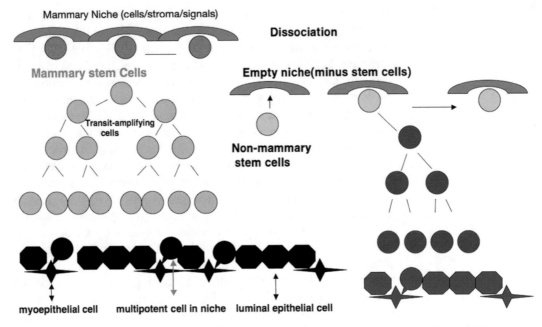

Fig. 1 Schematic presentation of niche and occupation by non-mammary stem cells. The *left side* shows a normal mammary microenvironment occupied by mammary stem cells. The *right side* shows the occupation of reforming mammary niches by non-mammary stem cell populations. We have found that PI-MECs cannot usually make cap cells

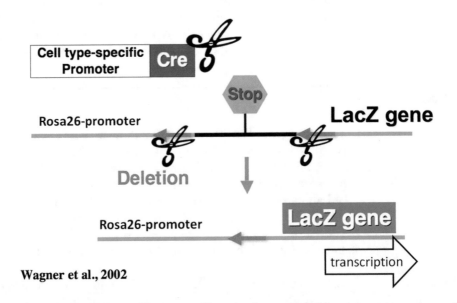

Fig. 2 Wap-Cre/Rosa 26stop model. The diagram illustrates the removal of the stop sequence upon activation of Cre driven by a Whey Acidic Protein promoter (WAP) in the mammary gland (Wagner et al. [12])

3. 4% paraformaldehyde: 40 g/l in 1× PBS. Start at 800 ml PBS. In laminar flow hood heat to 60 °C, adding 1 N NaOH until powder goes into solution. Cool and adjust pH to 7.4.

2.3 Tissue Culture Supplies

1. DMEM media for mixing with collagenase and pronase.
2. 250 ml disposable Erlenmeyer flasks for shaking cells in incubator.
3. 60 mm dishes for mincing cells.
4. Scalpels (any size blade).
5. Centrifuge.
6. 10 ml pipets.
7. 10 cc syringes.
8. 19 G needles.
9. T-75 flasks.
10. Trypsin.
11. DNase.
12. 40-μm filters.
13. Trypan blue stain.
14. Hemocytometer.

2.4 Surgical Instruments

1. Deweckers scissors.
2. Watchmakers forceps.
3. Bent angle forceps.
4. Micro dissection scissors.
5. Low temperature cautery.
6. Cork board.
7. Tie downs made from string to immobilize mouse during surgery (*see* **Note 2**).
8. Microscope for visualizing animal (or magnifying headset) (*see* **Note 3**).
9. Light source (*see* **Note 4**).
10. Wound clips (9 mm).
11. Wound clip applicator.

3 Methods

3.1 Single Cell Dissociation Procedure

1. Excise the two inguinal (#4) mammary glands from the donor mouse.
2. Mince the glands into 1–2 mm pieces in DMEM (20 ml/mg tissue), containing 0.1% collagenase and 0.1% hyaluronidase, using scalpels and sterile technique.

3. Shake the mixture at 100 rpm for approx. 90 min at 37 °C, until the pieces are dissipated.

4. Centrifuge the mixture at $100 \times g$ for 5 min at room temperature.

5. Save the fatty layer on top and the pellet.

6. Resuspend the pellet in 10 ml DMEM containing 1.25 % pronase, using 10 ml/g of original tissue.

7. Shake the cells at 100 rpm for 15 min at 37 °C.

8. Again, centrifuge the cells at $100 \times g$ for 5 min at room temperature.

9. Wash the cell pellet in 3× DMEM.

10. Resuspend the pellet in 10 ml DMEM.

11. Filter the cells through 40 µm cell strainer.

12. Use 5 µl for trypan blue viability stain.

13. Count the cells using a hemocytometer.

3.2 Clearing of Mammary Fat Pad

Mammary fat pad clearance and transplantation is performed on female mice between 3 and 4 weeks of age.

1. Anesthetize the mice with an intraperitoneal (IP) injection of ketamine–xylazine (120–130 µl at 10 mg/ml ketamine and 1 mg/ml xylazine).

2. Once the mouse is unconscious, secure it to a surgical table with string tie downs.

3. The ventral surface is further anesthetized with a topical solution such as Sensorcaine solution (*see* **Note 5**)

4. Expose the mammary fat pads by cutting and folding back the ventral skin. Use a cautery tool to block blood vessels in the inguinal (#4) fat pads, and also to disconnect these fat pads from nearby tissue.

5. "Clear" the fat pads off endogenous epithelium by surgically removing the proximal portion (from the nipple to the lymph nodes) (Fig. 3).

6. In order to transplant the tissue fragments (1–2 mm^2 of mammary tissue) create a small cavity in the fat pad with watchmaker forceps.

7. For cell innoculations, use a Hamilton syringe with a fine gauge (e.g., 30 G) needle to inject 10 µl of cell preparations into each cleared fat pad.

8. Once the transplant is in place, pull back the ventral skin over the exposed area and close the wound using metal clips.

9. Place the mouse on a heating pad in a "recovery cage" with nonstick bedding, and monitor them until they are fully recovered.

10. Remove the metal clips 7–10 days post-surgery.

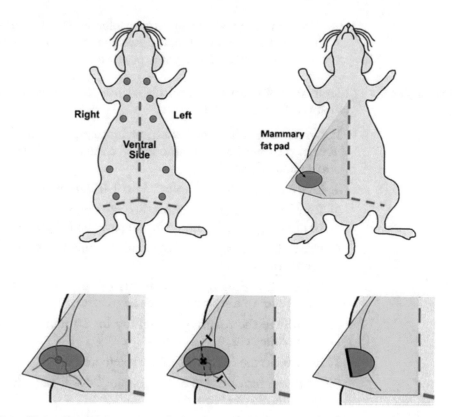

Fig. 3 Fat pad clearing. The mammary fat pad (*blue*) is exposed by cutting open the ventral skin. To clear endogenous epithelium, the lymph nodes and blood vessels (*red*) are severed with a cautery tool, and the proximal portion of the gland is cut away. Cells or tissue fragments can be transplanted into the remaining fat pad

3.3 Sperm/Germ Cell Dissociation Procedure

Excise the testes from WAP-Cre/Rosa26 R males following the protocol in Bellve et al. [11] with a few modifications:

1. De-capsulate the testes to remove the tunica albuginea.

2. Place them in DMEM containing 0.5 mg/ml collagenase and incubate at 33 °C in a shaking water bath for 15 min at 120 rpm.

3. Centrifuge the suspension at $120 \times g$ for 5 min to pellet the seminiferous chords.

4. Wash the top, fatty layer, and the pellet twice in DMEM to facilitate removal of the interstitial tissue.

5. Place the dispersed seminiferous chords in 20 ml DMEM containing 0.5 mg/ml trypsin and 1 g/ml DNase, and incubate as above.

6. Shear the remaining cell aggregates by pipetting 10–12 times.

7. Recover the cells by centrifugation and wash as above.

8. Resuspend the cells in 10 ml DMEM containing 0.5% BSA and filter them through a 40-μm filter to remove any remaining clumps.

9. Determine the viability by Trypan blue exclusion, and count the cells in a hemocytometer. Average yields of $15–20 \times 10^6$ cells are usually obtained per procedure.

3.4 Mammary Cell Isolation

Mammary epithelial cells are isolated using a standard protocol employed for primary cell culture:

1. On day one, place the excised mammary glands in a 60 mm petri dish with a small amount of collagenase (1 mg/ml in complete media).

2. Mince the glands with scalpels into 1–2 mm fragments.

3. Transfer the fragments into a 50 ml conical tube containing 10 ml collagenase (1 mg/ml in complete media) per two glands.

4. Place the tubes in a tissue culture incubator at 37 °C overnight.

5. On day 2, Triturate the fragments through a 10 ml pipette three times to dissociate the cell aggregates.

6. Centrifuge the resulting suspension for 10 min at $100 \times g$ to recover the cells.

7. Resuspend the cells in 10 ml complete media.

8. Shear the resuspended cell aggregate ONCE only through a 19-G needle (*see* **Note 6**).

9. Recover the cells by centrifugation for 10 min at $100 \times g$.

10. Resuspend the cells in 15 ml complete medium, transfer them into a T-75 flask and place it in a tissue culture incubator.

11. Carry out differential trypsinization to remove fibroblasts after 3–4 days (*see* **Note 7**).

12. Collect cells from primary mammary cultures after 4–7 days on plastic culture flasks.

3.5 Mixing Experiments

Setup the exogenous stem/progenitor cell populations from Wap-Cre/Rosa26 transgenic mice in the following manner (Fig. 4).

1. Mix 50,000 cells from normal FVB or Balb/C mice in a culture tube with 50,000 stem/progenitor cells from Wap-Cre/Rosa26 mice (*see* **Note 8**).

2. For ES cells use stem/progenitor cells at 1000 and 10,000 cells and mix with 50,000 mammary cells (*see* **Note 9**).

3.6 Cell and Tissue Transplantation

The surgical techniques used to clear the mammary epithelium from the no. 4 fat pad of 3-week old host mice, and the subsequent transplantation of tissue fragments or cell suspensions, have been described in detail [4, 6–8].

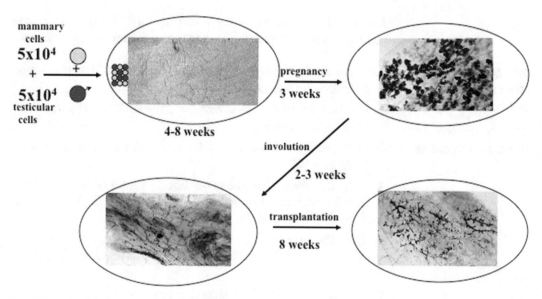

Fig. 4 Experimental design. A schematic presentation of the cell mixing experiment and subsequent second-generation transplantation. WAP-Cre/Rosa26-fl-stop-LacZ marked testicular cells from seminiferous tubules can combine with wild type mammary epithelial cells to regenerate a mammary gland

1. Anesthetize the mice and clear the gland as described above immediately before insertion of the transplanted fragment or cell suspension.

2. Inject the cell suspensions in 10 μl volumes with a Hamilton (Reno, NV) syringe equipped with a 30-G needle.

3. Using correct ratio (1:1 or less depending upon starting material) mix mammary and progenitor cell populations diluted to a final volume of 10 μl.

4. Mix cells well before inoculation into animal.

5. Draw cells into syringe.

6. While holding the cleared fat pad with forceps, gently insert needle at an angle and slowly inject 10 μl cell mixture.

7. Remove the needle and gently set the flap of skin over the gland.

8. Repeat this procedure for each gland.

9. Close the mouse with wound clips post inoculation.

10. Cut the fragments to be implanted into 1–2 mm pieces.

11. Use watchmaker forceps to create a small pocket in cleared gland (*see* **Note 10**).

12. Gently place the fragment into the hole.

13. Close the flap of skin over transplanted gland.

14. Staple the mouse closed upon completion of both inguinal glands.

15. Place the implanted females with males 6–8 weeks following implantation to initiate pregnancy and secretory development.

3.7 Mammary Gland Whole Mounts and X-Gal Staining

3.7.1 Paraformaldehyde Fixation and X-Gal Staining

1. Excise the cleared/implanted inguinal fat pad from the mouse and place on a glass slide.

2. Fix the gland with 4% paraformaldehyde (PFA) at room temperature for 1–1.5 h.

3. Permeabilize the gland in 0.01% Nonidet P-40 in PBS overnight at 4 °C.

4. Place the glands (on the glass slides) in 5 ml of X-gal mixture in a 50 ml conical tube, and add 125 μl (1× final concentration) of X-gal substrate.

5. Wrap the tubes in aluminum foil and incubate for 24–30 h at 37 °C in an incubator. After staining, wash the glands 2 × 30 min in 1× PBS.

6. Dehydrate the glands by stepwise treatment with ethanol, starting with 2×70% for 30 min, and then 2×100% for 60 min.

7. Once dehydrated, clear the glands in xylene for 2 h.

8. Whole mount the glands on slides using mounting medium and cover slips. They are now ready for image analysis (Fig. 5).

3.8 Preparation of Mammary Gland Whole Mounts for Immunostaining

1. Spread the whole inguinal gland on a glass slide and fix in 4% PFA overnight at 4 °C as above.

2. Process the gland and stain with X-Gal as above.

3. Rinse the stained glands repeatedly in PBS.

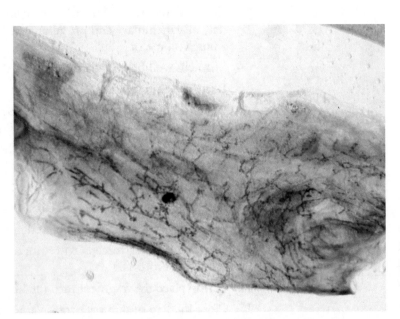

Fig. 5 Whole mount. Image of a cleared whole mount from a chimeric gland, illustrating how the testicular cells have been reprogrammed to a mammary fate

4. Fix the glands in Carnoy's fixative for 1–2 h at room temperature.

5. Clear the glands in 100% ethanol, 2×60 min.

6. Complete the clearing of the glands by placing them in xylene (before whole-mount analysis).

7. For histological examination, X-Gal-stained whole mounts are re-embedded in paraffin, sectioned at 6.0 μm, and counterstained with nuclear fast red.

8. Immunocytochemistry is carried out on deparaffinized sections.

3.9 DNA Extraction and PCR

PCR analysis is performed for identification of male DNA in chimeric reprogrammed populations.

1. Extract the genomic DNA from mammary tissue using an appropriate DNA isolation kit. Perform PCR analysis with WAP and Cre primers as described by Wagner et al. [10, 12], Y6 primers described by Peters et al. [13] and standard mouse GAPDH primers [14].

2. Reaction times are as follows:

3. 95 °C for 5 min.

4. 30 cycles:

   (a) 95 °C for 30 s.

   (b) 55 °C for 30 s.

   (c) 72 °C for 30 s.

   (d) 72 °C for 10 min.

   (e) 4 °C until removed.

5. Amplified DNA is loaded into 1.25% agarose gels containing ethidium bromide (0.5 μg/ml), electrophoresed at 60–100 V, and visualized under UV light (Fig. 6).

4 Notes

1. When making this solution always start with ethanol and add chloroform second, acetic acid is added last.

2. Tie downs can be made by looping string through a 200 μl pipet tip that has had the end cut off. The loop allows the hand or foot to be held and the remaining string out of the back of the pipet tip can be pinned down with a pushpin.

3. Using either a microscope of magnifying glasses allows a clearer surgical view and more precise cautery/incision/implantation. The use of these is strongly recommended.

4. An alternate light source is useful to make the surgical area bright enough to see without straining one's eyes. This is

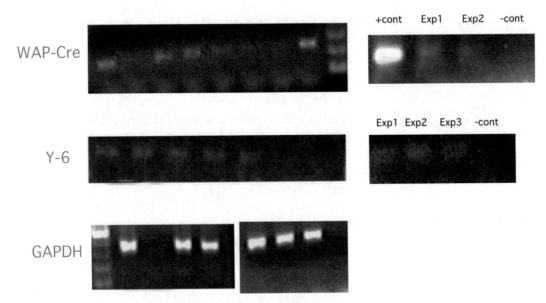

Fig. 6 Expression of male-specific Y6 transcripts in regenerated mammary gland post-transplantation of marked testicular cells. PCR detection of the chimeric outgrowth and control cell populations to identify the presence of male cell contribution in the mammary cell populations

    also strongly recommended, but care must be used as this can cause the mouse to dry out quickly if the light is too close to the animal.

5. These are available through a veterinary pharmacy.

6. This procedure is only performed ONCE.

7. Fibroblast levels should be reduced before collection of the epithelial cells by differential trypsinization.

8. This number represents a limiting dilution of mammary populations that allows for competition from alternative stem/progenitor cells in reforming mammary niches.

9. For ES cells, the number of stem/progenitor cells is greatly reduced due to the ES cells capacity for tumor formation when transplanted into the mammary fat pad.

10. Instruments must be kept sterile during the survival surgical procedure.

Acknowledgements

The work reported herein was supported in its entirety by the intermural research fund supplied by the Center for Cancer Research, NCI, Bethesda, MD 20892.

References

1. Daniel CW, DeOme KB, Young JT, Blair PB (1968) The in vivo life span of normal and pre-neoplastic mouse mammary glands: a serial transplantation study. Proc Natl Acad Sci U S A 61(1):53–60

2. Young LJ, Medina D, DeOme KB, Daniel CW (1971) The influence of host and tissue age on life span and growth rate of serially transplanted mouse mammary gland. Exp Gerontol 6:49–56

3. Daniel CW, Young LJ (1971) Influence of cell division on an aging process. Life span of mouse mammary epithelium during serial propagation in vivo. Exp Cell Res 65:27–32

4. DeOme KB, Faulkin LJ, Bern HA, Blair PB (1959) Development of mammary tumors from hyperplastic alveolar nodules transplanted into gland-free mammary fat pads of female C3H mice. Cancer Res 19:515–520

5. Li L, Xie T (2005) Stem cell niche: structure and function. Annu Rev Cell Dev Biol 21:605–631. doi:10.1146/annurev.cellbio.21.012704.131525

6. Daniel CW, Deome KB (1965) Growth of mouse mammary glands in vivo after monolayer culture. Science 149:634–636

7. Smith GH (1996) Experimental mammary epithelial morphogenesis in an in vivo model: evidence for distinct cellular progenitors of the ductal and lobular phenotype. Breast Cancer Res Treat 39:21–31

8. Smith GH, Gallahan D, Zwiebel JA et al (1991) Long-term in vivo expression of genes introduced by retrovirus-mediated transfer into mammary epithelial cells. J Virol 65:6365–6370

9. Kordon EC, Smith GH (1998) An entire functional mammary gland may comprise the progeny from a single cell. Development 125:1921–1930

10. Wagner K-U, Boulanger CA, Henry MD et al (2002) An adjunct mammary epithelial cell population in parous females: its role in functional adaptation and tissue renewal. Development 129:1377–1386

11. Bellve AR, Millette CF, Bhatnagar YM, O'Brien DA (1977) Dissociation of the mouse testis and characterization of isolated spermatogenic cells. J Histochem Cytochem 25:480–494

12. Wagner KU, Wall RJ, St-Onge L et al (1997) Cre-mediated gene deletion in the mammary gland. Nucleic Acids Res 25:4323–4330

13. Peters SO, Bauermeister K, Simon JP et al (2002) Quantitative polymerase chain reaction-based assay with fluorogenic Y-chromosome specific probes to measure bone marrow chimerism in mice. J Immunol Methods 260:109–116

14. Rasmussen SB, Kordon E, Callahan R, Smith GH (2001) Evidence for the transforming activity of a truncated Int6 gene, in vitro. Oncogene 20:5291–5301. doi:10.1038/sj.onc.1204624

Chapter 15

Lineage Tracing of Mammary Stem and Progenitor Cells

Anoeska A.A. van de Moosdijk, Nai Yang Fu, Anne C. Rios,
Jane E. Visvader, and Renée van Amerongen

Abstract

Lineage tracing analysis allows mammary epithelial cells to be tracked in their natural environment, thereby revealing cell fate and proliferation choices in the intact tissue. This technique is particularly informative for studying how stem cells build and maintain the mammary epithelium during development and pregnancy. Here we describe two experimental systems based on Cre/loxP technology (Cre^{ERT2}/loxP and rtTA/tetO-Cre/loxP), which allow the inducible, permanent labeling of mammary epithelial cells following the administration of either tamoxifen or doxycycline.

Key words Mouse, Mammary gland, Lineage tracing, Cre, Cre^{ERT2}, loxP, rtTA, tetO, Tamoxifen, Doxycycline

1 Introduction

The mammary epithelium is a remarkably dynamic tissue, which undergoes extensive changes during the lifespan of an organism. These include rapid and invasive branching morphogenesis during puberty, additional side branching during consecutive reproductive cycles, massive expansion and terminal differentiation during pregnancy and, finally, complete remodeling during involution. Furthermore, the epithelium harbors extraordinary regenerative potential: an entire branched and fully functional mammary epithelium can grow out following transplantation of a small piece of epithelial tissue or epithelial cell suspensions into the cleared fat pad [1, 2]. Even a single mammary stem cell was able to generate a fully functional mammary gland in this transplantation assay.

Over the past few decades it has become evident that in spite of its deceptively simple appearance, the bilayered mammary epithelium harbors multiple distinct cell populations in both the basal and the luminal layer. To fully understand how these different cell types are related, much effort has been dedicated towards unraveling the mammary gland stem and progenitor cell hierarchy [3].

Finian Martin et al. (eds.), *Mammary Gland Development: Methods and Protocols*, Methods in Molecular Biology, vol. 1501, DOI 10.1007/978-1-4939-6475-8_15, © Springer Science+Business Media New York 2017

A complete fate map of the mammary epithelium is not only of interest from a developmental perspective, but also for cancer research. Breast cancer is a heterogeneous disease comprising multiple distinct subtypes and this may at least in part be due to tumors arising from different cells of origin.

Much of our knowledge regarding the different cell populations in the mammary epithelium comes from cell sorting experiments using combinatorial cell surface markers, followed by either transplantation (to interrogate the regenerative potential of putative stem cell populations) or colony formation assays (to score the proliferative potential of progenitors) [2, 4–8]. Recently however, lineage tracing has emerged as a new gold standard for demonstrating stem cell activity within a tissue or organ and for tracking the fate of specific cells [9–15]. This technique relies on the lineage-specific expression of a DNA recombinase to activate expression of a reporter gene (Fig. 1). Indeed, inducible lineage tracing analyses are a more recent addition to the mammary gland biology toolbox. They provide both spatial and temporal control and offer the possibility of permanently marking a cell population of interest in the intact mammary gland in order to track its developmental fate in vivo. Importantly, this circumvents disrupting the tissue and taking cells out of their natural environment, which is unavoidable for most other experimental analyses.

This chapter focuses on the two leading methods for in vivo lineage tracing, namely Cre^{ERT2}/loxP and rtTA/tetO-Cre/loxP. Both of these methods use genetically engineered mice to label a cell population of choice and both are inducible, thereby providing the investigator with full experimental control over the time point at which the trace is initiated.

2 Materials

2.1 Mouse Strains

1. Mice carrying a Cre reporter allele (*see* **Note 1**), allowing the inducible expression of a marker gene (e.g., lacZ or a fluorescent protein), to be combined with either a Cre^{ERT2} driver or an rtTA/tetO-Cre system (**items 2** or **3** below, respectively, *see* **Note 2**).

2. Mice carrying a tamoxifen-inducible Cre^{ERT2} recombinase allele (*see* **Note 3**) under the control of a cell- or tissue-specific promoter (*see* **Note 4**).

3. Mice carrying both a reverse tetracycline-controlled transactivator allele (rtTA) under the control of a cell- or tissue-specific promoter (*see* **Note 4**) and a tetracycline-inducible Cre recombinase allele (tetO-Cre, *see* **Note 5**).

2.2 Administration of Tamoxifen to Activate the Reporter Gene

1. Gloves.

2. Microbalance.

3. Four 1.5 ml or 2 ml Eppendorf tubes.

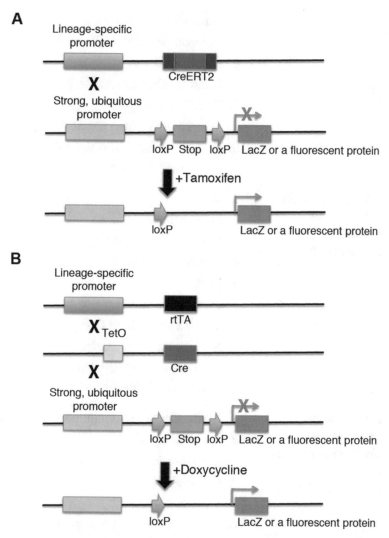

Fig. 1 Overview of the two-component Cre[ERT2]/loxP system and the three-component rtTA/tetO-Cre/LoxP system. (**a**) The Cre[ERT2]/LoxP system consists of two transgenic alleles: one has Cre[ERT2] expression under the control of a lineage-specific promoter and the other has a marker gene (fluorescent or LacZ) under the control of a strong, ubiquitous promoter. Prior to Cre-mediated recombination, the presence of a stop sequence prevents expression of the marker gene. When tamoxifen is administered, Cre[ERT2] becomes activated in cells where the lineage-specific promoter is active. It excises the stop cassette by recombining the flanking loxP sites, resulting in expression of the marker gene. (**b**) The rtTA/tetO-Cre/LoxP system consists of three alleles. Instead of the Cre-recombinase, the rtTA transactivator is expressed under the control of a lineage-specific promoter. When doxycycline is administered, rtTA can bind and activate the tetO promoter on the tetO-Cre allele. The resulting Cre expression again causes excision of the stop cassette on the reporter allele, causing expression of the marker gene. In both systems, all progeny of an activated cell will keep expressing the marker gene under the control of the ubiquitous promoter, even when the pulse of Cre recombinase activity itself is transient

4. Tamoxifen (*see* **Note 6**).

5. Corn oil (*see* **Note 7**).

6. Absolute ethanol.

7. Nutator mixer or rotator.

8. Two 3 ml syringes.

9. Two 22 μm syringe filters.

10. Two 1 ml BD™ slip-tip syringes with 26 G×5/8 in. subQ needles (*see* **Note 8**).

2.3   Administration of Doxycycline to Activate the Reporter Gene

1. Gloves.

2. Microbalance.

3. Four 1.5 ml or 2 ml Eppendorf tubes.

4. Doxycycline (*see* **Note 9**).

5. PBS.

6. Two 3 ml syringes.

7. Two 22 μm syringe filters.

8. Nutator mixer or rotator.

9. Two 1 ml BD™ slip-tip syringes with 26 G×5/8 in. subQ needles (*see* **Note 8**).

2.4   Components for Harvesting the Mammary Glands for Further Downstream Analyses

1. Euthanasia setup or equipment (*see* **Note 10**).

2. Dissection pad.

3. Spray bottle with 70% ethanol.

4. 6–8 Pushpins.

5. Tissues.

6. Two pairs of surgical scissors.

7. Two pairs of fine (Iris or Graefe) forceps.

8. One pair of Dumont No. 5 forceps.

9. Razor blade (optional).

2.5   Whole-Mount Confocal Analysis

1. PBS.

2. 4% Paraformaldehyde in PBS.

3. PBT (PBS + 0.1% Tween-20).

4. Alexa Fluor 647 Phalloidin.

5. Glycerol.

6. Cover slips No 1.5.

7. Microscopy slides.

8. Dissecting fluorescence microscope.

9. Micro-dissecting scissors and forceps.

10. Tape.

3   Methods

**3.1   Breeding
of the Mice**

1. Design the lineage-tracing experiment and determine the required time points for initiating (t_0) and analyzing (t_x) the trace (Fig. 2) (*see* **Note 11**).

2. For the Cre^{ERT2} system, cross heterozygous Cre^{ERT2} mice with homozygous reporter mice to generate double heterozygous transgenic mice. For the generation of rtTA/tetO-Cre/loxP mice, cross the double-heterozygous rtTA/tetO-Cre mice with homozygous reporter mice to generate triple heterozygous transgenic mice.

3. Genotype the double (for the Cre^{ERT2} system) or triple (for the rtTA/tetO-Cre/loxP system) transgenic mice by PCR (*see* **Note 12**).

4. Determine your experimental cohorts (*see* **Notes 13–15**).

**3.2   Induction
of the Reporter Allele
with Tamoxifen or
Doxycycline**

1. When using the Cre^{ERT2}/loxP system, proceed with **steps 2–4**. When using the rtTA/tetO-Cre/loxP system, proceed with **steps 5–7**.

2. Prepare the tamoxifen solution using the materials listed in Subheading 2.2. Remember to wear gloves. Weigh the required amount of tamoxifen and dissolve it at 5–20 mg/ml in 90 % corn oil and 10 % ethanol (for example, for 1 ml of 10 mg/ml tamoxifen solution, dissolve 10 mg of tamoxifen in 900 μl oil and 100 μl ethanol) (*see* **Note 16**).

3. Prepare the control oil solution. Mix oil and ethanol in a 90:10 ratio (for example, for 1 ml of control solution, mix 900 μl oil and 100 μl ethanol).

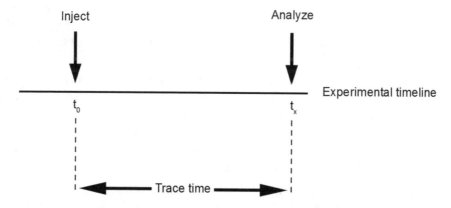

Fig. 2 Experimental setup of a lineage tracing experiment. Overview of the experimental setup for a lineage tracing experiment. The trace can be started at any desired time point, even before birth (tamoxifen or doxycycline is administrated to the pregnant mother in that case). t_0 depicts the start of the actual tracing experiment, when the cells are labeled through administration of tamoxifen (for the Cre^{ERT2}/loxP system) or doxycycline (for the rtTA/tetO-Cre/loxP system). Analysis follows at t_x, which is any time point 24 h or later after cell labeling

4. Incubate the solutions on a nutator or rotator to let the tamoxifen dissolve (*see* **Notes 17** and **18**). Continue with **step 8**.

5. Prepare the doxycycline solution using the materials listed in Subheading 2.3. Remember to wear gloves. Weigh the required amount of doxycycline and dissolve it at 20 mg/ml in PBS (for example, for 1 ml of 20 mg/ml doxycycline solution, dissolve 20 mg of doxycycline in 1 ml PBS).

6. Use PBS as control solution.

7. Incubate the solutions on a nutator or rotator to allow the doxycycline to dissolve. Continue with **step 8**.

8. Filter the solutions prepared in **steps 2–4** or **5–7** through a 22 μm syringe filter (*see* **Note 19**).

9. Aliquot the stock solution into 500 μl/vial and store at –20 °C.

10. Weigh the mice to determine how much of the stock solution should be injected (*see* **Notes 20** and **22**).

11. Fill a 1 ml syringe attached to a subQ needle with the tamoxifen, doxycycline, or control solution.

12. Hold the syringe vertically, with the needle pointing upwards, and remove air bubbles by flicking the syringe with your fingers, forcing the air to the top. Insert the plunger a little to let all air escape. Be careful not to spill any of the solution.

13. Put down the syringe and open the cage.

14. Lift the mouse that is to be injected out of the cage, holding it by its tail.

15. Check the identity of the mouse to be sure that you inject the correct one.

16. Place the mouse on the wire lid of the cage and let it grab the bars.

17. Fix the animal; make sure its head and tail are properly secured. Turn the mouse over and slightly tilt your hand, so that the head of the mouse is slightly lower than its abdomen. You now have one hand free to inject (*see* **Note 23**).

18. Pick up the syringe and gently insert the needle in the lower right or left quadrant at a low angle to prevent penetration of any organs (*see* **Notes 24** and **25**).

19. Inject the required amount of tamoxifen (for the Cre^{ERT2}/loxP system), doxycycline (for the rtTA/tetO-Cre/loxP) or appropriate control solution by pushing the plunger.

20. Carefully pull back the needle and put it down (*see* **Note 26**).

21. Transfer the mouse to a clean cage.

22. Repeat **steps 13–20** with the remainder of the mice that need to be injected (*see* **Note 27**).

23. Clean the workspace to remove any traces of tamoxifen or doxycycline (*see* **Note 28**).

24. Check the welfare of your animals daily in the first week following injection.

3.3 Dissection of Mammary Glands

1. At the required analytical time-point (t_x, Fig. 2), euthanize the animal according to your institutional or national guidelines.

2. Secure the animal on a dissection pad with the belly facing up.

3. Spray the animal with 70% ethanol (*see* **Note 29**).

4. Using a pair of scissors, make an incision in the skin along the vertical midline, moving from the groin up to the top of the sternum. Cut superficially, leaving the peritoneum intact and make sure not to hit any blood vessels.

5. Make a second incision towards the right knee of the animal. Start your incision at the point you began the incision from **step 4** and try to prevent hitting any blood vessels in the lower limb.

6. Make a third incision towards the left knee of the animal. Again, start your incision at the point you began the incision from **step 4** and try to prevent hitting any blood vessels in the lower limb (*see* **Note 30**).

7. Make a fourth incision towards the right shoulder of the animal. Start your incision at the point you ended the incision from **step 4** and try to prevent hitting any blood vessels in the upper neck and limbs.

8. Make a fifth incision towards the left shoulder of the animal. Start your incision at the point you ended the incision from **step 4** and try to prevent hitting any blood vessels in the upper neck and limbs.

9. Using two pairs of tweezers, gently peel the skin sideways on either side of the animal, separating it from the peritoneum (which should still be intact).

10. Using pushpins, secure the skin flaps onto the dissecting pad (Fig. 3).

11. At this point, you should be able to dissect the third and fourth mammary gland (*see* **Note 31**).

12. To remove the fourth mammary gland, gently grab the distal tip of the fat pad using a pair of forceps (*see* **Note 32**).

13. Hold onto the distal tip, gently pulling it up. Using a pair of scissors or a razor blade, gently remove the ligaments that attach the fat pad to the body wall, working your way towards the proximal end, where the nipple is located.

14. Process the mammary gland for further downstream analysis as required (*see* **Notes 33** and **34**).

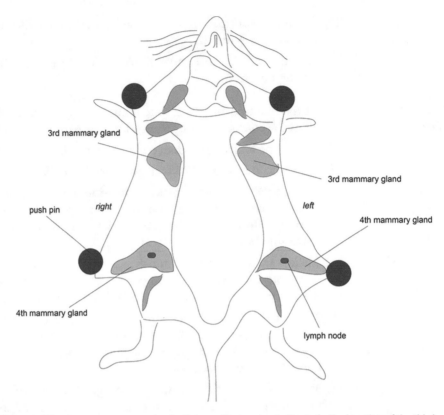

Fig. 3 Anatomy of the mouse mammary glands. Schematic drawing to illustrate the location of the third and fourth mammary glands at the time of dissection. At this stage, the skin has been peeled away and has been pinned down onto the dissection pad. The peritoneum is still intact, hiding the abdominal and thoracic organs from view. The fourth mammary gland is clearly visible, as is the third mammary gland and both can be readily dissected

3.4 Whole-Mount Confocal Analysis

1. Prepare cold fixation medium: heat 100 ml PBS to approximately 60 °C. Add 4 g paraformaldehyde powder to the solution and let dissolve on a heat stirring plate. After the solution becomes clear, put the solution on ice or store at –20 °C (*see* **Note 35**).

2. Rapidly dissect the third and fourth mammary glands as described in Subheading 3.3 and transfer directly into cold fixation medium (4% PFA in PBS).

3. Incubate the tissues on ice for 30 min (*see* **Note 36**).

4. Prepare PBT: PBS + 0.1% Tween-20. (v/v; for example, dissolve 1 ml Tween-20 in 1000 ml PBS). Put this solution on ice.

5. Transfer the tissues to a 15 ml tube containing 10 ml of cold PBT.

6. Wash the tissues for 30 min on a rotator to remove residual PFA and allow gentle permeabilization.

7. Incubate the tissues in 500 μl PBT containing phalloidin (1:50; 500 μl PBT + 10 μl phalloidin). (*see* **Note 37**). Store them at 4 °C overnight.

8. On the next day, wash the tissues three times with ice cold PBS (20 min each time).

9. Prepare a clearing solution of 80% glycerol in PBT (for example 8 ml glycerol plus 2 ml PBT).

10. Transfer the tissues to a 15 ml tube containing 10 ml clearing solution and incubate at room temperature for 3 h (*see* **Note 38**).

11. Dissect the fluorescent ducts, using a dissecting microscope with a fluorescent light source. To obtain a good sample for confocal imaging, you need to trim the fat on the surface to reveal the ductal tree.

12. Mount the dissected tissues on slides and cover them with a cover slip. Tightly press down on the cover slip to flatten the tissue and let all air escape. Remove excess liquid (*see* **Note 39**).

13. Tape the cover slip to the slide on both sides of the sample.

14. Mount the slide on the stage of a confocal microscope and record Z-stacks (*see* **Note 40**) (Fig. 4).

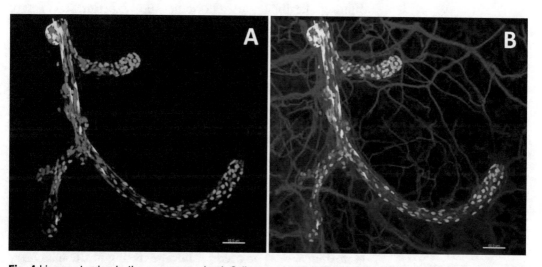

Fig. 4 Lineage tracing in the mammary gland. Cells were labeled ("pulse") in puberty and traced for 8 weeks ("chase"). (**a**) Composite image showing a branched epithelium containing GFP+, RFP+ and YFP+ cells. (**b**) The same picture as in (**a**), now including an F-actin staining in blue to help visualise the overall structure of the tissue. Image by Anne Rios

4   Notes

1. Although alternatives exist (e.g., FLP/FRT technology), Cre/loxP technology remains the most commonly used tool for in vivo lineage tracing experiments. As a result, multiple Cre reporter strains are available to the scientific community (see http://jaxmice.jax.org/list/xprs_creRT1805.html).   The reporter lines for lineage tracing in the mammary gland generally express a fluorescent protein gene under control of a strong, ubiquitous promoter once a so-called stop sequence is removed by Cre-mediated recombination (Fig. 1). The choice of reporter strain is up to the investigator and depends on the experimental question and/or preferred mode of analysis. For instance, expression of a fluorescent reporter can be visualized directly, but this endogenous signal decreases over time in both whole-mount preparations (as described in this chapter) and tissue sections. In contrast, enzymatic detection of lacZ activity provides a permanent record, but lacZ detection is not easily compatible with fluorescence-activated cell sorting (FACS) or confocal microscopy.

2. Both the Cre^{ERT2}/loxP and the rtTA/tetO-Cre/loxP systems can be used for inducible lineage tracing analyses. The Cre^{ERT2} system involves less breeding as it relies on double-transgenic mice, and a wide variety of Cre^{ERT2} driver strains are available, offering flexibility in choosing the experimental setup. However, one needs to be very cautious about the toxicity of tamoxifen, which is required to induce Cre activity in this system. For example, Bmi1-expressing stem cells in the intestinal crypt have been shown to undergo apoptosis in response to tamoxifen, resulting in unintentional bias during lineage tracing [16]. It remains unclear whether mammary stem cells are also sensitive to tamoxifen, but the hormone sensitivity of this population and their susceptibility to the inhibitor of estrogen biosynthesis letrozole [17] suggest that they will be. Furthermore, haematopoietic stem cells are highly sensitive to estrogen and hence may be impacted by tamoxifen [18]. In the mammary gland, where estrogen signaling plays a critical role, tamoxifen can have adverse effects on development, especially when higher doses are administered [9, 19]. A low dose of tamoxifen should be used for lineage tracing in the mammary gland. In this respect, the doxycycline-inducible rtTA driver is more favorable. This three-component system necessarily incorporates a tetO-Cre allele and thus requires more complex breeding than the two-component Cre^{ERT2}/loxP system.

3. In theory, one could opt to perform lineage-tracing experiments with a non-inducible Cre driver (e.g., Wap-Cre [20, 21]). While offering tissue specificity, this eliminates all experi-

mental control over the time point of Cre-mediated recombination and thus complicates the interpretation of the experimental data.

4. Both the Cre^{ERT2}/loxP and rtTA/tetO-Cre/loxP systems rely on either a knock-in or transgenic mouse strain expressing Cre^{ERT2} or rtTA under a lineage-specific promoter. Regarding tissue specificity, any cell- or tissue-specific promoter is only as good as tested. It might not faithfully recapitulate endogenous gene expression, even in the context of a knock-in construct. Therefore, use a promoter that is well characterized or, better yet, carefully characterize the Cre- or rtTA-driver yourself.

5. The tetO-Cre line obtained from Jax can be leaky in the germline, and thus mice carrying both the tetO-Cre and reporter alleles should be avoided as the breeder.

6. Tamoxifen can function as an estrogen agonist or antagonist, depending on the target tissue. It is classified as carcinogenic and toxic. Always wear gloves when handling this product. Tamoxifen is light sensitive.

7. Alternatively, you can use regular sunflower oil fit for human consumption.

8. Tamoxifen and doxycycline are both administered by intraperitoneal injection in this protocol. Another option is oral gavage, or, in the case of doxycycline, administration via the drinking water and the feed. Total uptake might differ for these options [22, 23], so choose one for all traces within one study.

9. Doxycycline is a tetracycline antibiotic. It is classified as harmful and is light sensitive.

10. Always follow your institutional or national guidelines for euthanizing animals.

11. You will need to analyze several time points. The first one is 24–48 h after labeling, depending on how long it takes for the marker gene to be expressed. For example, LacZ can easily be detected at 24 h after injection of tamoxifen. GFP is also detectable from 24 h onwards, but robust levels can often only be detected after 48 h. Checking these early time points will show you what population(s) of cells were initially labeled. Any time point(s) after that will likely show the offspring of these cells as well. To follow the trace really well, consider including some intervening time points in addition to the early (24–28 h) and final analytical time point. This is particularly informative if the gland will dramatically change in appearance (for instance, when the trace is initiated during puberty and the glands are analyzed during pregnancy, it could be informative to include experimental animals in which the trace is analyzed in adult virgins). Importantly, if you want to provide evidence

that you are tracing a stem cell population, the mice will have to complete at least one round (and preferably multiple rounds) of pregnancy, lactation, and involution. This ensures complete turnover of the mammary epithelium, since massive turnover of cells takes place during these processes. Stem cells can self-renew and therefore should survive this turnover. Next to that, they should give rise to differentiated offspring.

12. Depending on the parental genotype (homozygous or heterozygous for each of the alleles), only part of the offspring may carry both alleles. Always check all of your animals by genotyping, even if you are sure what the genotype should be! There is always a chance that animals (or their labels) get mixed up, especially in big animal facilities with large numbers of animals and multiple strains. Make sure that your animals have a unique identifier (for instance an ear clip) when you genotype them. Collect a piece of tissue (ear, toe, or tail can all be used) according to your institutional or national guidelines. It is easiest to do this when the animals are weaned (around postnatal day 21, P21). Lyse the tissue, for example in Viagen tail lysis buffer (approx. 100 µl per sample) supplemented with proteinase K (100 µg/ml) and incubate overnight at 55 °C. Next day, inactivate the proteinase K at 85 °C for 15–45 min. This lysate can now directly be used in a genotyping PCR (use 1–10 µl). Another option is to use a homemade lysis buffer [24]. This will take more time, but using homemade buffer is cheaper and will lead to a similar quality of DNA.

13. It is important to think well about your controls. It will usually be difficult to include a proper positive technical control, because this requires an independent Cre driver that is known to work in the mammary gland and therefore, potentially, importing an additional transgenic mouse line. If you do not have access to such a line, you can opt to use a different tissue from the double-transgenic mice in which you initiate the trace as a positive control (provided that there is another tissue in which Cre-mediated recombination should take place very effectively using this driver). For instance, when using the Wnt//β-catenin responsive $Axin2^{CreERT2}$ allele in combination with the $Rosa^{26mTmG}$ reporter [10], we often quickly inspect the intestine (which also contains Wnt/β-catenin responsive stem cells) for successful recombination (i.e., the presence of a GFP signal) using an inverted fluorescence microscope.

14. As a negative control, inject double-transgenic mice with the control solution (i.e., the solution without tamoxifen or doxycycline). Should your Cre driver be "leaky" (i.e., activated by endogenous estrogen or not properly shielded by the ER-moiety), the negative control will reveal this. Of note, leakiness will be far less likely with Cre^{ERT2}, since it contains three point

mutations in the ER moiety (compared to Cre^{ER}; another version, Cre^{ERT}, contains one point mutation compared to Cre^{ER}), which make it less sensitive to estrogen and more sensitive to tamoxifen. Because animal numbers are often limiting, it may be difficult to take along a negative control for every single experiment. As an alternative, analyze a negative control at least once per Cre strain and protocol (i.e., time point of tamoxifen administration). Another negative control is to inject a mouse that carries the reporter allele, but not the Cre^{ERT2} driver, with tamoxifen and analyze the mammary glands from this animal alongside the glands from double-transgenic mice to determine the background levels (either following enzymatic lacZ detection or "straight up" for fluorescent reporter alleles). In practice however, most of the mammary gland tissue will serve as a negative control in itself, because you should aim to label only a subset of the cells in order to be able to perform clonal analyses (*see* **Note 20**).

15. Biological controls are important as well for the correct interpretation of your experiment. Try to use littermates whenever possible and house animals in the same cage.

16. When designing your experiments, be aware that high doses of tamoxifen can have adverse effects on mammary gland development, as mentioned in **Note 2** [9, 19]. Set up pilot experiments to carefully determine the minimum dose you can use with your particular Cre^{ERT2} driver and labeling time point. We have noticed that the efficiency of labeling cells with tamoxifen in the mammary gland can be much lower than the efficiency in other tissues, such as intestine, using the same Cre^{ERT2} driver.

17. It is advisable to always prepare fresh solutions for each experiment, although it is possible to make the solution the day before and store it at 4 °C overnight for use the next morning.

18. Tamoxifen is notoriously difficult to dissolve. Adding ethanol to the oil solution helps (tamoxifen is soluble in pure ethanol), as will heating to 37 °C, but it will still take some time. Tamoxifen is light sensitive, so wrap the tube in aluminum foil.

19. The solution can be very viscous, so be careful when applying pressure to the syringe.

20. When determining the right concentration of tamoxifen or doxycycline to use, keep in mind that, ideally, lineage tracing analyses are performed at clonal density (*see* **Note 14**). This means that only a small part of the population you are interested in is labeled at the time of tamoxifen administration, so as to be able to robustly detect the clonal offspring of the initially labeled cells. To be absolutely sure that a labeled patch of cells is indeed a clone (i.e., comes from one cell), one would

want to label only one cell per mammary gland. However, this is practically unfeasible, as it would require large numbers of animals to reach sufficient experimental coverage. Therefore it is advisable to optimize the concentration of tamoxifen such that each labeled cell (clone) is separated by large pieces of non-labeled tissue.

21. The absolute amount of tamoxifen (and thus the injection volume) is dependent on the age and weight of the mice, as well as on the aim of your experiment. When injecting a low dose of tamoxifen (i.e., no more than 0.5 mg), it is helpful to prepare a solution that has a low concentration (2–5 mg/ml) so you have good control about the volume you are injecting.

22. Not much is known about the half-life of tamoxifen in the mammary gland of mice and, related to this, about the duration of Cre activity following a single pulse of tamoxifen. It may take several hours for tamoxifen to be converted into its active metabolite (4-hydroxytamoxifen). Generally speaking, the half-life of tamoxifen will differ depending on the tissue, the age of the mice, the administered dose, and the route of administration [25–27].

23. Find out on beforehand which hand to use for injecting and which one for fixing the mouse. This is a personal preference.

24. IP injections can either be administered in the lower right or the lower left quadrant of the mouse abdomen, depending on which hand you are using to hold the syringe. If you hold the syringe in your right hand, it is easiest to inject the left quadrant of the mouse and vice versa. However, injecting on the right side will ensure not puncturing the caecum and is the safest option.

25. Insert the needle at an angle of approximately 30° to penetrate the skin, while preventing damage to any internal organs such as liver, bladder, small intestine or caecum. Insertion of about 0.5 cm of the needle should be enough. To check if your needle penetrated any organs, you can aspirate a small volume before injecting. The aspirate should appear clear and not colored red (blood), green/brown (intestine), or yellow (bladder).

26. When injecting multiple animals housed in the same cage, it is possible to re-use the same needle, provided that you check the sharpness of the needle between injections. A blunt needle causes discomfort for the animal. If you want to be absolutely safe and minimize the chance of infections, use a clean needle (i.e., fill a new syringe) for each animal.

27. Never pool injected and non-injected mice. Similarly, do not put mice injected with different drugs in the same cage.

28. Under normal circumstances, no tamoxifen should be spilt when following this procedure.

29. This will prevent hair from sticking to your instruments as well as the mammary tissue.

30. At this point, the incisions should look like an inverted Y.

31. The mammary glands should now be facing up. Usually, the fourth mammary gland is isolated for further analysis. The fifth mammary gland may still be attached to the fourth mammary gland, but the two are readily distinguishable: the fourth mammary gland is the largest gland and has a distinct boomerang shape with a clearly visible lymph node. The third mammary gland is also relatively easy to excise and can be taken along for analysis as well. However, you should take care to prevent isolating muscle in the process (the muscle is a bit more brownish, whereas the mammary gland is more pinkish in color).

32. The distal tip is the part closest to the peritoneum and may even extend below the peritoneum towards the back of the animal. Push back the peritoneum to find the end of the fat pad.

33. There are multiple ways to analyze a tracing experiment. First, fluorescence-activated cell sorting (FACS) using combinatorial cell surface markers allows the overall contribution of the labeled cell lineage to the basal and luminal cell populations to be assessed [9, 10]. These analyses are suitable for fluorescent Cre reporter alleles, but less so for lacZ reporter alleles. In addition, FACS offers a population based analysis and individual cell clones cannot be analyzed. Second, individual cell clones can be analyzed by whole mount confocal microscopy [9, 10], provided that the labeling was performed at clonal density such that individual cell clones can be readily distinguished. Finally, mammary glands can be processed for paraffin embedding and analyzed by immunohistochemical or immunofluorescence staining. While this allows co-staining of labeled cells with structural markers, it again complicates the analysis of entire cell clones, although this can be performed using serial sections.

34. In Subheading 3.4 we present one of the strategies used to further investigate the kinetics of stem and progenitor cells in the mammary gland by performing a clonal analysis using the multi-color Cre-reporter Confetti. Two challenges need to be tackled to perform a non-biased analysis using this Cre-reporter: strong detection of all four fluorescent proteins, and covering a large area of tissue in order to view many clones. The R26R-confetti allele consists of the strong CAGGS promoter, a LoxP-flanked NeoR cassette serving as a transcriptional roadblock, and the original Brainbow-2.1 cassette [28, 29]. After Cre-mediated recombination, the roadblock is removed and one of the four fluorescent marker genes (membrane-bound CFP, nuclear GFP, YFP, or RFP) is stochastically placed

under control of the CAGGS promoter, allowing one to trace the contribution of individual cells in the same population in a given tissue. Detection of signals from the native fluorescent proteins requires fresh tissue as there are no antibodies available to distinguish CFP, GFP, and YFP.

An additional obstacle to performing a non-biased clonal analysis is the mode of imaging used. Sectioning (a few micrometers of thickness) has been used for many decades but has some limitations as it does not precisely account for cell morphology and their positioning within the ductal mammary epithelium. Indeed, myoepithelial cells are highly elongated (average of 100 μm) while the cuboidal luminal cells are organized in an inner layer that lies perpendicular to the myoepithelial cell layer. This organization and the differing morphologies of the two cell types imposes challenges for scoring cell clones in 2D sections. Different strategies for whole mount fluorescence analysis have been described [10, 30]. The Visvader lab recently developed 3D confocal imaging to view entire regions of the mammary ductal tree down to the single cell level [9]. Here, we provide a brief protocol that allows strong detection of the native fluorescent proteins in regions up to 1 cm of tissue at cellular resolution, thus allowing analysis of both clonal localization and composition.

35. Paraformaldehyde is classified as toxic and carcinogenic. Work in the fume hood while heating and dissolving it.

36. Placing the tissues on ice is necessary to preserve fluorescence.

37. Phalloidin is used to label and visualize F-actin containing structures. It is the easiest way to visualize multiple different cell types in a complex tissue. The staining reveals structural information and provides a global overview of the tissue architecture during image acquisition and analysis.

38. A clearing step is required for ex vivo optical imaging to reduce background of surrounding tissue.

39. If the tissue is too large (for example from a pregnant female) you might need to chop it in several pieces before it fits on a slide.

40. Confocal imaging (Fig. 4) can be performed using a Leica SP5 confocal microscope equipped with a 40× oil lens NA 1.2. Tile scans with Z-stacks must be acquired to cover a large ductal tree area, 1024×1024 pixels and 12 bits. Use sequential scanning for XFP excitations, to prevent spectral bleed-through. Nuclear GFP can be excited using an argon laser 488 nm line; for EYFP 514 nm line; for RFP a red diode laser emitting at 561 nm, and blue membrane-bound CFP can be excited using a laser line at 458 nm. In general, GFP fluorescence can be collected between

~498 and 510 nm; YFP fluorescence collected between ~521 and 560 nm; RFP fluorescence collected between ~590 and 620 nm; CFP fluorescence collected between ~466 and 495 nm. Phalloidin-647 is excited with the laser 633 and the fluorescence collected after 650 nm.

References

1. Deome KB, Faulkin LJ Jr, Bern HA, Blair PB (1959) Development of mammary tumors from hyperplastic alveolar nodules transplanted into gland-free mammary fat pads of female C3H mice. Cancer Res 19(5):515–520

2. Shackleton M, Vaillant F, Simpson KJ, Stingl J, Smyth GK, Asselin-Labat ML, Wu L, Lindeman GJ, Visvader JE (2006) Generation of a functional mammary gland from a single stem cell. Nature 439(7072):84–88. doi:10.1038/nature04372

3. Visvader JE, Stingl J (2014) Mammary stem cells and the differentiation hierarchy: current status and perspectives. Genes Dev 28(11):1143–1158. doi:10.1101/gad.242511.114

4. Stingl J, Eirew P, Ricketson I, Shackleton M, Vaillant F, Choi D, Li HI, Eaves CJ (2006) Purification and unique properties of mammary epithelial stem cells. Nature 439(7079):993–997. doi:10.1038/nature04496

5. Smalley MJ, Kendrick H, Sheridan JM, Regan JL, Prater MD, Lindeman GJ, Watson CJ, Visvader JE, Stingl J (2012) Isolation of mouse mammary epithelial subpopulations: a comparison of leading methods. J Mammary Gland Biol Neoplasia 17(2):91–97. doi:10.1007/s10911-012-9257-1

6. Smalley MJ (2010) Isolation, culture and analysis of mouse mammary epithelial cells. Methods Mol Biol 633:139–170. doi:10.1007/978-1-59745-019-5_11

7. Sleeman KE, Kendrick H, Ashworth A, Isacke CM, Smalley MJ (2006) CD24 staining of mouse mammary gland cells defines luminal epithelial, myoepithelial/basal and non-epithelial cells. Breast Cancer Res 8(1):R7. doi:10.1186/bcr1371

8. Prater M, Shehata M, Watson CJ, Stingl J (2013) Enzymatic dissociation, flow cytometric analysis, and culture of normal mouse mammary tissue. Methods Mol Biol 946:395–409. doi:10.1007/978-1-62703-128-8_25

9. Rios AC, Fu NY, Lindeman GJ, Visvader JE (2014) In situ identification of bipotent stem cells in the mammary gland. Nature 506(7488):322–327. doi:10.1038/nature12948

10. van Amerongen R, Bowman AN, Nusse R (2012) Developmental stage and time dictate the fate of Wnt/beta-catenin-responsive stem cells in the mammary gland. Cell Stem Cell 11(3):387–400. doi:10.1016/j.stem.2012.05.023

11. Wang D, Cai C, Dong X, Yu QC, Zhang XO, Yang L, Zeng YA (2014) Identification of multipotent mammary stem cells by protein C receptor expression. Nature. doi:10.1038/nature13851

12. Prater MD, Petit V, Alasdair Russell I, Giraddi RR, Shehata M, Menon S, Schulte R, Kalajzic I, Rath N, Olson MF, Metzger D, Faraldo MM, Deugnier MA, Glukhova MA, Stingl J (2014) Mammary stem cells have myoepithelial cell properties. Nat Cell Biol 16(10):942–950. doi:10.1038/ncb3025, 941-947

13. Sale S, Lafkas D, Artavanis-Tsakonas S (2013) Notch2 genetic fate mapping reveals two previously unrecognized mammary epithelial lineages. Nat Cell Biol 15(5):451–460. doi:10.1038/ncb2725

14. de Visser KE, Ciampricotti M, Michalak EM, Tan DW, Speksnijder EN, Hau CS, Clevers H, Barker N, Jonkers J (2012) Developmental stage-specific contribution of LGR5(+) cells to basal and luminal epithelial lineages in the postnatal mammary gland. J Pathol 228(3):300–309. doi:10.1002/path.4096

15. Van Keymeulen A, Rocha AS, Ousset M, Beck B, Bouvencourt G, Rock J, Sharma N, Dekoninck S, Blanpain C (2011) Distinct stem cells contribute to mammary gland development and maintenance. Nature 479(7372):189–193. doi:10.1038/nature10573

16. Zhu Y, Huang YF, Kek C, Bulavin DV (2013) Apoptosis differently affects lineage tracing of Lgr5 and Bmi1 intestinal stem cell populations. Cell Stem Cell 12(3):298–303. doi:10.1016/j.stem.2013.01.003

17. Asselin-Labat ML, Vaillant F, Sheridan JM, Pal B, Wu D, Simpson ER, Yasuda H, Smyth GK, Martin TJ, Lindeman GJ, Visvader JE (2010) Control of mammary stem cell function by steroid hormone signalling. Nature 465(7299):798–802. doi:10.1038/nature09027

18. Nakada D, Oguro H, Levi BP, Ryan N, Kitano A, Saitoh Y, Takeichi M, Wendt GR, Morrison SJ (2014) Oestrogen increases haematopoietic

stem-cell self-renewal in females and during pregnancy. Nature 505(7484):555–558. doi:10.1038/nature12932

19. Shehata M, van Amerongen R, Zeeman AL, Giraddi RR, Stingl J (2014) The influence of tamoxifen on normal mouse mammary gland homeostasis. Breast Cancer Res 16(4):411. doi:10.1186/s13058-014-0411-0

20. Chang TH, Kunasegaran K, Tarulli GA, De Silva D, Voorhoeve PM, Pietersen AM (2014) New insights into lineage restriction of mammary gland epithelium using parity-identified mammary epithelial cells. Breast Cancer Res 16(1):R1. doi:10.1186/bcr3593

21. Wagner KU, Wall RJ, St-Onge L, Gruss P, Wynshaw-Boris A, Garrett L, Li M, Furth PA, Hennighausen L (1997) Cre-mediated gene deletion in the mammary gland. Nucleic Acids Res 25(21):4323–4330

22. Hayashi M, Sutou S, Shimada H, Sato S, Sasaki YF, Wakata A (1989) Difference between intraperitoneal and oral gavage application in the micronucleus test. The 3rd collaborative study by CSGMT/JEMS.MMS. Collaborative Study Group for the Micronucleus Test/Mammalian Mutagenesis Study Group of the Environmental Mutagen Society of Japan. Mutat Res 223(4):329–344

23. Cawthorne C, Swindell R, Stratford IJ, Dive C, Welman A (2007) Comparison of doxycycline delivery methods for Tet-inducible gene expression in a subcutaneous xenograft model. J Biomol Tech 18(2):120–123

24. Laird PW, Zijderveld A, Linders K, Rudnicki MA, Jaenisch R, Berns A (1991) Simplified mammalian DNA isolation procedure. Nucleic Acids Res 19(15):4293

25. Nakamura E, Nguyen MT, Mackem S (2006) Kinetics of tamoxifen-regulated Cre activity in mice using a cartilage-specific CreER(T) to assay temporal activity windows along the proximodistal limb skeleton. Dev Dyn 235(9):2603–2612. doi:10.1002/dvdy.20892

26. Reinert RB, Kantz J, Misfeldt AA, Poffenberger G, Gannon M, Brissova M, Powers AC (2012) Tamoxifen-induced Cre-loxP recombination is prolonged in pancreatic islets of adult mice. PLoS One 7(3), e33529. doi:10.1371/journal.pone.0033529

27. Reid JM, Goetz MP, Buhrow SA, Walden C, Safgren SL, Kuffel MJ, Reinicke KE, Suman V, Haluska P, Hou X, Ames MM (2014) Pharmacokinetics of endoxifen and tamoxifen in female mice: implications for comparative in vivo activity studies. Cancer Chemother Pharmacol. doi:10.1007/s00280-014-2605-7

28. Livet J, Weissman TA, Kang H, Draft RW, Lu J, Bennis RA, Sanes JR, Lichtman JW (2007) Transgenic strategies for combinatorial expression of fluorescent proteins in the nervous system. Nature 450(7166):56–62. doi:10.1038/nature06293

29. Snippert HJ, van der Flier LG, Sato T, van Es JH, van den Born M, Kroon-Veenboer C, Barker N, Klein AM, van Rheenen J, Simons BD, Clevers H (2010) Intestinal crypt homeostasis results from neutral competition between symmetrically dividing Lgr5 stem cells. Cell 143(1):134–144. doi:10.1016/j.cell.2010.09.016

30. van Amerongen R (2015) Lineage Tracing in the Mammary Gland Using Cre/lox Technology and Fluorescent Reporter Alleles. Methods Mol Biol. 1293:187–211. doi:10.1007/978-1-4939-2519-3_11

Part IV

Translational Applications

Chapter 16

Assessment of Significance of Novel Proteins in Breast Cancer Using Tissue Microarray Technology

Laoighse Mulrane, William M. Gallagher, and Darran P. O'Connor

Abstract

The arraying of formalin-fixed paraffin-embedded (FFPE) tissue, or less commonly frozen tissue, in tissue microarrays (TMAs) is an invaluable method with which to assess the association of novel proteins with a myriad of diseases in large cohorts of patients allowing high throughput evaluation as potential biomarkers. TMAs are most frequently used in cancer studies although they are not limited to this application. The most common method of evaluation of TMAs is via immunohistochemistry (IHC) which is an antibody-based protein localisation method routinely used in the clinical laboratory. However, significant issues still exist with respect to the validation of antibodies for use on TMA sections, with a large number of published studies failing to do so correctly [O'Hurley et al. Mol Oncol, doi:10.1016/j.molonc.2014.03.008, 2014]. Here, we present a method to determine the antibody specificity for use in immunohistochemistry (IHC), as well as the analysis and interpretation of results from an IHC-stained TMA.

Key words Tissue microarray, Immunohistochemistry, High throughput, Cancer

1 Introduction

TMAs were first introduced in 1986 by Battifora [1] and have gained extensive recognition following a seminal publication in *Nature Medicine* by Kononen and colleagues [2]. They are constructed from circular cores extracted from blocks of FFPE tissue and set in an empty paraffin block, thus providing an "array" of up to 500 tissue samples from individual patients in a single block. Multiple sections of the array may be harvested and subjected to a number of independent tests including IHC and in situ hybridization (ISH). With individual samples it is well documented that experimental results can vary based on the time from the cutting of FFPE sections to their use [3], as well as between different experimental runs [4]. The advent of TMA technology revolutionized biomarker validation by allowing putative markers to be assayed under standard conditions in a large set of samples simultaneously, leading to a marked reduction in staining variability. Additionally,

Finian Martin et al. (eds.), *Mammary Gland Development: Methods and Protocols*, Methods in Molecular Biology, vol. 1501, DOI 10.1007/978-1-4939-6475-8_16, © Springer Science+Business Media New York 2017

with tumors routinely processed into FFPE blocks in hospital histology labs, TMA technology offers a cost-effective high-throughput method of extracting further benefit from archival material and has been validated for use in breast cancer biomarker studies [5]. Moreover, as the detection of protein biomarkers by IHC is currently the "gold standard" in breast cancer assessment, the use of TMAs coupled with IHC provides a directly translatable method for the validation of clinically relevant diagnostic, prognostic, or predictive biomarkers [6].

IHC was first introduced in 1941 by Coons and Jones [7] but did not become a well-used analytical tool in the pathology laboratory for many years. Today, it is a standard technique used to assess biomarker expression in tissue in both an experimental and clinical setting, for example, to assess estrogen receptor (ER), progesterone receptor (PR) and HER2/neu expression in breast tumor tissue. [Amplification of the HER2/neu gene is subsequently confirmed with fluorescent in situ hybridisation (FISH).] However, despite the success of these biomarkers in the clinic, the use of many others [e.g., CA 15-3, CA 27-29, carcinoembryonic antigen (CEA), cathepsin D, cyclin E] has yet to be fully validated and their utility confirmed [8]. Breast tumor TMAs will significantly facilitate the speed with which the assessment of new biomarkers can be carried out. But it is clear that their successful use depends on standardisation of tissue collection, processing and staining but most importantly the full validation of antibodies to be used in this setting. Here we provide protocols for validation of antibodies to be used in IHC analysis of TMAs.

2   Materials

2.1   Construction of FFPE Cell Pellets

1. Chloroform.
2. Dulbecco's phosphate-buffered saline (PBS).
3. Ethanol.
4. Paraffin-embedding station.
5. Low-gelling-point agarose.
6. 10 % v/v Neutral-buffered formalin.
7. Paraffin wax.
8. Tissue cassettes.
9. Tissue processor.
10. 0.05 % w/v Trypsin-0.5 mM EDTA (1×).

2.2   Construction of Cell Pellet Arrays/ TMAs

1. Paraffin-embedding station.
2. Microtome.
3. Paraffin wax.
4. SuperFrost® Plus slides.

5. Manual tissue micro-arrayer [9].

6. TMA punches 0.6 mm/1 mm/1.5 mm.

2.3 Immuno-histochemistry

1. Citrate buffer (10 mM Sodium citrate, 0.05 % Tween 20), pH 6.0.

2. Dulbecco's phosphate-buffered saline, pH 7.3.

3. 1 mM EDTA, buffer pH 8.0 or 9.0 with NaOH.

4. Ethanol.

5. 3 % w/v H_2O_2.

6. Tris–HCl buffer (10 mM Tris Base, 0.05 % Tween 20), pH 10.

7. Automated dewaxing and epitope recovery device or pressure cooker.

8. Automated stainer.

9. Automated glass coverslipper.

10. Mayer's Hematoxylin.

11. Pertex®

12. Tween 20.

13. Detection system for specific mouse IgG or rabbit IgG antibody bound in tissue sections. (The specific antibody is located by a universal secondary antibody formulation conjugated to an enzyme-labeled polymer. The polymer complex is then visualized with a chromogen (e.g., diaminobenzidine (DAB).)®

14. Xylene.

3 Methods

3.1 Selection of Antibodies and Determination of Antibody Specificity

Appropriate antibody selection and validation is critical to the success of IHC. Selection of an antibody from a reputable company that has been validated for IHC use and has been reported on many times is desirable. However, despite thousands of studies being published using antibodies to immunohistochemically stain clinical material, relatively few have fully validated the antibody used as being suitable for IHC [10].

It is of the utmost importance that antibody specificity is determined through a rigorous validation process prior to utilising a particular antibody to evaluate protein expression in large cohorts of patients, as failure to do so may result in the generation of erroneous results. The most common method to achieve this is through the evaluation of cell lines containing varying levels of the specific protein of interest using Western blotting and IHC on cell pellets. However, not all antibodies will work for both techniques. For example, formalin fixation (which remains the standard method of tissue fixation in the clinic) can cause epitope masking [11], leaving such epitopes inaccessible to antibody binding, while a reduced protein evaluated on a Western blot is not in the same natural conformation as it would be in tissue.

Furthermore, although an antibody may pass validation at the stage of Western blot and IHC analysis of FFPE cell pellets, it may still bind nonspecifically in tissue (given its more complex cellular milieu). As such, determining the predicted localisation of the protein in the tissue being assayed is key. This can usually be determined using literature/bioinformatic searches, as well as having knowledge of the function of the respective protein. If this predicted localization does not match with the tissue localization seen upon staining, this can indicate a potential problem with the antibody in use. However, it is wise to take into account the variability in the localization of some proteins, which may be trafficked to different locations depending on cellular context or intracellular conditions.

Another matter to consider in the choice of antibody is whether it is polyclonal or monoclonal. Monoclonal antibodies are clonally produced from a single immune cell and specifically recognize one particular epitope within a protein. Thus, they are considered more specific. Polyclonal antibodies may produce a stronger signal owing to the fact that they bind to multiple epitopes within the protein but may also result in a higher level of non-specific staining, as well as exhibiting batch effects. However, they have the advantage that even if some epitopes are masked, others may still be accessible.

3.2 Antibody Validation by Western Blotting

The first step in the antibody validation process for IHC use is to confirm binding of the antibody to the correct sized protein of interest on a Western blot. As this is a standard procedure we would like to refer the reader to the many published protocols. In brief, proteins are transferred to membranes which can be probed with specific antibodies. Multiple bands are acceptable if different isoforms of the protein exist and are at the expected molecular weight. However, non-specific bands on a Western blot generally indicate that there may be at least some non-specific binding on tissue. Testing of antibody specificity using Western blotting is usually coupled with siRNA/shRNA knockdown of the protein of interest in cell lines to ensure that the band being detected is the correct one. Cells with knockdown of a specific protein can then be used to create cell pellet arrays, as detailed below, to determine of antibody specificity for IHC use.

3.3 Construction of FFPE Cell Pellets

FFPE cell pellets can be used to determine specificity of the antibody for use in immunohistochemical staining (of TMAs) in conjunction with Western blotting. The process results in paraffin blocks containing plugs of cells suspended in agarose which can be sectioned and stained in the same way as FFPE tissue. As with Western blotting, knockdown of the specific protein of interest using siRNA/shRNA can be used to determine specific binding of the antibody (Fig. 1). Cell pellets also allow the user to staining quality before moving on to stain precious tissue sections.

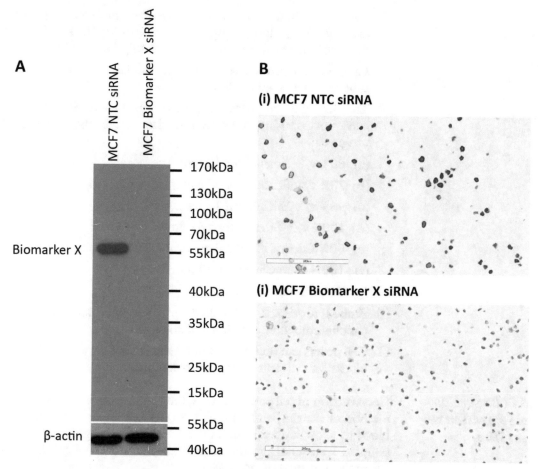

Fig. 1 Validation of antibody against Biomarker X using western blotting and IHC on cell pellets. (**a**) Western blot of Biomarker X on lysates from MCF7 cells transfected with either a non-targeting control (NTC) siRNA or siRNA against Biomarker X. One clear band is seen at the correct size. (**b**) Cell pellets from the same cells as stained with antibody against Biomarker X. The same pattern of expression seen in the Western blot is observed

1. Grow adherent cells to approximately 70% confluency in a 175 cm^2 flask. Include lines transfected with siRNA/shRNA against the protein of interest as well as lines transfected with a non-targeting control.

2. Aspirate media and wash cells with PBS.

3. Add 4 ml w/v 0.05% w/v trypsin-0.5 mM EDTA (1× solution) to the flask and place flask at 37 °C until cells have detached. Transfer detached cell suspension to sterile centrifuge tube.

4. Inactivate trypsin with 8 ml media and centrifuge at $250 \times g$ for 3 min to pellet cells.

5. Aspirate media and wash cells with PBS. Centrifuge tubes at $250 \times g$ for 3 min to pellet cells.

6. Resuspend the cell pellet in 10% neutral-buffered formalin and incubate at room temperature for 4 h.

7. Following fixation, wash cells in PBS and resuspend pellet quickly in 200 μl 1% w/v low-gelling-point agarose cooled to 40 °C. Take care not to allow agarose to set in pipette tip (*see* **Note 1**).

8. Pipette cell-agarose suspension into the lid of an Eppendorf tube or other mould.

9. Transfer solidified pellets to tissue cassettes lined with sponges and process using a tissue processor using the following program:

   (a) 50% v/v Ethanol for 15 min.

   (b) 70% v/v Ethanol for 15 min.

   (c) 95% v/v Ethanol for 15 min.

   (d) 100% v/v Ethanol for 15 min X3.

   (e) 50% Ethanol/50% chloroform for 20 min.

   (f) 100% Chloroform for 30 min X2.

   (g) Paraffin wax for 2 h X2.

10. Embed processed cell pellets into paraffin blocks using a paraffin-embedding station and allow to set.

11. Cell pellets can then be sectioned and stained or used as donor blocks for cell pellet array construction.

3.4 Construction of Cell Pellet Arrays and TMAs

Construction of cell pellet arrays (CPAs) and TMAs is a convenient method of arraying multiple cell lines/tissue samples into a single block to allow high-throughput evaluation of a large number of samples. Instructions in this section assume the use of an MTA-1 (or similar) manual tissue arrayer

1. Prepare empty recipient paraffin blocks using a paraffin-embedding station.

2. Allow to set on cooling plate for approximately 30 min.

3. Prior to construction, prepare a map for the TMA/CPA to be constructed. It is essential to include reference cores to allow for correct orientation of the slide when sectioned (*see* **Note 2**).

4. Affix needles of desired size (0.6 mm/1 mm/1.5 mm to the MTA-1 arrayer, red needle to punch hole in recipient block to the left, blue needle to extract tissue/cell pellet core from donor block to the right. These needles are of a slightly different size to allow the donor core to fit securely into the recipient block (*see* **Note 3**).

5. Screw the paraffin block into the indent, ensuring that it is fully immobilized.

6. Move the punches to the starting position (usually the left-hand corner of the recipient block) and set the levers to zero. Ensure sufficient space is left at each of the edges of the recipient block (at least 2–3 mm). If cores are placed too close to the edge of the block, they are liable to fall off during sectioning.

7. Set the depth that the needle will punch to using the stopper. This is set based on the depth of the tissue/cell pellets to be arrayed. Setting too great a depth will result in cores being potentially pushed too far into the block, while too small a depth will result in sample loss at the top of the block.

8. Punch a hole in the recipient block using the red needle.

9. Punch a similar hole in the donor block (cell pellet or tissue) using the blue punch. Twist needle back and forth to remove core from block. If making a tissue microarray, it is essential to do this in conjunction with hematoxylin and eosin-stained slides which have been marked by a pathologist. This will allow the correct areas to be selected for inclusion in the TMA.

10. Slowly push the donor core into the recipient block, taking care that the top of the core is level with the block.

11. Move the handle to the position of the next core. It is generally sufficient to leave 0.8–1.5 mm between cores. Leaving an appropriate amount of space will ensure that cores do not become joined or mixed up during sectioning.

12. Samples are arrayed in triplicate or quadruplicate. This is essential to allow representation of tissue in a section as well as to allow for core loss during sectioning.

13. Seal the block by placing upside-down on a clean slide at 60 °C for approximately 10 min. Allow to cool on ice and remove slide.

14. Cut 3–7 μm sections from the CPA/TMA using a microtome and adhere to SuperFrost® Plus slides.

15. Bake sections at 60 °C for 1 h to melt away excess paraffin wax before use.

16. Store cut sections in a dessicator.

3.5 Immuno-histochemistry

Bright-field IHC is a method commonly used in both the laboratory and the clinic for the detection and localisation of proteins in frozen or FFPE tissue using an antibody-based technique not unlike a Western blot. Following heat-induced epitope retrieval, a primary antibody against a protein of interest is bound to that protein in the tissue. Secondary antibodies either directly or indirectly coupled to an enzyme (usually a peroxidase or alkaline phosphatase) are used to detect specific binding of the primary antibody. The enzyme then catalyses a colorimetric reaction using a substrate (e.g., 3′-diamino-benzidinetrahydrochloride (DAB) or 3-amino-9-ethylcarbazole (AEC)) allowing visualization of antibodies bound to the tissue.

1. Deparaffinize and rehydrate sections as follows:

    (a) 60 °C for 10 min

    (b) Xylene for 5 min X2

    (c) 100 % v/v Ethanol for 5 min X2

    (d) 95 % v/v Ethanol for 5 min

    (e) 80 % v/v Ethanol for 5 min

    (f) Deionized water for 5 min

2. Formalin fixation can mask or reduce immunoreactivity of epitopes in tissue sections. Therefore, it is usually necessary to perform heat-induced epitope retrieval (HIER). This can be carried out in a commercially available dewaxing and antigen retrieval module or a pressure cooker for 95 °C for 15 min in an appropriate buffer. Buffers suitable for antigen retrieval include 10 mM citrate, 10 mM Tris–HCl and 1 mM EDTA at a range of different pH levels. It is essential to optimize this component of the protocol for each primary antibody.

3. Include appropriate negative (isotype-matched and no antibody) and positive controls in each experiment.

4. Transfer slides to an automated staining device and conduct immunohistochemical staining using a HRP polymer system as follows:

    (a) Rinse slides with PBS 0.1 % v/v Tween 20 (PBS-T).

    (b) Dispense 300 μl of 3 % w/v H_2O_2 onto each slide. This functions as an enzyme block for endogenous peroxidases which can be found in many tissues. Volumes can be altered to reflect the size of tissue on the slide.

    (c) Rinse slides with PBS-T.

    (d) Add UV block to slides for 5 min.

    (e) Blow excess solution from slides and dispense primary antibody solution for 1 h at room temperature. Antibody is diluted to appropriate concentration in PBS-T (*see* **Note 4**).

    (f) Rinse slides with PBS-T X3.

    (g) Add secondary reagent (primary antibody enhancer) for 10 min. This contains secondary antibodies against both rabbit and mouse species.

    (h) Rinse slides with PBS-T X3.

    (i) Dispense labeled HRP polymer onto slides for 15 min. Ensure bottle is protected from light as this kit component is light sensitive.

    (j) Rinse slides with PBS-T X3.

    (k) Incubate slides with colorimetric substrate for 10 min. Working solution of DAB is prepared fresh by adding one drop of DAB Chromagen to 2 ml DAB substrate.

    (l) Rinse slides with deionized water.

5. Counterstain slides with Mayer's hematoxylin for 1–3 min followed by 2–10 min in tepid water with agitation (*see* **Note 5**).

6. Dehydrate slides as follows:

  (a) 80 % v/v Ethanol for 3 min

  (b) 95 % v/v Ethanol for 3 min

  (c) 100 % v/v Ethanol for 3 min X2

  (d) Xylene for 3 min X2

7. Coverslip slides manually using Pertex® or using an automated glass coverslipper.

8. Allow to dry overnight.

3.6 Evaluation of Staining

Evaluation of staining can be carried out either manually or using automated image analysis algorithms to produce a score. Automated systems are fast becoming popular owing to the relatively high-throughput nature of the analysis along with the reduction of the variability seen with manual scoring due to observer fatigue and the subjective nature of the analysis [6]. However, manual scoring carried out by an experienced pathologist remains the gold standard. TMAs should be scored by two independent observers and inter-observer variability noted. Scores generated using an automated algorithm are generally compared to a manual score to ensure that the algorithm is detecting the staining correctly. However, as automated scores are generated on a continuous scale, they can provide additional information over and above the manual scores.

3.6.1 Manual Scoring Systems

1. Staining can be cytoplasmic, nuclear, membranous, or a combination of the three depending on the cellular localisation of the protein and can also be present in different areas of the tissue on the TMA. For example, in cancer, particular proteins may be expressed in the cancerous tissue, as well as the stroma or infiltrating immune cells.

2. A numerical value is assigned based on the intensity of the staining. Typically, this could be from 0 to 3 with 0 being negative, 1 being weak staining, 2 being moderate staining and 3 being strong staining.

3. Staining results may also be reported as a binary yes/no (for example, >10 % stained cells) or positive versus negative.

4. Unless 100 % positivity is seen in the cell type being evaluated, percentage positivity should also be recorded. This can be combined with intensity score to produce a weighted histoscore, e.g., (% of cells with an intensity of 1) + (% of cells with an intensity of 2) + (% of cells with an intensity of 3).

5. Additionally, another scoring method combining a proportion score with intensity score was described in 1998 and has been adopted by some laboratories [12].

6. Scoring should be carried out in a blinded manner by two independent observers, at least one of whom should be an

experienced pathologist. If scoring is discordant for a core, it may be reviewed by both observers and a consensus reached or a max value taken for this core. As described in Subheading 3.7, a correlation coefficient of at least 0.7 is required for scores from two independent observers.

7. In order to maintain the 0–3 scoring system without introducing scores with decimal places, the max value is usually taken for samples which are arrayed in duplicate if the same score is not assigned to both cores. If discordant scores are noted for triplicate or quadruplicate scores, staining should be reviewed by both observers and a median value assigned. However, if a large range of staining is seen across replicate cores, this sample should potentially be removed from the analysis or a full face section from the block from which these cores were taken reviewed.

8. Always ensure that the scores are correctly lined up with the TMA map, bearing in mind that cores may move during sectioning.

3.6.2 Automated Scoring Systems

A number of image analysis solutions exist to automatically quantify IHC staining, including those from Aperio (now part of Leica Biosystems, Nussloch, Germany), Definiens (Carlsbad, CA, USA), and 3D Histech (Budapest, Hungary) [13]. These can analyse the amount of staining in a given section, without reference to subcellular localization, or can quantify specific staining of the nucleus, cytoplasm, or membrane. Outputs include percentage positivity, as well as staining intensity and sometimes a combinatory score comprised of the two measurements (*see* **Note 6**).

1. Analysis of a TMA using an automated image analysis algorithm requires that a high-resolution digital scan is created from the slides using a digital slide scanner [e.g., Aperio ScanScope XT slide scanner (Leica Biosystems, Nussloch, Germany) or 3DHistech Panoramic slide scanner (3D Histech, Budapest, Hungary)].

2. Ensure that the TMA is manually quality controlled to eliminate cores not suitable for analysis. Failure to do so will result in the inclusion of scores for cores which should not have been evaluated (e.g., core missing, core folded over, cores which include staining artifacts).

3. Optimize color parameters within the algorithm (positive stain and nuclear counterstain) using slides that have been stained with that reagent alone.

4. Optimize other parameters such as nuclear size, roundness, and elongation to ensure that the algorithm detects the correct area on the tissue.

5. Run the algorithm and extract automated scores. Ensure that automated scores are correctly aligned with TMA map.

6. When calculating a single score from multiple replicate cores on an array, the median is used.

7. Correlate percentage positivity scores or intensity scores to manually generated scores and conduct statistical analysis as detailed below.

3.7 Statistical Analysis and Survival Estimations

Statistical analysis of TMA data can be performed using multiple statistical packages including R (http://www.r-project.org/) and SPSS (SPSS Inc., Chicago, IL, USA).

1. Spearman's Rho or Intraclass correlation analysis can be used to estimate the relationship between duplicate cores from individual tumors, between manual scores from two independent observers and between manual score and automated score. Cohen's kappa coefficient can also be used to determine the correlation between independent observer's scores. A correlation coefficient of >0.7 is considered the minimum acceptable standard but higher coefficients are always desirable [14].

2. Kaplan-Meier as well as univariate Cox regression analyses are used to analyze correlations between a survival endpoint and a biomarker of interest. Endpoints can include recurrence-free survival (RFS), disease-free survival (DFS), metastasis-free survival (MFS), and overall survival (OS) depending on the dataset [15] (*see* **Note 7**).

3. Cox regression analysis can also be used to correct for various clinicopathological characteristics of patients within the cohort. This is conducted by building a multivariate Cox regression model and can correct for many confounding factors such as age, tumor size, disease grade or stage, or the expression of other biomarkers [16]. If a biomarker is seen to still be significantly associated with survival following this correction, it can be said to be independent of these clinical features.

4. These analyses can be stratified for various groups such as high grade versus low grade, disease or specific treatment versus treatment naïve. The latter is useful to evaluate biomarkers which are predictive of treatment response but material from clinical trials which is generally used to test such a theory can be difficult to obtain.

5. Pearson's X^2 test is used to evaluate associations between biomarker expression and clinicopathological characteristics of the patients.

3.8 Presentation of Results

Results of a TMA analysis are usually presented in the form of graphs and tables summarizing the key statistics for the cohort of patients.

1. Kaplan-Meier analyses are generally represented in graphical format. Figure 2 contains example graphs illustrating that high

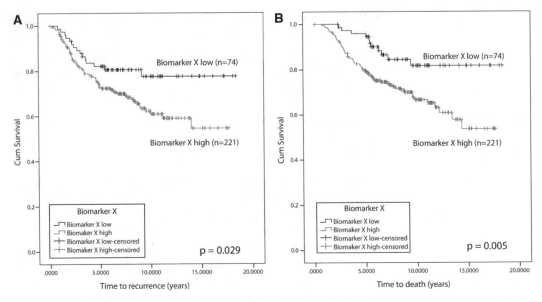

Fig. 2 Biomarker X expression is associated with decreased recurrence-free and overall survival in a cohort of breast cancer patients. Kaplan-Meier survival analysis of Biomarker X expression in breast cancer patients. Association of Biomarker X expression with (**a**) recurrence-free survival in (**b**) overall survival in a cohort of 295 breast cancer patients. *P* values were estimated using a log rank test

expression of Biomarker X is associated with decreased RFS and OS. The p value is included in the graph and numbers of patients in each group can be optionally included.

2. Cox regression analysis produces a hazard ratio or numerical measurement of risk associated with a particular biomarker along with a 95 % confidence interval. This can be presented in text as *HR 0.59, 95 % CI 0.39–0.90, p = 0.013* or in a format. Table 1 displays representative data showing that Biomarker X is significantly associated with decreased RFS and OS, both in a univariate and multivariate Cox regression model correcting for age, tumor size, Nottingham histological grade (NHG), lymph node status, and ER status.

3. For multivariate Cox regression models, all variables used to build the model and their associations with survival may be listed in the table or a footnote may be included detailing which variables were significant contributing factors in the analysis.

4. Correlations with particular clinicopathological characteristics of patients within the data set are presented in a tabular format. Table 2 demonstrates that expression of Biomarker X is associated with higher grade tumors and ER negativity.

5. It is imperative to validate findings in more than one cohort of patients.

Table 1
Univariate and multivariate Cox regression analysis of Biomarker X in a cohort of 295 breast cancer patients

	Univariate			Multivariate[a]		
	HR	95 % CI	p value	HR	95 % CI	p value
			RFS			
All patients (*n* = 295)	1.84	1.06–3.21	0.031	1.633[a]	1.08–2.88	0.040
			OS			
All patients (*n* = 295)	2.41	1.27–4.55	0.007	1.939[b]	1.01–3.71	0.046

[a]Corrected for age, tumor size, Nottingham histological grade (NHG), lymph node status, and ER status
Significant contributing factors in multivariate Cox regression analysis were [a]NHG, ER, and lymph node status, [b]NHG, ER status, and size

Table 2
Example table depicting association of clinicopathological characteristics of tumors with expression of Biomarker X in a cohort of 295 breast cancer patients using Pearson's X^2 test

	Low biomarker X, n (%)	High biomarker X, n (%)	P value
Age (years)			
<=50	8 (10.8)	23 (10.4)	0.922
>50	66 (89.2)	198 (89.6)	
Tumor size (mm)			
<=20	33 (44.6)	107 (48.4)	0.569
>20	41 (55.4)	114 (51.6)	
NHG			
Grade I	28 (37.8)	47 (21.3)	0.003
Grade II	27 (36.5)	74 (33.5)	
Grade III	19 (25.7)	100 (45.2)	
Nodal status			
Negative	38 (51.4)	113 (51.1)	0.974
Positive	36 (48.6)	108 (48.9)	
ER status			
ER–	7 (9.5)	62 (28.1)	0.001
ER+	67 (90.5)	159 (71.9)	

ER Estrogen receptor

NHG Nottingham histological grade

4   Notes

1. When resuspending cells in 200 µl 1% w/v agarose (using a 1 ml pipette tip), allow both pelleted cells and agarose to equilibrate in a water bath set to 40 °C prior to mixing. This will minimise the chance of the agarose solidifying in the pipette tip.

2. Ensure that reference cores are positioned in such a way as to allow correct identification of cores on the slides in any orientation. For example, one row of reference cores along the top of the array may allow orientation along the x/y-axis but will not allow determination of the position of cores along the z-axis (i.e., if the section is flipped 180° along that axis).

3. Precious tissue may be conserved by using smaller needles but note that a greater number of punches may be required for full representation of the tissue section.

4. If an antibody is specific at the Western level but showing non-specific staining on tissue, blocking buffer can be added to the antibody solution to reduce non-specific binding. Number and duration of washes can also be altered or blocking time increased.

5. This is a progressive haematoxylin, so the depth of staining depends on the length of incubation in water. As such, care must be taken to optimise this incubation time. Overstaining will result in positive DAB staining being more difficult to evaluate.

6. Expression in multiple cell types is expected in the case of the majority of proteins. However, this can prove challenging when using automated algorithms to evaluate staining. In this case, care must be taken to ensure that the algorithm is selecting the correct areas to analyse.

7. For Kaplan-Meier survival analysis, patients must be stratified into groups (e.g. manual scores 0-3, quartiles for automated scores etc.). However, the Cox proportional hazards model allows analysis of continuous variables such as continuous automated scores.

Acknowledgements

Funding is acknowledged from the Irish Research Council for Science, Engineering and Technology (IRCSET), Science Foundation Ireland through the *Molecular Therapeutics for Cancer, Ireland* Strategic Research Cluster (award 08/SRC/B1410; http://www.mtci.ie) and the Irish Cancer Society Collaborative Cancer Research Centre BREAST-PREDICT grant, CCRC13GAL (http://www.breastpredict.com). No conflicts of interest are declared by the authors.

References

1. Battifora H (1986) The multitumor (sausage) tissue block: novel method for immunohistochemical antibody testing. Lab Invest 55(2):244–248

2. Kononen J, Bubendorf L, Kallioniemi A, Barlund M, Schraml P, Leighton S, Torhorst J, Mihatsch MJ, Sauter G, Kallioniemi OP (1998) Tissue microarrays for high-throughput molecular profiling of tumor specimens. Nat Med 4(7):844–847

3. Fergenbaum JH, Garcia-Closas M, Hewitt SM, Lissowska J, Sakoda LC, Sherman ME (2004) Loss of antigenicity in stored sections of breast cancer tissue microarrays. Cancer Epidemiol Biomarkers Prev 13(4):667–672

4. Pozner-Moulis S, Cregger M, Camp RL, Rimm DL (2007) Antibody validation by quantitative analysis of protein expression using expression of Met in breast cancer as a model. Lab Invest 87(3):251–260. doi:10.1038/labinvest.3700515

5. Camp RL, Charette LA, Rimm DL (2000) Validation of tissue microarray technology in breast carcinoma. Lab Invest 80(12):1943–1949

6. Brennan DJ, O'Connor DP, Rexhepaj E, Ponten F, Gallagher WM (2010) Antibody-based proteomics: fast-tracking molecular diagnostics in oncology. Nat Rev Cancer 10(9):605–617. doi:10.1038/nrc2902

7. Coons AH, Jones RN (1941) Immunological properties of an antibody containing a fluorescent group. Proc Soc Exp Biol Med 47:200–202

8. Harris L, Fritsche H, Mennel R, Norton L, Ravdin P, Taube S, Somerfield MR, Hayes DF, Bast RC Jr (2007) American Society of Clinical Oncology 2007 update of recommendations for the use of tumor markers in breast cancer. J Clin Oncol 25(33):5287–5312. doi:10.1200/JCO.2007.14.2364

9. Springall RJ, Gillett CE (2006) Breast tissue microarrays. Methods Mol Med 120:43–50

10. O'Hurley G, Sjostedt E, Rahman A, Li B, Kampf C, Ponten F, Gallagher WM, Lindskog C (2014) Garbage in, garbage out: a critical evaluation of strategies used for validation of immunohistochemical biomarkers. Mol Oncol. doi:10.1016/j.molonc.2014.03.008

11. Arnold MM, Srivastava S, Fredenburgh J, Stockard CR, Myers RB, Grizzle WE (1996) Effects of fixation and tissue processing on immunohistochemical demonstration of specific antigens. Biotech Histochem 71(5):224–230

12. Allred DC, Harvey JM, Berardo M, Clark GM (1998) Prognostic and predictive factors in breast cancer by immunohistochemical analysis. Mod pathol 11(2):155–168

13. Mulrane L, Rexhepaj E, Penney S, Callanan JJ, Gallagher WM (2008) Automated image analysis in histopathology: a valuable tool in medical diagnostics. Expert Rev Mol Diagn 8(6):707–725

14. Kirkegaard T, Edwards J, Tovey S, McGlynn LM, Krishna SN, Mukherjee R, Tam L, Munro AF, Dunne B, Bartlett JM (2006) Observer variation in immunohistochemical analysis of protein expression, time for a change? Histopathology 48(7):787–794. doi:10.1111/j.1365-2559.2006.02412.x

15. Clark TG, Bradburn MJ, Love SB, Altman DG (2003) Survival analysis part I: basic concepts and first analyses. Br J Cancer 89(2):232–238. doi:10.1038/sj.bjc.6601118

16. Bradburn MJ, Clark TG, Love SB, Altman DG (2003) Survival analysis part II: multivariate data analysis—an introduction to concepts and methods. Br J Cancer 89(3):431–436. doi:10.1038/sj.bjc.6601119

Chapter 17

Patient-Derived Xenografts of Breast Cancer

Damir Varešlija, Sinead Cocchiglia, Christopher Byrne, and Leonie Young

Abstract

With the advancement of translational research, particularly in the field of cancer, it is now imperative to have models which more clearly reflect patient heterogeneity. Patient derived xenograft (PDX) models, which involve the orthotopic implantation of breast tumors into immune-compromised mice, recapitulate the native tumor biology. Despite the considerable challenges that establishing PDX models present, they are the ultimate model to study tumorigenesis of refractory disease and for assessing the efficacy of new pharmaceutical compounds.

Key words Patient-derived xenografts, PDX, Primary tissue, Tumor fragments, Translational research

1 Introduction

The use of cell line models, coupled with cell line xenograft models, to study cancer has greatly enhanced our understanding of biological processes regarding the tumorigenicity of cells. However, preclinical data has not always translated into substantial improvements in the clinical setting. A chief limitation of traditional in vitro cell culture models is that the culture conditions used to propagate these cells create an environment that diverges distinctly from the normal breast tumor microenvironment [1]. The cells that have adapted to thrive in plastic flasks are selected for phenotypically homogeneous populations that are not indicative of the natural tumor state.

In recent times, the cancer research community is turning to a novel and more accurate model approach to study tumors. This involves transplanting a freshly resected patient tumor into immunocompromised mouse host. These aptly named patient-derived xenografts (PDXs) can potentially recapitulate the tumor heterogeneity observed in human tumors. The PDXs maintain the clinical and histopathological features observed in the patient (Fig. 1). Similarly, they remain true to the biology and disease outcomes of

Finian Martin et al. (eds.), *Mammary Gland Development: Methods and Protocols*, Methods in Molecular Biology, vol. 1501, DOI 10.1007/978-1-4939-6475-8_17, © Springer Science+Business Media New York 2017

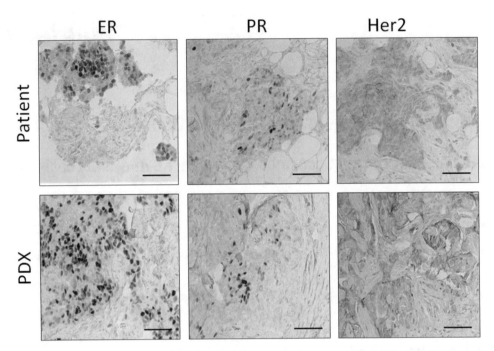

Fig. 1 Retention of ER, PR, and Her2 expression in hormone receptor positive patient-derived xenografts. Patient-derived xenografts were established in NSG mice, and subsequently propagated via direct transplantation of solid tumor pieces into recipient mice. Tumors were grown under continuous estrogen supplementation. Sections of xenograft tumor were stained by IHC for ER (Ventana 790-4324 Rabbit monoclonal), PR (Ventana 790-2223) and Her2 (Ventana 790-2991). Magnification 20×. Scale bars are 100 μM

the patient tumors from which they were derived as the PDX tumors respond to treatment and metastasize to the same sites as the original tumor [2–5].

One of the main issues facing PDX models is the percentage of successful engraftments and the rates of growth of the tumors once implanted into a mouse host. Efforts to develop PDX models of hormone-driven tumors such as breast and prostate cancer have been met with only partial success, compared with ovarian or colorectal cancer [6]. Reported engraftment rates of breast cancer PDX models vary between 5 and 35%, with triple negative tumors being most successful which is consistent with the aggressive nature of this particular subtype [2–4, 7].

ER positive tumors appear to be have the lowest engraftment rates but several groups have successfully improved the engraftment efficacy and their studies have shown great mechanistic and genomic insight into therapy-resistant ER positive disease [2, 3, 8].

One of the biggest drawbacks with establishing PDX models is the associated cost as the establishment of a PDX line requires genetically engineered mice and the other expenses related to keeping large number of animals. Similarly, unlike with other traditional models, the time for a PDX tumor to manifest itself can be anywhere between 4 months to a year which can often impact significantly on experimental planning [6]. Nevertheless, PDX models offer the true recapitulation of the tumor biology in its natural state and are the ultimate model to study tumorigenesis of refractory disease and for assessing the efficacy of new pharmaceutical compounds.

2   Materials

2.1   Collecting of Surgical Breast Tumor Tissue Samples

1. 70 % ethanol.
2. Sterile scalpels.
3. Sterile forceps.
4. Sterile 10-cm petri dishes.
5. 15-ml conical vials.
6. Sterile cryovials.
7. Small liquid nitrogen container.
8. RNA*later*.
9. MEGM medium.

2.2   Processing of the Tissue Samples for Implantation

1. Tumor tissue for implantation.
2. MEGM medium.
3. Tumor-freezing medium (95 % (v/v) fetal bovine serum (FBS)/5 % (v/v) DMSO).
4. Sterile 10-cm petri dishes.
5. Disposable scalpels.
6. Sterile razor blades.
7. Sterile 1.5-ml cryovials.
8. Cryo-freezing container.
9. Isopropanol.

2.3   Implantation of Tissue Fragments into the Mice and Estrogen Pellet Implantation

1. Specific-pathogen free room in the animal house.
2. Anesthesia induction chamber.
3. NOD/SCID/IL2Rγ−/− (NSG) female mice.
4. 70 % ethanol.
5. Disinfectant.
6. Vetergesic (Buprenorphine 0.3 mg/ml).
7. Surgical blades.
8. Sterile dissecting scissors.
9. Sterile tweezers.
10. Sterile scalpels.

11. Sterile forceps.

12. 90-day release estrogen pellets (0.36 mg/pellet).

13. Tumor samples for implantation.

14. Basement membrane matrix (Matrigel).

15. 100% w/w isofluorane.

16. Electric shaver.

17. Heat source for the mice during recovery.

18. Precision trochar.

19. MEGM media.

20. Absorbable surgical sutures.

21. Cotton swabs, sterile.

22. Microlane 3 syringe (26G-1/2″).

2.4 Harvesting and Processing of Tumors from Mice

1. 70% ethanol.

2. Tumor-bearing xenografts.

3. Sterile 10-cm petri dishes.

4. Surgical blades.

5. Sterile dissecting scissors.

6. Sterile tweezers.

7. Calipers for measurement of tumor size.

8. Small liquid nitrogen container.

9. Tumor-freezing medium (95% (v/v) FBS/ 5% (v/v) DMSO).

10. Sterile 1.5-ml cryovials.

11. RNA*later*.

12. 10% v/v neutral-buffered formalin.

13. Molecular grade ethanol.

14. Paraffin wax.

15. Paraffin embedding station.

16. Sterile 15-ml conical tubes.

3 Methods

3.1 Sourcing, Collection, and Storage of Surgical Breast Tumor Tissue Samples

All human patient tissue samples must be collected from informed, consenting patients under an approved ethics and following an approved research protocol. It is imperative to have a well coordinated system and a member of a research staff on site to ensure fast and reliable patient recruitment and tissue collection. Speed of processing is vital in ensuring viable breast cancer tissue and it increases the chance of successful engraftment into a mice.

1. The research person in charge of patient recruitment and tissue collection is responsible for tracking surgery schedules in order to identify suitable candidates. It is imperative that the research person works closely in conjunction with the hospital staff involved so that a streamlined system for collection exists.

2. It is very important that prior to the tissue collection the patient is recruited to the study and the informed consent is obtained by a research nurse.

3. Once a possible collection has been identified, alert everyone involved in the process and inform the lab by email or by phone at the earliest possible time. Note the patient's clinical diagnosis.

4. Inform the researcher responsible for the tissue collection prior to the removal of the tumor tissue by the operating room staff.

5. When the breast tumor tissue is removed, place the tissue on a petri dish and keep it on ice; record the time of removal and transport the sample to pathology for processing (*see* **Note 1**).

6. When the pathologist has processed the patient sample and has removed excess tumor and adjacent normal tissue for research, label cryovials with a hospital number only so that the patients cannot be identified. Record information regarding the pathology of the tumor, such as hormone receptor status, grade and type. Flash-freeze sections of the tissue in the cryovials and store these at –80 °C. Place additional tissue pieces in 1.5 ml reaction tubes containing 400 µl of RNAlater, which are kept at 4 °C overnight and then stored long-term at –80 °C.

7. Process the remaining tissue for implantation into the mice as per protocol in Subheading 3.2.

8. Record patient identification, tissue, and storage information into a database.

3.2 Processing of the Tissue Samples for Implantation

The tumor tissue obtained from patient surgery and designated for animal implantation must be processed quickly under sterile conditions and kept on ice. The tumor tissue can either be freshly implanted or frozen in cryovials to be implanted at a later stage. Choosing whether to carry out fresh implantation or implantation from frozen stocks will be dependent on the proximity of the hospital site to the lab. Logistically implanting from frozen allows for greater control in experimental planning. For instance, the mice can be specifically sourced at a young age (28–35 days) and can be implanted at the same time. It also allows the researcher time to choose which tumor subtype would be best suited for implantation. Each technique has been reported to work as effectively as the other [2].

1. Place the tumor sample on a petri dish in media to keep it from drying out. Depending on the size the tumor cut it into small fragments of ~2 mm using a sterile scalpel. If possible, remove as much of the necrotic tissue present which differs to the tumor tissue in color and consistency (*see* **Note 2**).

2. If implanting from fresh tissue proceed to Subheading 3.3. If the tumor samples are being frozen in cryovials proceed to **step 3**.

3. If the tumor tissue pieces are frozen to be implanted at a later stage place them in a sterile cryovial containing tumor-freezing medium. Freeze the tissue slowly in a cryo container stored at –80 °C. Keep the samples here overnight after which they are placed in liquid nitrogen for long-term storage.

3.3 Implantation of Tumor Tissue Fragments and Estrogen Pellets into the Mice

Implanting the tumor tissue fragments into the mice is a minimally invasive subcutaneous procedure. The procedure is normally carried out on young NSG female mice. Other models have been used elsewhere such as NOD-SCIDs or Swiss nude mice [4, 7]. If desirable, the procedure can be coupled with the clearing of the immature endogenous mouse mammary tissue at 3–4 weeks of age. This method has been reported to increase the success of tumor transplantation [9]. When mice are ordered they must be allowed to acclimatize for 1 week prior to experimentation.

Estrogen receptor (ER) positive breast cancers are estrogen dependent and require estrogen for growth in vivo [10]. The ER status of a patient is determined at diagnosis by the pathologist and the information on each patient is available in the pathology reports. If the tumor is classified as ER positive, the mice will require estrogen supplementation. Estrogen pellets can be ordered and they provide sustained release for 3 months after which they have to be replenished. It is also possible to make your own estrogen pellets which will be more cost effective [9] (*see* **Note 3**).

1. Disinfect the area designated for the surgery and sterilize all the surgical tools prior to use. Prepare the surgical area so that there are three distinct areas for preparation, surgery, and recovery (*see* **Note 4**).

2. Administer a preemptive analgesic before surgery begins (*see* **Note 5**).

3. After administering analgesic anesthetize the mouse by placing it into an induction chamber and exposing it to 3–3.5% isofluorane. Lower the isofluorane to 2.5% when the mouse is fully anesthetized.

4. Transfer the mouse to the surgical area and place it under the anesthesia unit. Gently stabilize the animal and position it for surgery, making sure that the nose cone is safely secured for the continuous delivery of isofluorane.

5. Shave the incision site area of animal fur (*see* **Note 6**).

6. Use a Steri-Wipe to disinfect the incision site prior to surgery.

7. Pinch the foot of the mouse before making any incisions to ensure that the animal is completely under anesthetic.

8. If the tumor sample receptor status is ER positive, it is essential to carry out the implantation of an estrogen pellet first prior to tumor implantation. If the tumor sample receptor status is ER negative, proceed to **step 9**. Implant the pellet subcutaneously in the area around the shoulder (*see* Fig. 2). Shave the area around the incision site and make a small incision measuring 2–3 mm. Deliver one estrogen pellet subcutaneously using a sterile trochar and close and disinfect the skin at the incision.

9. The tumor tissue fragments are now ready to be implanted. Implant the tumor fragments into the inguinal (fourth) mammary fat pad (*see* **Note 7**). Make a small upward subcutaneous incision 3-4 mm as shown in Fig. 2. Utilizing sterile tweezers carefully make a pocket within the inguinal fat pad (*see* **Note 8**).

10. Utilizing sterile forceps mix the tumor fragments with the Matrigel solution and place in the pocket made within the fat pad (*see* **Note 9**).

11. Close the wound using absorbable sutures and place the mouse in a recovery cage with a warm lamp to aid the recovery (*see* **Note 10**). Once the mouse recovers fully, place it in its home cage, and closely monitor the mouse post-surgery to ensure it is healthy.

12. Monitor growth of successful tumor implantations and palpate weekly.

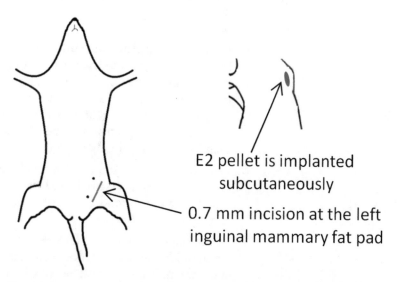

E2 pellet is implanted subcutaneously

0.7 mm incision at the left inguinal mammary fat pad

Fig. 2 Schematic showing the location of surgical incisions and sites for tissue implants. The area of implantation is the fourth inguinal mammary fat pad. The image also shows the location of the estrogen pellet subcutaneous implantation

3.4 Harvesting and Processing Successful Tumors from Mice

Once a successful implantation has been identified it is left to proliferate in the mice. When ready for harvesting, it is important to have a clear plan of what is to be done with the tumor sample and how it is going to be processed. At the early stages of tumor engraftment the major focus is on re-implantation into subsequent generations of mice and building up the tumor bio-bank. However, the resulting tumor is generally divided into a flash-frozen piece, formalin-fixed paraffin embedded (FFPE) piece for immunohistochemistry (IHC) and remaining tumor fragments for transplantations into subsequent generations.

1. Prepare the area as before. Disinfect the area designated for the surgery and sterilize all the surgical tools prior to use. Prepare the surgical area so that there that there are three distinct areas for preparation, surgery, and recovery.

2. Euthanize the tumor-bearing mouse with an isoflurane overdose following the relevant regulatory body approved procedures. The euthanized mouse can then be harvested.

3. Wipe the area around the inguinal fat pad with 70% ethanol and make an incision to expose the tumor. Carefully expose the tumor utilizing surgical scissors and tweezers. Measure the tumor at this stage utilizing calipers, and completely remove it from the body. Once the tumor is removed from the body it needs to be processed.

4. Keep the tumor in a 10-cm petri dish on ice and cover it with medium to keep it moist. At this stage, divide it appropriately for either transplantation, flash freezing for nucleic acid experiments and also for FFPE. The size of the tumor fragments will depend on the application as some require more material than others.

5. In case of transplants into other mice cut the tumor into small fragments using a surgical scalpel and place it in a 1.5 ml cryovial containing tumor freezing media. Freeze the tumor fragments in a cryo-container and store it at –80 °C. Keep the samples here overnight after which they can be stored in liquid nitrogen for long-term storage. These tumor fragments can then be used for reimplantation.

6. Place the tumor fragments that will be used in nucleic acid and protein studies in cryovials and flash-freeze these in a Dewar flask containing liquid nitrogen. These can then be used for subsequent experiments.

7. As the tumor cells are not fluorescent, we rely on IHC and necropsy to identify any potential metastasis sites. IHC and necroscopy are utilized to track potential metastasis sites. Surgically remove organs typical of breast cancer metastasis (liver, lung, bone) and place them in 10% formalin. Leave the organs rotating overnight at room temperature for fixation and then incubate them in 70% ethanol at 4 °C overnight. After ethanol incubation

embed the organs in paraffin wax and section the tissue as needed. A small fragment of the primary tumor may also be processed for IHC in order to track changes between the patient primary tumor, the xenografts and the potential metastatic cells found.

4   Notes

1. This has to be done in a prompt manner in order to minimize the processing time.

2. The tumor size available for the implantation will vary depending on how much extra tumor tissue is available after surgery. If the tumor is to be transported to the lab from the hospital site, it must be placed in a sterile 15-ml conical tube filled with MEGM medium and transported on ice

3. Important! All surgical procedures must be approved by the local ethical or legal authority. It is your responsibility that the approval is in place before beginning any animal experimentation. Should there be any doubt regarding the approval for surgery or use of the animals, you should not proceed with surgery!!

4. As with any surgical procedure the implantation has to be carried out using aseptic techniques with the animal welfare being the utmost priority.

5. We utilize Vetergesic which must be administered 30 min prior to surgery.

6. A shaved area of approximately three times the size of the planned incision site should be adequate.

7. The tumor tissue fragments should be kept in the media until implantation in order to keep the tissue moist.

8. The aim is to have the tumor fragment implanted at the center of the fat pad.

9. The Matrigel provides an extracellular matrix for the tumor fragments and can augment the tumors.

10. If carried out correctly, there should be very little bleeding and the wound itself should be relatively small.

Acknowledgements

Funding is acknowledged from the Irish Cancer Society Collaborative Cancer Research Centre BREAST-PREDICT grant, CCRC13GAL and Breast Cancer Ireland. No conflict of interest is declared by the authors.

References

1. Vargo-Gogola T, Rosen JM (2007) Modelling breast cancer: one size does not fit all. Nat Rev Cancer 7(9):659–672, doi:10.1038/nrc2193 nrc2193 [pii]

2. DeRose YS, Wang G, Lin YC, Bernard PS, Buys SS, Ebbert MT, Factor R, Matsen C, Milash BA, Nelson E, Neumayer L, Randall RL, Stijleman IJ, Welm BE, Welm AL (2011) Tumor grafts derived from women with breast cancer authentically reflect tumor pathology, growth, metastasis and disease outcomes. Nat Med 17(11):1514–1520, doi:10.1038/nm.2454 nm.2454 [pii]

3. Li S, Shen D, Shao J, Crowder R, Liu W, Prat A, He X, Liu S, Hoog J, Lu C, Ding L, Griffith OL, Miller C, Larson D, Fulton RS, Harrison M, Mooney T, McMichael JF, Luo J, Tao Y, Goncalves R, Schlosberg C, Hiken JF, Saied L, Sanchez C, Giuntoli T, Bumb C, Cooper C, Kitchens RT, Lin A, Phommaly C, Davies SR, Zhang J, Kavuri MS, McEachern D, Dong YY, Ma C, Pluard T, Naughton M, Bose R, Suresh R, McDowell R, Michel L, Aft R, Gillanders W, DeSchryver K, Wilson RK, Wang S, Mills GB, Gonzalez-Angulo A, Edwards JR, Maher C, Perou CM, Mardis ER, Ellis MJ (2013) Endocrine-therapy-resistant ESR1 variants revealed by genomic characterization of breast-cancer-derived xenografts. Cell Rep 4(6):1116–1130, doi:10.1016/j.celrep.2013.08.022 S2211-1247(13)00463-4 [pii]

4. Marangoni E, Vincent-Salomon A, Auger N, Degeorges A, Assayag F, de Cremoux P, de Plater L, Guyader C, De Pinieux G, Judde JG, Rebucci M, Tran-Perennou C, Sastre-Garau X, Sigal-Zafrani B, Delattre O, Dieras V, Poupon MF (2007) A new model of patient tumor-derived breast cancer xenografts for preclinical assays. Clinical Cancer Res 13(13):3989–3998. doi:10.1158/1078-0432.CCR-07-0078

5. Siolas D, Hannon GJ (2013) Patient-derived tumor xenografts: transforming clinical samples into mouse models. Cancer Res 73(17):5315–5319, doi:10.1158/0008-5472.CAN-13-1069 0008-5472.CAN-13-1069 [pii]

6. Williams SA, Anderson WC, Santaguida MT, Dylla SJ (2013) Patient-derived xenografts, the cancer stem cell paradigm, and cancer pathobiology in the 21st century. Lab Invest 93(9):970–982, doi:10.1038/labinvest.2013.92 labinvest201392 [pii]

7. Zhang X, Claerhout S, Prat A, Dobrolecki LE, Petrovic I, Lai Q, Landis MD, Wiechmann L, Schiff R, Giuliano M, Wong H, Fuqua SW, Contreras A, Gutierrez C, Huang J, Mao S, Pavlick AC, Froehlich AM, Wu MF, Tsimelzon A, Hilsenbeck SG, Chen ES, Zuloaga P, Shaw CA, Rimawi MF, Perou CM, Mills GB, Chang JC, Lewis MT (2013) A renewable tissue resource of phenotypically stable, biologically and ethnically diverse, patient-derived human breast cancer xenograft models. Cancer Res 73(15):4885–4897. doi:10.1158/0008-5472.CAN-12-4081

8. Landis MD, Lehmann BD, Pietenpol JA, Chang JC (2013) Patient-derived breast tumor xenografts facilitating personalized cancer therapy. Breast Cancer Res 15(1):201, doi:10.1186/bcr3355 bcr3355 [pii]

9. DeRose YS, Gligorich KM, Wang G, Georgelas A, Bowman P, Courdy SJ, Welm AL, Welm BE (2013) Patient-derived models of human breast cancer: protocols for in vitro and in vivo applications in tumor biology and translational medicine. Curr Protoc Pharmacol Chapter 14:Unit 14:23. doi:10.1002/0471141755.ph1423s60

10. Clarke R, Leonessa F, Welch JN, Skaar TC (2001) Cellular and molecular pharmacology of antiestrogen action and resistance. Pharmacol Rev 53(1):25–71

Conclusion

Our aim with this collection of chapters is to provide protocols that have been applied at first hand to the molecular analysis or modeling of key phases of mammary gland development and function. For various reasons there have been restrictions to the number and range of protocols we could include in this work; however, our limited compendium is complemented by a range of recent books, reviews, and original research that describe many of the other methodologies that have contributed to our understanding of mammary gland development [1–6]. In this, the concluding chapter, for the sake of completeness we wish to supplement the protocols we have described by highlighting some particularly seminal work so that the reader can have a comprehensive and full overview of the techniques that have been instrumental in our progress to date. Readers may also like to access a comprehensive video guide to the mouse mammary gland and its surgical manipulation released by Daniel and Strickland in 2012 [7]. In addition, it should be noted that the Journal of Visualized Experiments (JoVE) has over 100 instructional videos relating to techniques used for the study of mammary gland biology and breast cancer [8].

Patient derived xenograft (PDX) techniques are fast becoming a key preclinical approach to propagating, and analyzing, human breast tumor tissue in vivo. Varešlija and colleagues in Chapter 17 have described their PDX approach where human breast tumor tissue is grafted and allowed to establish itself within the confines of the mouse mammary fat pad. Graft survival is often dependent on the addition to parallel implantation of hormone releasing pellets. Non-orthotopic sites may also be used, but the mouse mammary fat pad has proven to be the most suitable microenvironment for the human epithelium to engraft successfully [9]. While such PDX techniques hold the promise of significantly advancing new drug development and novel treatments for breast cancer they tell us little about the developmental biology of the human gland, which for obvious reasons is difficult to capture, manipulate, and study in real time. In this regard, the work of Kupperwasser should be noted for its pioneering approach to the human-in-mouse concept, and humanization model [10]. Instead of relying on the mouse stroma to support the human epithelial cell growth, as not all human and mouse growth factors are interchangeable, Kupperwasser and col-

Finian Martin et al. (eds.), *Mammary Gland Development: Methods and Protocols*, Methods in Molecular Biology, vol. 1501, DOI 10.1007/978-1-4939-6475-8, © Springer Science+Business Media New York 2017

leagues developed a technique for transplanting immortalized human stromal cells into a cleared mouse fat pad so as to have a model where both the epithelial and stromal compartment were human in origin [10]. These "humanization" approaches have also allowed the establishment of human breast out-growths in vivo, from reduction mammoplasties, which are functionally and structurally normal in their development, and formation of tumors. Naturally, the humanization procedure can also be used for PDX experiments or grafting of tumor organoids, as well as primary HMECs. A detailed protocol for the technique can also be found in the book *Mammary Stem Cells: Methods and Protocols* [11].

One notable finding from the Kupperwasser studies was that tumors could develop from ostensibly normal epithelial cells when transplanted into growth factor activated humanized stroma. However, the tumor forming capacity of these cells, which had likely already undergone premalignant changes in vivo, was not apparent within the context of a normal stromal environment [12]. This observation, among many others, nicely illustrates the central tenet put forward in the early 1980s by Bissell and colleagues, that phenotype is capable of dominating genotype, and that the role of the microenvironment, most significantly the ECM, ultimately determines the cell behavior and function [13, 14]. Indeed the Bissell laboratory pioneered the mammary gland 3D culture model that has influenced and been further developed by so many other researchers (for example the Brugge, Barcellos-Hoff, Weaver, Werb, Kenny, and Martin groups, to name but a few) [15]. In Chapter 9, Gajewska and McNally present the basic protocol for generating functional acini from primary (mid-pregnancy) mammary epithelial cells. However, numerous adaptations of the technique have been used since its inception in 1980s, such as the fibroblast coculture model described by *Koledova* and *Lu* in Chapter 10 [16]. The Bissell laboratory continues to innovate and among other things has developed microfabricated organotypic tissue arrays. These arrays consist of defined cavities molded into the surface of collagen I gels. Mammary epithelial cells, when added, form hollow tubules capable of responding to additional biological cues by outgrowth [17, 18]. These experiments highlighted the influences of tissue geometry in determining tubule behavior such as branch initiation, and confirmed the previously hypothesized role of TGFβ proposed almost 25 years earlier by Daniel and Silberstein using slow release Elvax pellets implanted in vivo [3, 19]. Impressively, these relatively simple structures are capable of undergoing bifurcation and branch patterning in a manner identical to that of the mammary epithelial tree in vivo [17, 20]. Owing to the importance of not only spatial and geometric cues, but also mechanical pressure, a further microfabrication approach is described in Chapter 12 contributed by Nelson, the author of the original paper from the Bissell lab. This technique elegantly describes a method to examine and quantify how changes in fluid pressure and flow affect cell dynamics. In this case however it is not tubules that are formed but rather epithelial cell aggregates whose collective migratory behavior is monitored. However, the role of mechanical stress in phenotype determination was long considered important to understand prior to the work of Nelson. In this regard Valerie Weaver and her colleagues, also formerly of the Bissell laboratory, have been crucial in advancing the field of mechano-biology as it relates to mammary epithelial cell behavior. Importantly, her work has linked both endogenous (cytoskeletal) and exogenous (stromal/matrix-derived) mechanical stresses to determining the malignant phenotype [21].

The main non-cellular contributor to the physical microenvironment of the mammary epithelium that is sensitive to mechanical stress, is the ECM and, in addition to the Weaver group, Zena Werb, from her work both with the Bissell group, and independently, is undoubtedly at the forefront of ECM research today as it relates to both the biology and pathobiology of epithelial tissues [22]. We have provided detailed protocols for the 3D culture and analysis of primary and immortalized mammary epithelial cells (Chapters 9–12); including a protocol for generating alveolus-like structures from primary mammary epithelial cells or MCF10A cells cultured "on top" of Matrigel. However, another variation, mammary organoid culture (see Chapter 10), is distinguished by the use not of primary epithelial cells from the mid pregnant mouse gland, or from MCF10A cells, but by harvesting of less mature ductal structures (from 6- to 10-week-old virgin mice). These ductal elements are cultured, completely embedded within Matrigel, making it possible to study branching morphogenesis in vitro. This technique (developed in the Bissell laboratory [23]) proved important as it enabled, for instance, the investigation of ductal elongation: Ewald and Werb could demonstrate that pubertal ductal elongation from terminal end bud structures is driven by collective epithelial cell migration and not by creation of protrusions or cellular extensions [24, 25]. Ewald and Werb also used this technique of primary organoid culture to show that the type of surrounding ECM determined the migration phenotype and invasion of mammary epithelium. Thus, stromal-like primarily collagen I basement membrane facilitated invasion with protrusions and dissemination of normal or malignant mammary epithelium. In contrast, however, Matrigel, which recapitulates the epithelial basement membrane, facilitated multilayered epithelial cell collective migration without protrusions, and local dissemination was rarely observed [26]. Of note too are the strikingly beautiful images that these authors produced using long-term time-lapse microscopy, which provide such clear visualization of the morphological process ex vivo [25, 27]. The crucial role of integrins and integrin-associated proteins in supporting ECM-directed cell signaling in mammary epithelial cells has been comprehensively demonstrated by the ground-breaking studies of the Streuli laboratory [see also Chapter 3]: Using in vitro and in vivo gene knockout models, they did not just reveal, for example, that integrin β1 is required for mammary alveolar development and differentiation but further elucidated the molecular mechanisms by which integrin β1 signaling determines lumen formation, cell polarity, cell cycle progression and even nuclear architecture. Their work has therefore been crucial for our understanding of how the cellular microenvironment functions to influence integrated morphogenesis during mammary gland development and provides a potential target for disruption during mammary tumor progression [28–31].

Mammary epithelial cell–ECM interactions are reciprocal and dynamic but the tightly controlled remodeling as seen in branching morphogenesis during development is in stark contrast to the abnormal ECM dynamics that are a hallmark of pathological mammary epithelium. The ECM is also an essential component of the adult mammary stem cell niche and, by extension, potentially facilitates a tumor favorable microenvironment or cancer stem cell niche, when deregulated [22]. Accordingly, various ECM receptors (integrins and cell adhesion molecules) have been found to be useful markers with which to enrich for adult mammary stem cells. In 2006, the Visvader laboratory were first to identify cell surface markers that could be used in flow cytometry to isolate mammary epithelial cell populations that were capable of repopulating a mammary fat pad with a fully functional mammary tree, even from a single cell [32]. Several groups (including the Smalley, Stingl, and Lindeman groups) have since further developed cell sorting protocols for isolating various luminal progenitor cell populations from mouse and the human breast tissue, and so, have dissected the mammary stem cell hierarchy. An excellent comparison of these different protocols can be found elsewhere [33]. In Chapter 13, of this volume, Shehata and Stingl describe their method for the isolation of human breast epithelial progenitor cell populations; the protocol for isolating these cell populations from mouse mammary gland has previously been described in the MiMB series [34].

Protocols for mammary tissue explant culture were developed in the 1970s in an attempt to bridge the gap between the in vitro and in vivo experimental environment [35]. It proved possible to culture whole mouse mammary glands intact for several days and, for example, to induce alveolar epithelial cell differentiation by hormonal stimulation. This ex vivo technique facilitated the study of hormones and local growth factors on epithelial cell differentiation. In contrast with single cell or monolayer mammary epithelial cell culture models it retains all cellular compartments: structurally intact and biologically viable [36]. A further advantage explant and whole organ cultures have over monolayer and 3D epithelial cell culture models is of course the presence of the intact stroma. The ideal 3D in vitro model would include a stromal compartment, and the epithelium supported by an

ECM, in order to most accurately reflect the microenvironment. To this end, the Watson group, who contributed Chapter 7 of this volume, has recently described an impressive 3D model comprising an in vitro fad pad derived from pre-adipocytes: porous scaffolds were fabricated from collagen and hyaluronic acid and consistent with the gland in vivo allowed for the interaction of epithelial (KIM-2), stromal (3T3-L1), and immune cells (RAW-264.7) in coculture. Using this unique model it was possible, for the first time, to recapitulate the remodeling that occurs in the involuting gland [37].

One important aspect of Watson's model that is overlooked in others is the inclusion of immune cells in the extracellular milieu. The immune cells that normally populate the stroma throughout mammary gland development consist of leukocytes both of innate and adaptive origin, although the role of the adaptive immune system appears to be restricted to lactation and involution. The absolute requirement for these cells in vivo during development of the mammary gland was most elegantly demonstrated by Pollard and colleagues. They used various KO mouse models (see Chapter 3) and ablation of the bone marrow by irradiation to demonstrate the block in pubertal mammary gland development resulting from the absence of macrophages or eosinophils. This work highlighted the fact that during mammary gland development the role of these innate immune cells is not immunological, in the sense that there is no pathogenic tissue entity, but rather they function primarily, for example in the case of macrophages, to promote epithelial cell growth, engulf apoptotic cells, and remodel the ECM [38].

Interestingly, breast cancer too is defined by immune cell infiltration, with macrophages being the most abundant leukocyte in the tumor stroma. The intravital multiphoton imaging techniques developed by Condeelis and Segall and their colleagues to study the tumor microenvironment led to the discovery of the role of macrophages in the migration and invasion by tumor cells. Breast tumor associated macrophages (TAMs) adopt the properties of the macrophages utilized during ductal development, such as their matrix remodeling ability, for pathological gains and are therefore key determinants of tumor cell intravasation and metastasis [39–42]. It should be noted, however, that the rather less sophisticated method of zymography, prior to the advent of intravital imaging, has also been very useful in studying the activity of the matrix proteases in the mammary gland, often derived from macrophages, particularly during the second remodeling phase of involution [43, 44]. The role of the matrix proteases in determining the bioavailability of growth factors, particularly the ligands of the EGF receptor family has been highlighted by the work of Paraic Kenny [45] and this, along with the discovery of the existence of distinct paracrine loops between TAMs and tumor cells, underscores the multifaceted role of these immune cells in mammary gland development and cancer [46, 47].

How the ECM, normally associated with a particular stage of mammary gland development, can directly influence tumor progression has been elegantly demonstrated in the Schedin laboratory: They could demonstrate that a breast cancer cell line injected into mammary fat pads of nude mice together with ECM isolated from an involuting mammary gland showed significantly enhanced metastasis formation [48, 49]. Interestingly, however, targeting the cytokine IL-10 with a neutralizing antibody selectively inhibited tumor growth in the involution microenvironment but not in a nulliparous-derived one, suggesting that the transient immunosuppressive effect of IL-10 plays a key role in promoting tumor growth post-lactation [50].

The study of mammary gland development and investigation into the origins of breast cancer have always gone hand-in-hand. Many of the researchers that have stood at the forefront of mammary gland research have also sought to understand the organ in terms of oncogenic transformation, cancer development and progression. It follows the idea that tumor and tumor-associated cells abide by instructions that parallel those used to drive development. A better understanding of mammary gland development can therefore have a significant impact on our understanding of breast cancer development and progression. In agreement, basic biological studies have not only shed light on the fundamentals, allowing us to ask questions such as: how does the mechanism of morphogenesis relate to that of pathological invasion, but also helped us utilize this information, even clinically. A key example is our understanding of extracellular microenvironment, particularly in relation to stiffness or mechanical pressure: this is clearly reflected in clinical studies where high mammographic density (MD), indicative of high cellularity, high stromal ECM, and presence of immune cells, is linked with higher breast cancer risk. Conversely a marked reduction in MD, which can be seen for example following anti-hormone therapy, improves prognosis in relation to patients showing no change in MD [51, 52]. Another example is the work, discussed above, that went some way towards resolving the paradox that although early full term pregnancy is generally considered protective for lifetime breast cancer risk, the post-partum years hold considerably higher risk and in fact show a transient increase in breast cancer incidence, owing to the fact that the involution-associated pro-inflammatory microenvironment is permissive for tumor promotion and progression. This, we understand now, is one of the key reasons why women with pregnancy-associated breast cancer have a higher incidence of metastatic disease and a poorer prognosis [53, 54]. Our understanding of the basic biology underlying clinical and epidemiological observations such as these, allows for the design of medical intervention that would have not been imagined otherwise. The mouse study demonstrating that administration of anti-IL10 neutralizing antibodies succeeded in making the involution-associated stroma less immunosuppressive and permissive for tumor growth, is an example of this, as it suggests the possibility of developing a similar prophylactic treatment for humans [50].

The Future: Integrated Omics and Gene Editing

The future of experimental mammary gland biology is undoubtedly bright and some of the novel approaches currently developed are contributing and will continue to contribute new and exciting data to this field. It is foreseeable that the full integration of omics technologies (whole genome, transcriptome, epigenome, and proteomics) will provide a comprehensive integrated molecular explanation to firstly, how the mammary gland develops; and secondly, in combination, how mammary tumors of different cellular origin develop, are sustained, and disseminate. We highlighted here the application of some omics techniques such as the gene expression analysis of TEBs in Chapter 5 and RNA-seq analysis that defined an EMT profile during 3D organization of MEC in vitro in Chapter 11. However, a fully integrated overview will necessitate an inclusive model, which explains not only how an initial genome sequence dictates cell behavior, but also how "out-put" from an individual's genome is continuously changing over time in response to environmental cues, epigenetically, by means of DNA methylation, histone modification, the influence of noncoding RNAs, etc. In that vein too it is worth mentioning that the biology of exosome or microvesicular-mediated cell signaling has come to the forefront of cancer biology in more recent times. Exosomes likely have a fundamental role in mammary gland development and are readily found in human milk where they have some immunomodulatory functions [55, 56]. However, there is a new appreciation of their role in tumor cell communication via trafficking of various signaling molecules. In particular, cell specific integrins, which appear to act as a homing signal in organotypic metastasis [57]. In addition to the rapid development of several types of liquid biopsy for the detection of circulating tumor cells or cell free DNA in patient blood samples, the isolation of tumor cell derived exosomes from a liquid biopsy allows for the exciting possibility that the specific exosome-integrin expression profile could predict sites of future metastasis [57, 58]. This would provide a significant advance over the simple diagnostic readouts possible from liquid biopsies today. Experimentally, the methodology that will perhaps have most influence over the study of mammary gland biology and cancer in the coming years is the CRISPR/Cas9 technology [59], briefly discussed in our Chapter 3. It should rapidly advance our ability to introduce more precise, more complex and more extensive genetic changes than was heretofore possible, and determine their influence on mammary gland development. In addition, one can also envisage the use of virally delivered gene-editing to exactly define the genetic requirement to reverse the pathological profile in implanted tumor samples and so define a specific therapeutic regimen. Exciting times!.

References

1. Bissell MJ (2011) The mammary gland as an experimental model. In: Bissell MJ, Polyak K, Rosen JM (eds) Cold Spring Harbor perspectives in biology. Cold Spring Harbor, NY

2. Gjorevski N, Nelson CM (2011) Integrated morphodynamic signalling of the mammary gland. Nat Rev Mol Cell Biol 12:581–593. doi:10.1038/nrm3168

3. Ip MM, Asch BB (2000) Methods in mammary gland biology and breast cancer research. Springer, New York, NY. doi:10.1007/978-1-4615-4295-7

4. Smith BA, Welm AL, Welm BE (2012) On the shoulders of giants: a historical perspective of unique experimental methods in mammary gland research. Semin Cell Dev Biol 23:583–590. doi:10.1016/j.semcdb.2012.03.005

5. Macias H, Hinck L (2012) Mammary gland development. WIREs Dev Biol 1:533–557. doi:10.1002/wdev.35

6. McNally S, Martin F (2011) Molecular regulators of pubertal mammary gland development. Ann Med 43:212–234. doi:10.3109/07853890.2011.554425

7. Daniel CW, Strickland P (2012) The mouse mammary video: an introduction to the mouse mammary gland and instruction in its surgical manipulation. J Mammary Gland Biol Neoplasia 17:165. doi:10.1007/s10911-012-9262-4

8. JoVE Journal of Visualized Experiments

9. Whittle JR, Lewis MT, Lindeman GJ, Visvader JE (2015) Patient-derived xenograft models of breast cancer and their predictive power. Breast Cancer Res 17:17. doi:10.1186/s13058-015-0523-1

10. Proia DA, Kuperwasser C (2006) Reconstruction of human mammary tissues in a mouse model. Nat Protoc 1:206–214. doi:10.1038/nprot.2006.31

11. Wronski A, Arendt LM, Kuperwasser C (2015) Humanization of the mouse mammary gland. Mammary Stem Cells 1293:173–186. doi:10.1007/978-1-4939-2519-3_10

12. Kuperwasser C, Chavarria T, Wu M et al (2004) Reconstruction of functionally normal and malignant human breast tissues in mice. Proc Natl Acad Sci U S A 101:4966–4971. doi:10.1073/pnas.0401064101

13. Boudreau N, Bissell MJ (1998) Extracellular matrix signaling: integration of form and function in normal and malignant cells. Curr Opin Cell Biol 10:640–646. doi:10.1016/S0955-0674(98)80040-9

14. Weaver VM, Petersen OW, Wang F et al (1997) Reversion of the malignant phenotype of human breast cells in three-dimensional culture and in vivo by integrin blocking antibodies. J Cell Biol 137:231–245.

15. Lee GY, Kenny PA, Lee EH, Bissell MJ (2007) Three-dimensional culture models of normal and malignant breast epithelial cells. Nat Methods 4:359–365. doi:10.1038/nmeth1015

16. Campbell JJ, Hume RD, Watson CJ (2014) Engineering mammary gland in vitro models for cancer diagnostics and therapy. Mol Pharm 11:1971–1981. doi:10.1021/mp500121c

17. Nelson CM, Vanduijn MM, Inman JL et al (2006) Tissue geometry determines sites of mammary branching morphogenesis in organotypic cultures. Science 314:298–300. doi:10.1126/science.1131000

18. Nelson CM, Inman JL, Bissell MJ (2008) Three-dimensional lithographically defined organotypic tissue arrays for quantitative analysis of morphogenesis and neoplastic progression. Nat Protoc 3:674–678. doi:10.1038/nprot.2008.35

19. Silberstein G, Daniel C (1987) Reversible inhibition of mammary gland growth by transforming growth factor-beta. Science 237:291–293. doi:10.1126/science.3474783

20. Pavlovich A, Boghaert E, Nelson CM (2011) Mammary branch initiation and extension are inhibited by separate pathways downstream of TGFβ in culture. Exp Cell Res. doi:10.1016/j.yexcr.2011.03.017

21. Paszek MJ, Zahir N, Johnson KR et al (2005) Tensional homeostasis and the malignant phenotype. Cancer Cell 8:241–254. doi:10.1016/j.ccr.2005.08.010

22. Lu P, Weaver VM, Werb Z (2012) The extracellular matrix: a dynamic niche in cancer progression. J Cell Biol 196:395–406. doi:10.1083/jcb.201102147

23. Fata JE, Mori H, Ewald AJ et al (2007) The MAPKERK-1,2 pathway integrates distinct and antagonistic signals from TGFα and FGF7 in morphogenesis of mouse mammary epithelium. Dev Biol 306:193–207. doi:10.1016/j.ydbio.2007.03.013

24. Ewald AJ, Brenot A, Duong M et al (2008) Collective epithelial migration and cell rearrangements drive mammary branching morphogenesis. Dev Cell 14:570–581. doi:10.1016/j.devcel.2008.03.003

25. Ewald AJ (2013) Isolation of mouse mammary organoids for long-term time-lapse imaging. Cold Spring Harbor Protocols 2013:pdb.prot072892–pdb.prot072892. doi:10.1101/pdb.prot072892

26. Nguyen-Ngoc K-V, Cheung KJ, Brenot A et al (2012) ECM microenvironment regulates collective migration and local dissemination in normal and malignant mammary epithelium. Proc Natl Acad Sci U S A 109:E2595–E2604. doi:10.1073/pnas.1212834109

27. Ewald AJ (2013) Practical considerations for long-term time-lapse imaging of epithelial morphogenesis in three-dimensional organotypic cultures. Cold Spring Harbor Protoc 2013:pdb.top072884–pdb.top072884. doi:10.1101/pdb.top072884

28. Akhtar N, Streuli CH (2013) An integrin-ILK-microtubule network orients cell polarity and lumen formation in glandular epithelium. Nat Cell Biol 15(1):17–27. doi:10.1038/ncb2646

29. Jeanes AI, Wang P, Moreno-Layseca P, Paul N, Cheung J, Tsang R, Akhtar N, Foster FM, Brennan K, Streuli CH (2012) Specific beta-containing integrins exert differential control on proliferation and two-dimensional collective cell migration in mammary epithelial cells. J Biol Chem 287(29):24103–24112. doi:10.1074/jbc.M112.360834

30. Maya-Mendoza A, Bartek J, Jackson DA, Streuli CH (2016) Cellular microenvironment controls the nuclear architecture of breast epithelia through beta1-integrin. Cell Cycle 15(3):345–356. doi:10.1080/15384101.2015.1121354

31. Naylor MJ, Li N, Cheung J, Lowe ET, Lambert E, Marlow R, Wang P, Schatzmann F, Wintermantel T, Schuetz G, Clarke AR, Mueller U, Hynes NE, Streuli CH (2005) Ablation of beta1 integrin in mammary epithelium reveals a key role for integrin in glandular morphogenesis and differentiation. J Cell Biol 171(4):717–728. doi: 10.1083/jcb.200503144

32. Shackleton M, Vaillant F, Simpson KJ et al (2006) Generation of a functional mammary gland from a single stem cell. Nature 439:84–88. doi:doi:10.1038/nature04372

33. Smalley MJ, Kendrick H, Sheridan JM et al (2012) Isolation of mouse mammary epithelial subpopulations: a comparison of leading methods. J Mammary Gland Biol Neoplasia 17:91–97. doi:10.1007/s10911-012-9257-1

34. Smalley MJ (2010) Isolation, culture and analysis of mouse mammary epithelial cells. Methods Mol Biol 633:139–170. doi:10.1007/978-1-59745-019-5_11

35. Banerjee MR, Wood BG, Lin FK, Crump LR (1976) Organ culture of whole mammary gland of the mouse. Tca Manual 2:457–462. doi:10.1007/BF00918341

36. Casey TM, Boecker A, Chiu JF, Plaut K (2000) Glucocorticoids maintain the extracellular matrix of differentiated mammary tissue during explant and whole organ culture. Proc Soc Exp Biol Med 224:76–86. doi:10.1111/j.1525-1373.2000.22404.x

37. Campbell JJ, Botos L-A, Sargeant TJ, et al (2014) A 3-D in vitro co-culture model of mammary gland involution. Integr Biol (Camb) 6:618–626. doi:10.1039/C3IB40257F

38. Coussens LM, Pollard JW (2011) Leukocytes in mammary development and cancer. Cold Spring Harb Perspect Biol 3:a003285–a003285. doi:10.1101/cshperspect.a003285

39. Condeelis J, Segall JE (2003) Intravital imaging of cell movement in tumours. Nat Rev Cancer 3:921–930. doi:10.1038/nrc1231

40. Entenberg D, Wyckoff J, Gligorijevic B et al (2011) Setup and use of a two-laser multiphoton microscope for multichannel intravital fluorescence imaging. Nat Protoc 6:1500–1520. doi:10.1038/nprot.2011.376

41. Condeelis J, Pollard JW (2006) Macrophages: obligate partners for tumor cell migration, invasion, and metastasis. Cell 124:263–266. doi:10.1016/j.cell.2006.01.007

42. Noy R, Pollard JW (2014) Tumor-associated macrophages: from mechanisms to therapy. Immunity 41:49–61. doi:10.1016/j.immuni.2014.06.010

43. Lund LR, Rømer J, Thomasset N et al (1996) Two distinct phases of apoptosis in mammary gland involution: proteinase-independent and -dependent pathways. Development 122:181–193.

44. Kupai K, Szucs G, Cseh S et al (2010) Matrix metalloproteinase activity assays: importance of zymography. J Pharmacol Toxicol Meth 61:205–209. doi:10.1016/j.vascn.2010.02.011

45. Kenny PA (2007) Tackling EGFR signaling with TACE antagonists: a rational target for metalloprotease inhibitors in cancer. Expert Opin Therap Targets 11:1287–1298. doi:10.1517/14728222.11.10.1287

46. Goswami S, Sahai E, Wyckoff JB et al (2005) Macrophages promote the invasion of breast carcinoma cells via a colony-stimulating factor-1/epidermal growth factor paracrine loop. Cancer Res 65:5278–5283. doi:10.1158/0008-5472.CAN-04-1853

47. Su S, Liu Q, Chen J et al (2014) A positive feedback loop between mesenchymal-like cancer cells and macrophages is essential to breast cancer metastasis. Cancer Cell 25:605–620. doi:10.1016/j.ccr.2014.03.021

48. McDaniel SM, Rumer KK, Biroc SL et al (2006) Remodeling of the mammary microenvironment after lactation promotes breast tumor cell metastasis. Am J Pathol 168:608–620. doi:10.2353/ajpath.2006.050677

49. O'Brien J, Martinson H, Durand-Rougely C, Schedin P (2012) Macrophages are crucial for epithelial cell death and adipocyte repopulation during mammary gland involution. Development 139:269–275. doi:10.1242/dev.071696

50. Martinson HA, Jindal S, Durand-Rougely C et al (2015) Wound healing-like immune program facilitates postpartum mammary gland involution and tumor progression. Int J Cancer 136:1803–1813. doi:10.1002/ijc.29181

51. Huo CW, Chew GL, Britt KL et al (2014) Mammographic density—a review on the current understanding of its association with breast cancer. Breast Cancer Res Treat 144:479–502. doi:10.1007/s10549-014-2901-2

52. Nyante SJ, Sherman ME, Pfeiffer RM et al (2015) Prognostic significance of mammographic density change after initiation of tamoxifen for ER-positive breast cancer. J Natl Cancer Inst 107:dju425–dju425. doi:10.1093/jnci/dju425

53. Schedin P (2006) Pregnancy-associated breast cancer and metastasis. Nat Rev Cancer 6:281–291. doi:10.1038/nrc1839

54. J O, P S (2009) Macrophages in breast cancer: do involution macrophages account for the poor prognosis of pregnancy-associated breast cancer? J Mammary Gland Biol Neoplasia 14:145–157. doi:10.1007/s10911-009-9118-8

55. Admyre C, Johansson SM, Qazi KR et al (2007) Exosomes with immune modulatory features are present in human breast milk. J Immunol 179:1969–1978. doi:10.4049/jimmunol.179.3.1969

56. Hendrix A, Hume AN (2011) Exosome signaling in mammary gland development and cancer. Int J Dev Biol 55:879–887.

57. Hoshino A, Costa-Silva B, Shen T-L et al (2015) Tumour exosome integrins determine organotropic metastasis. Nature 527:329–335. doi:10.1038/nature15756

58. Ignatiadis M, Lee M, Jeffrey SS (2015) Circulating tumor cells and circulating tumor DNA: challenges and opportunities on the path to clinical utility. Clin Cancer Res 21:4786–4800. doi:10.1158/1078-0432.CCR-14-1190

59. Mou H, Kennedy Z, Anderson DG et al (2015) Precision cancer mouse models through genome editing with CRISPR-Cas9. Genome Med 7:53. doi:10.1186/s13073-015-0178-7

INDEX

Finian Martin et al. (eds.), *Mammary Gland Development: Methods and Protocols*, Methods in Molecular Biology, vol. 1501,
DOI 10.1007/978-1-4939-6475-8, © Springer Science+Business Media New York 2017

Printed in the United States
By Bookmasters